INTERNAL KINEMATICS AND DYNAMICS OF GALAXIES

INTERNATIONAL ASTRONOMICAL UNION
UNION ASTRONOMIQUE INTERNATIONALE

SYMPOSIUM No. 100

HELD IN BESANÇON, FRANCE, AUGUST 9–13, 1982

INTERNAL KINEMATICS
AND
DYNAMICS OF GALAXIES

EDITED BY

E. ATHANASSOULA

Observatoire de Besançon, France

D. REIDEL PUBLISHING COMPANY

DORDRECHT : HOLLAND / BOSTON : U.S.A. / LONDON : ENGLAND

Library of Congress Cataloging in Publication Data

Main entry under title:

Internal kinematics and dynamics of galaxies.

    Includes index.
    1. Galaxies–Congresses. I. Athanassoula, E. II. Inter-
national Astronomical Union.
QB857.I56   1983   523.1'12   82-24151
ISBN 90-277-1546-7
ISBN 90-277-1547-5 (pbk.)

Published on behalf of
the International Astronomical Union
by
D. Reidel Publishing Company, P. O. Box 17, 3300 AA Dordrecht, Holland

Sold and distributed in the U.S.A. and Canada
by Kluwer Boston, Inc.,
190 Old Derby Street, Hingham, MA 02043, U.S.A.

In all other countries, sold and distributed
by Kluwer Academic Publishers Group,
P. O. Box 322, 3300 AH Dordrecht, Holland

D. Reidel Publishing Company is a member of the Kluwer Group

Printed in The Netherlands

TABLE OF CONTENTS

Titles of reviews are given in capital letters.

II. SPIRAL STRUCTURE

III. WARPS

IV. BARRED GALAXIES

V. SPHEROIDAL SYSTEMS

VI. MERGERS

VII. GALAXY FORMATION

VIII. SUMMARIES

FOREWORD

It was particulary appropriate that the IAU Symposium N° 100 be held in Besançon in 1982 (August 9-13), since this coincided with the 100th anniversary of the Observatoire de Besançon. The meeting was held on the campus of the Université de Franche-Comté and was sponsored by IAU commission 28 and co-sponsored by commissions 33 and 40. It was organized under the auspices of the Ministère de l'Education Nationale, the Centre National de la Recherche Scientifique and the Institut National d'Astronomie et de Geophysique. It was attended by 166 scientists from 22 countries.

The subject of the symposium was internal kinematics and dynamics of galaxies and it was aimed to confront recent theoretical developments in this field with the wealth of new observational material obtained during the last few years. Barred galaxies and spiral structure were two major topics, though unfortunately at the last minute the main proponent of one of the spiral formation theories was unable to attend. The program included 23 reviews, 53 contributions and 39 poster papers, and there was ample time for discussion.

This book contains the proceedings of the symposium and was produced from camera-ready manuscripts prepared by the authors. I believe that in rapidly evolving fields a fast publication of meeting proceedings is a must and this has greatly influenced my editorial policy.Reviews are in the order presented at the meeting, but contributions have been rearranged and posters interlaced,as seemed appropriate. Several contributions were shortened or reworded. When somebody gave both a presentation and a poster on the same or similar subjects the manuscripts were merged into one longer unit. Discussion was included only after the reviews. It was edited partly from the discussion sheets filled by particicipants and partly from tape recordings.

It is a pleasure to thank all members of the scientific and local organizing committees,and in particular their chairmen P.O. Lindblad and M. Crézé, for their contributions to the success of the symposium. Generous grants from the Ministère de l'Education Nationale, the Centre National de la Recherche Scientifique and the Institut National d'Astronomie et de Geophysique are gratefully acknowledged. Danielle Chabod very efficiently assisted the local organizing committee in all stages of preparation and running of the symposium. Annika Johansson assisted the scientific organizing committee. J. Mulhauser, E. Rundo, D. Chabod and C. Cachoz helped with the preparation of the indexes and the final typing. Last but not least, thanks are due to

E. Athanassoula (ed.), Internal Kinematics and Dynamics of Galaxies, xi–xii.
Copyright © 1983 by the IAU.

A. Bosma for his interest and continual help-from the preparation of the meeting to the editing of this book.

Besançon, October 1982   E. Athanassoula

SCIENTIFIC ORGANIZING COMMITTEE

E. Athanassoula
A. Blaauw
J.J. Binney
G. Contopoulos
J. Delhaye
J. Einasto
K.C. Freeman
J. Kormendy
P.O. Lindblad (Chairman)
M.S. Roberts
G. de Vaucouleurs
R. Wielen
H. van Woerden

LOCAL ORGANIZING COMMITTEE

E. Athanassoula
D. Chabod
J. Colin
S. Considère
M. Crézé (Chairman)
E. Davoust
E. Oblak
F. Puel

LIST OF PARTICIPANTS

AGUILAR, C.L.University of California, Berkeley, U.S.A.
ALLEN, R.J. Kapteyn Astronomical Institute, Groningen, NL
AOKI, S. Tokyo Astronomical Observatory, Tokyo, Japan.
APPLETON, P.N. N.R.A.L., Jodrell Bank, Macclesfield, UK.
ATHANASSOULA, E. Observatoire, Besançon, France.
ATHERTON, P. Imperial College, London, UK.
BAILEY, M.E. Astronomy Centre University of Sussex, UK.
BAJAJA, E. Sterrewacht, Leiden, NL.
BARGE, P. Laboratoire d'Astronomie Spatiale, Marseille, France.
BASH, F. University of Texas, Austin, U.S.A.
BATTANER, J.J. University of Granada, Spain.
BECK, R. Max-Plank Institut Für Radioastronomie, Bonn, FRG.
BEKENSTEIN, J.D. University Ben Gurion, Beer Sheva, Israel.
BERTIN, G. Scuola Normale Superiore, Pisa, Italy.
BERTOLA, F. Osservatorio Astronomico, Padova, Italy.
BINNEY, J.J. Department of Theoretical Physics, Oxford, UK.
BLAAUW, A. Zuidvelde, NL.
BLACKMAN, C.P. University of Edinburgh,UK.
BLITZ, L. University of Maryland, U.S.A.
BOSMA, A. Sterrewacht, Leiden, NL.
BOULESTEIX, J. Observatoire, Marseille, France.
BRINKS, E. Sterrewacht, Leiden, NL.
BROSCHE, P. Sternwarte, Bonn, FRG.
BURTON, W.B. Sterrewacht, Leiden, NL.
CARLBERG, R. Institute of Astronomy, Cambridge, UK.
CARRASCO, L. Instituto de Astronomia, Mexico, Mexico.
CASERTANO, S. Scuola Normale Superiore, Pisa, Italy.
CHRISTIAN, C.A. CFHT Corporation, Kamuela, Hawaii, U.S.A.
CLUBE, S.V.M. Royal Observatory, Edinburgh,UK.
COHEN, R.J. N.R.A.L. , Jodrell Bank, Macclesfield, UK.
COHEN, R.S. Institute for Space Studies, New York, U.S.A.
COLIN, J. Observatoire, Besançon, France.
COMBES, F. Observatoire, Meudon, France.
COMTE, G. Observatoire, Marseille, France.
CONSIDERE, S. Observatoire, Besançon, France.
COURTES, G. Laboratoire d'Astronomie Spatiale, Marseille, France.
CREZE, M. Observatoire, Besançon, France.
DAVOUST, E. Observatoire, Besançon, France.
DEJONGHE, H. Gent, Belgium.
DEKEL, A. Yale University, Connecticut, U.S.A.
DELHAYE, J. Observatoire, Paris, France.
DESPOIS, D. Observatoire,Bordeaux, France.

De VAUCOULEURS, G. University of Texas, Austin, U.S.A.
De ZEEUW, T. Sterrewacht, Leiden, NL.
DJORGOVSKI, S. University of California, Berkeley, U.S.A.
DOOM, C. Université de Bruxelles, Belgium.
DRESSLER, A. Mount Wilson and Las Campanas Observatories,Pasadena, U.S.A.
DUVAL, M.F. Observatoire, Marseille, France.
DWORAK, Z. Institute of Meteorology and Water Management, Krakow, Poland
EVANS II N.J. University of Texas, Austin, U.S.A.
FALL, S.M. Institute of Astronomy, Cambridge, UK.
FEITZINGER, J.V.Astronomisches Institut der Universität, Bochum, FRG.
FREEMAN, K.C. Mt. Stromlo Observatory, Canberra, Australia
FRENCK, C.S. University of California, Berkeley, U.S.A.
FRIED, J. Max-Plank Institut für Astronomie, Heidelberg, FRG.
GOTTESMAN, S.T. University of Florida, Gainesville, U.S.A.
GRAYZECK, E. University of Nevada, Las Vegas, U.S.A.
GUNN, J.E. Princeton University Observatory, Princeton, U.S.A.
HAASS J. M.I.T. Cambridge, U.S.A.
HARDY, F. University Laval, Quebec, Canada.
HAUD, U. Tartu Astrophysical Observatory, Toravere, U.S.S.R.
HJALMARSON, A. Onsala Rymdobservatorium, Onsala, Sweden.
HRON, J. Institut für Astronomie, Wien, Austria
HUNTER, J.H. University of Florida, Gainesville, U.S.A.
ILLINGWORTH, G.D. Kitt Peak National Observatory, Tucson, U.S.A.
INAGAKI, S. Institute of Astronomy, Cambridge, UK.
IYE, M. Tokyo Astronomical Observatory, Mitaka, Japan.
JAMES, R.A. University of Manchester, Manchester, UK.
JORSATER, S. Stockholms Observatorium, Saltsjöbaden,Sweden.
KALNAJS, A. Mt. Stromlo Observatory, Canberra, Australia.
KENNICUTT,R. University of Minnesota, Minneapolis, U.S.A.
KENT, S.M. M.I.T Cambridge, U.S.A.
KNAPP, G.R. Princeton University Observatory, Princeton, U.S.A.
KORMENDY, J. Dominion Astrophysical Observatory, Victoria, Canada.
LACEY, C. Institute for Advanced Studies, Princeton, U.S.A.
LENTES, F.T. Observatorium Bonn, Daun, FRG.
LIN, C.C. M.I.T. Cambridge, U.S.A.
LINDBLAD, P.O. Stockholms Observatorium, Saltsjöbaden, Sweden.
LITZ, H.S. University of Virgina, Charlottesville, U.S.A.
LYNDEN-BELL, D. Institute of Astronomy, Cambridge, UK.
MADORE, B.F. University of Toronto, Toronto, Canada.
MAGNENAT, P. Observatoire, Genève, Switzerland.
MAMON, G. Princeton University Observatory, Princeton, U.S.A.
MANCHESTER, R.N. CSIRO Division of Radiophysics, Epping, Australia.
MARCELIN , M. Observatoire, Marseille, France.
MARTINET, L. Observatoire, Genève, Switzerland.
MENEGUZZI, M.M. DPEH Saclay, Gif sur Yvette, France.
MERRIT, D. N.R.A.O. Charlottesville, U.S.A.
MIKKOLA, S. Turku University Observatory, Turku, Finland.
MILLER, R.H. ESO, München, FRG.
MONNET, G. Observatoire, Lyon, France.
MULDER, W.A. Sterrewacht, Leiden, NL.
NELSON, A.H. University College Cardiff, Cardiff, UK.

NORMAN, C. Sterrewacht, Leiden, NL.
NOTNI, P. Zentralinstitut für Astrophysik, Babelsberg, FRG.
OBLAK, E. Observatoire, Besançon, France.
OKAMURA, S. Royal Observatory, Edinburgh,UK.
OLOFSSON, H. Onsala Space Observatory, Onsala, Sweden.
OORT, J. Sterrewacht, Leiden, NL.
PALMER, P.L. University of Athens, Greece.
PEDLAR, A. N.R.A.L. Jodrell Bank, Macclesfield, UK.
PENCE, W. A.A.O. Epping, Australia.
PETERSON, C.J. University of Missouri, Columbia, U.S.A.
PETROU, M. University of Athens, Greece.
PFENNIGER, D. Observatoire, Genève, Switzerland.
PICKLES, A.J. Mt. Stromlo Observatory, Canberra, Australia.
PRENDERGAST, K.H. Columbia University, New York, U.S.A.
PUEL, F. Observatoire, Besançon, France.
QUINN, P. Mt. Stromlo Observatory, Canberra, Australia.
RENZ, W. Institut für Theoretische Physik,Aachen FRG.
RICHSTONE, D.O. University of Michigan, Ann Arbor, U.S.A.
ROBE, H. Institut d'Astrophysique, Cointe Ougree, Belgium.
ROBERTS, M.S. N.R.A.O. Charlottesville, U.S.A.
ROBERTS, W.W. University of Virginia, Charlottesville, U.S.A.
ROHLFS, K. Astronomisches Institut, Bochum, FRG.
ROY, J.R. Université Laval, Québec, Canada.
RUBIN, V.C. DTM, Carnegie Institution, Washington, U.S.A.
RYDBECK, G. Onsala, Space Observatory, Onsala, Sweden.
SANCHEZ-SAAVEDRA, M.L. University of Granada, Spain.
SANCISI, R. Kapteyn Astronomical Institute, Groningen, NL.
SANDAGE, A. Mount Wilson and Las Campanas Observatories,Pasadena, U.S.A.
SANDERS, R. Kapteyn Astronomical Institute, Groningen, NL.
SASTRY, K.S. Osmania University, Hyderabad, India.
SCHWEIZER, F. DTM, Carnegie Institution, Washington, U.S.A.
SEITZER, P. Cerro Tololo Interamerican Observatory, La Serena, Chile.
SELLWOOD, J. Institute of Astronomy, Cambridge, UK.
SHANE, W.W. University of Nijmegen, NL.
SHARPLES, R.M. University of Durham, Durham, UK.
SHAYA, E. Institute for Astronomy, Honolulu, U.S.A.
SHOSTAK, G.S. Kapteyn Astronomical Institute, Groningen, NL.
SIMIEN, F. Observatoire, Lyon, France.
SIMKIN, S.M. Michigan State University, East Lansing, U.S.A.
SMITH, B.F. NASA/Ames Reseach Center, Moffet Field, U.S.A.
SOLOMON, P.M. Department of Earth and Space Science, Stony Brook, U.S.A.
SORENSEN, S.A. University College, London, UK.
TAMMANN, G.A. Astronomy Institute, Binningen, Switzerland.
TAYLOR, K. AAO, Epping, Australia.
TEUBEN, P.J. Kapteyn Astronomical Institute, Groningen, NL.
THIELHEIM, K.O. University of Kiel, FRG.
THONNARD, N. DTM, Carnegie Institution, Washington, U.S.A.
TOHLINE, J.E. Louisiana, State University, Lousiana, U.S.A.
TOOMRE, A., M.I.T. Cambridge, U.S.A.
TREMAINE, S., M.I.T. Cambridge, U.S.A.

TRINH X THUAN, Section d'Astrophysique, Gif sur Yvette, France.
TULLY, R.B. Institute for Astronomy, Honolulu, U.S.A.
ULRICH, M.H. ESO, München, FRG.
VADER, J.P. University of Amsterdam, NL.
VAGNETTI, F. Istituto Astronomico, Roma, Italy.
VAN ALBADA, G.D. Kapteyn Astronomical Institute, Groningen, NL.
VAN ALBADA,.T.S. Kapteyn Astronimical Institute, Groningen, NL.
VAN DER HULST, J.M. Radiosterrenwacht Westerbork, NL.
VAN DER KRUIT, P.C. Kapteyn Astronomical Institute, Groningen, NL.
VAN GORKOM, J. NRAO, New Mexico, U.S.A.
VAN MOORSEL, G. Kapteyn Astronomical Institute, Groningen, NL.
VAN WOERDEN, H. Kapteyn Astronomical Institute, Groningen, NL.
VIALLEFOND, F. Observatoire, Meudon, France.
VILLUMSEN, J. Yale University Observatory, New Haven, U.S.A.
WAMSTEKER, W. ESA-ESTEC Astronomy Division, Madrid, Spain.
WEVERS, B.M. Kapteyn Astronomical Institute, Groningen, NL.
WHITE, S. University of California, Berkeley, U.S.A.
WIELEN, R. Technische Universität Berlin, Berlin, FRG.
WILKINSON, A. University of Manchester, Manchester, UK.
WYSE, R.M. Institute of Astronomy, Cambridge, UK.
YOUNG, J.S. University of Massachusetts, U.S.A.

I

KINEMATICS OF GAS
AND THE UNDERLYING MASS DISTRIBUTION

SYSTEMATICS OF HII ROTATION CURVES

Vera C. Rubin
Department of Terrestrial Magnetism
Carnegie Institution of Washington

ABSTRACT. Systematic rotational properties of field spiral galaxies are presented, as a function of Hubble type and of luminosity. Within a Hubble type, radius, nuclear velocity gradient, rotation velocity, mass, and mass density increase with luminosity, while $\mathcal{M}/L$ is constant. At fixed luminosity, V(max), mass, density, and $\mathcal{M}/L$ decrease from Sa through Sc. These variations are so systematic that it is possible to display them on a single diagram.

1. SYSTEMATIC ROTATIONAL PROPERTIES OF FIELD SPIRALS

During the several years since the last conference on galaxy dynamics, we have increased our knowledge about the dynamical properties of spirals and ellipticals, bulges and disks. Progress has been due to systematic observational programs which have exploited the avilability of modern detectors on large telescopes. This paper will describe the

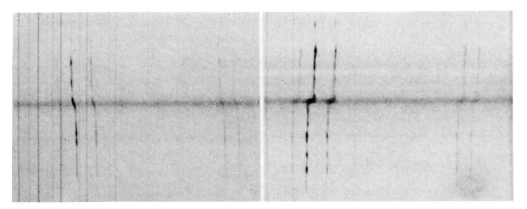

Fig. 1. Spectra near Hα of Sc galaxies of moderate and high luminosity. (left) NGC 2742, M = -20.5. (right) NGC 7541, M = -22.3. Kitt Peak 4-m spectrograph + Carnegie image tube; exposures 120 and 114 min. Note steeper nuclear velocity gradient in higher luminosity Sc.

3

E. Athanassoula (ed.), Internal Kinematics and Dynamics of Galaxies, 3–10.
Copyright © 1983 by the IAU.

results which my colleagues, W. Kent Ford, Jr., and Norbert Thonnard, and I have obtained from a study of rotations of field spirals.

Optical spectra for about 60 Sa, Sb, and Sc galaxies have been obtained with the 4-m telescopes at Kitt Peak and Cerro Tololo Observatories at high dispersion (25A/mm) and high spatial scale (25"/mm). A few spectra were also taken with the 100-in Las Campanas telescope. The spectra are centered at Hα. Fig. 1 shows spectra of two of the Sc galaxies. Program galaxies are relatively isolated, of high inclination, non barred, and have diameters of a few seconds, so that a single spectrograph slit subtends the optical image out to R(25), the isophote of 25th mag (arc sec)^2. For each Hubble type we have made every effort to observe galaxies with as large a range of luminosities as we could identify. Details and results are available elsewhere (Sc: Rubin, Ford and Thonnard 1980; Burstein et al. 1982; Sb: Rubin et al. 1982; comparison of Sa, Sb, and Sc: Rubin, Thonnard, and Ford 1982). Throughout the discussion, we use H = 50 km s^{-1}Mpc^{-1}.

Velocities from the advancing and receding major axes, measured from the Hα and [N II] lines, are smoothed to form the rotation curves (Fig. 2). Within each Hubble type, small low luminosity galaxies have velocities which rise gradually from the nucleus, and reach a low maximum velocity only at the galaxy isophotal radius. Large, high luminosity galaxies have velocities which rise steeply from the nucleus, reach the "nearly flat" portion in a smaller fraction of the galaxy radius, and reach a higher maximum

<u>Fig. 2.</u> Rotation velocity as a function of nuclear distance for 16 Sa, 23 Sb, and 21 Sc galaxies. Within each Hubble type, galaxies are arranged by increasing luminosity. Vertical lines: isophotal radius. Dashed lines: no measurements.

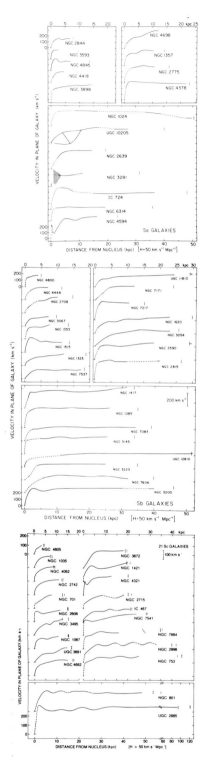

velocity. Virtually all rotation curves continue to rise with distance from the nucleus. For both Sb's and Sc's, in the mean, $V \propto R^{0.1}$ or $\propto R^{0.2}$. Sa's exhibit more individuality, due to increasing bulge importance. Because of the weaker emission, we have less radial coverage for the Sa galaxies. In the correlations we discuss below, we include only Sa galaxies with observations over at least 67% of R(25).

From these rotation velocities, we form "synthetic rotation curves" for galaxies of a given Hubble type, which show rotational velocity as a function of fractional isophotal radius. The procedure for forming these curves is described elsewhere (Thonnard and Rubin 1981). Figure 3 plots such a set for Sb galaxies. There is a neat progression with luminosity from a small nuclear velocity gradient, low rotational velocity to a steep nuclear gradient, high rotational velocity. Within a Hubble type, the form of the rotation curve is a clear luminosity indicator. A comparison of an Sb rotation curve with the set shown in Fig. 3 will permit the absolute magnitude to be deduced, even if only a fraction of the rotation curve is available.

The variation of rotational properties from one Hubble type to another is equally systematic. At a fixed luminosity, the rotation curve of an Sa is higher than that of an Sb, which in turn is higher than that of an Sc. This is illustrated in Fig. 4, where I plot synthetic rotation curves for Sa and Sc galaxies.

The increase in rotational velocity with luminosity produces the Tully-Fisher (TF) relation. However, note the significant displacement of galaxies of different types. This is illustrated in Fig. 5, a plot of M_B versus V(max) for program galaxies. Each Hubble type alone produces a TF line with a slope near 10, but the offset from type Sa to type Sc is over 2 magnitudes at any V(max). Or at fixed blue luminosity, the rotational velocity of an Sa is 1.6 times that of an Sc. A corresponding

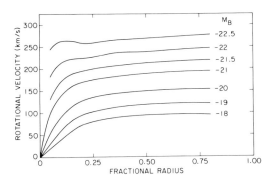

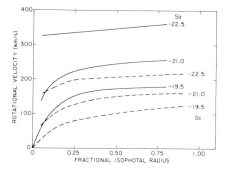

FIG 3. Synthetic Sb rotation curves, formed from observed velocities. Rotational velocity as a function of fractional isophotal radius is shown.

Fig. 4. Synthetic rotation curves for Sa and Sc galaxies. Curves are calulated with H=50 km s Mpc . For H=100, M is fainter by 1.5 mag.

correlation between V(max) and R(25) exists, but with slightly larger scatter. The agreement of these correlations with those based on red magnitudes (Aaronson, Huchra, and Mould 1979) is still under study.

For all of the program galaxies, luminosity is correlated with isophotal radius, with no type dependence (Fig. 6). Although Hubble (1936) knew this relation, it continues to be rediscovered annually.

2. SYSTEMATIC MASS PROPERTIES FOR FIELD SPIRALS

In the most simple spherical mass models, the mass $\mathcal{M}$ interior to R(25) is given by $G \cdot R(25) V^2(R25)$; G is the constant of gravitation. For flat rotation curves, integral mass rises linearly with R; for velocities rising as $R^{0.15}$, mass increases faster then R, $M \propto R^{1.3}$. Here I discuss the mass interior to R(25).

Within a Hubble type, velocity and hence mass and mass density increase with increasing luminosity. At a fixed radius and luminosity, early type galaxies have higher rotational velocity and hence higher density. But note the interplay of luminosity and Hubble type. An extremely luminous Sc galaxy can mimic the velocity, mass, and mass density of a mid-luminosity Sa, but the radius of the Sc must then be enormous.

For each Hubble type, there is a fairly tight correlation of mass with luminosity, i.e., $\mathcal{M}/L$ is independent of luminosity. This is

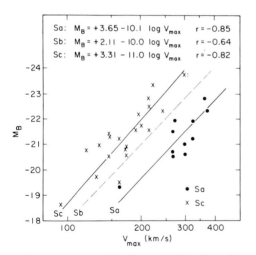

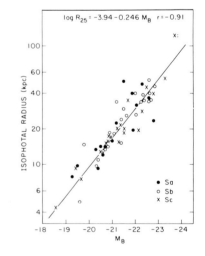

Fig. 5. Correlation of luminosity with V(max) for Sa, Sb, and Sc galaxies. Note slope near 10 for each type, but the displacement between Hubble classes.

Fig. 6. Correlation of radius with luminosity, for Sa, Sb, and Sc galaxies. The slope of the relation means that luminosity increases slower than surface area, $L \propto R^{\sim 1.7}$.

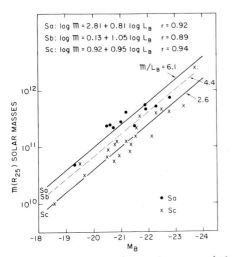

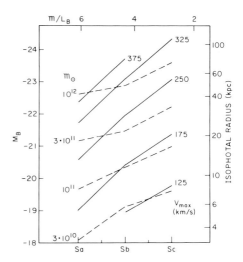

Fig. 7. Correlation of mass with luminosity for Sa, Sb, and Sc galaxies. At fixed Hubble type, $\mathcal{M}/L$ is independent of luminosity, as shown by the lines of constant $\mathcal{M}/L$.

Fig. 8. Interrelation of 6 parameters; lines of constant V(max) and constant mass are shown. For H = 100, M is brighter by 1.5 mag, radius and mass decrease 1/2, $\mathcal{M}/L$ increases x 2.

displayed in Fig. 7, where lines of constant $\mathcal{M}/L$ = 6.1 (Sa), 4.4 (Sb), and 2.6 (Sc) are shown. At a fixed Hubble type, the velocity, radius, mass, and mass density all increase with luminosity. The ratio of mass to luminosity is unique, being one important galaxy parameter which does not vary with luminosity, but does vary with Hubble type.

The correlations between luminosity, radius, mass, V(max), $\mathcal{M}/L$, and Hubble type can be combined into a single diagram, which illustrates the systematic interrelations of these parameters. I show such a diagram in Fig. 8. It contains, in extremely compact form, much of what we now know concerning the dynamics of galaxy disks. It also indicates that at least two parameters are necessary to locate a galaxy on this plane.

Finally, I point out that the curves of integral mass with radius of most Sc and Sb galaxies are remarkably similar, and can be generated from a single mass form, with suitable radial scaling. This follows from the rotation curve forms, which themselves can be transformed one into another with radial scaling (see discussion in Burstein et al. 1982).

3. ROTATIONAL PROPERTIES OF CLUSTER SPIRALS

The results discussed above come from observations of field spirals. We are presently engaged in an observing program to determine rotation curves for cluster spirals. Our aim is to see if the structure of galaxies is altered by the denser cluster environment, and if such altered structure leads to altered dynamical properties. For example, the high rotational velocities at large radial distances observed for field galaxies is one piece of evidence that spirals have massive halos extending to large radial distances. In the denser cluster environment, extended halos might be disturbed by the neighboring galaxies. A comparison of the rotational properties of field and cluster galaxies should enable us to infer the importance of environmental effects in galaxy evolution and dynamics.

To date, we have derived rotation curves for 8 galaxies in the Cancer and Pegasus I clusters, of Hubble types Sab through Sc. The dynamical properties of these galaxies are in no way different from those of the field spirals. These clusters, chosen because they were observed in the infrared by Aaronson et al. (1980), are not especially dense ones. We are now observing galaxies in two denser cluster (Hercules and DC 1842-63; Dressler 1980). If the remaining observations and analysis confirm the preliminary conclusion that field and cluster spirals are indistinguishable by their rotational properties, then we will have placed significant constraints on the properties of spiral halos.

ACKNOWLEDGMENTS

I thank the observatory Directors for observing time, numerous colleagues for valuable assistance, and Dr. R.J. Rubin for comments on the manuscript.

REFERENCES

Aaronson, M., Huchra, J., and Mould, J. 1979, Astrophys. J. 229, p.1.
Aaronson, M., Huchra, J., Mould, J., Sullivan, W.T. III, Schommer, R.A., and Bothun, G.D. 1980, Astrophys. J. 239, p.12.
Burstein, D., Rubin, V.C., Thonnard, N., and Ford, W. K. Jr. 1982, Astrophys. J. 253, p.70.
Dressler, A. 1980, Astrophys. J. Suppl. 42, p.569.
Hubble, E. 1936, in The Realm of the Nebulae, (New Haven: Yale University Press), p.48.
Rubin, V. C., Ford, W.K. Jr., and Thonnard, N. 1980, Astrophys. J. 238, p.471.
Rubin, V.C., Ford, W.K., Jr., Thonnard, N., and Burstein, D. 1982, Astrophys. J. 261 (in press).
Rubin, V.C., Thonnard, N., and Ford, W.K.,Jr. 1982, in preparation.
Thonnard, N., and Rubin, V.C. 1981, Carnegie Yrbk 80,, p.551.

DISCUSSION

INAGAKI : Is it possible to separate the disk mass from the halo mass
observationally, for example using statistical methods ? Are there any
rotation curves from absorption lines measurements i.e. stellar rotation
curves ?

RUBIN : For the simple analysis we employ, the total mass interior to
any radial distance R is derived, but not the distribution of that mass.
However F. Schweizer, B. Whitmore and I have observed a galaxy with a
polar ring, A0136-080, and have velocities in the S0 disk galaxy (from
absorption lines) and in the polar ring (from emission lines). The polar
ring extends to radial distances about three times as great as the S0 disk.
So for this case we are able to sample the gravitational potential beyond
the optical disk. The results indicate that the distant polar material is
feeling a spheroidal, rather than a disk potential. We have no stellar
rotation curves for our program galaxies, but absorption line rotation
curves for E's and S0's will be discussed later in this program.

GOTTESMAN : In A0136-080, what is the effect of tidal phenomena on the
halo of the galaxy ?

RUBIN : This depends on the mass ratio between the halo and the disk in
the S0 galaxy. From the flatness of the rotation curve in the outer ring,
we infer that the halo is massive and that it probably determines the
dynamics of the system. The outer ring itself contains so little material
that it probably reacts to the combined halo-disk potential like an
ensemble of test particles,and it seems unlikely that it can have any
appreciable effect on the massive halo.

SANDAGE : You have shown so convincingly here that the Tully-Fisher
relation is very dependent on Hubble type in M_B magnitudes, despite
certain contrary infrared data sets. Have you ever looked to see if this
dependence exists in the infrared observers' apparently magic H magnitudes,
or whether it disappears for your galaxies when H is used ? I think that
Tully-Fisher diagrams should be type dependent. Use of H magnitudes,
although reducing internal absorption problems, cannot conceal the large
difference in mass distribution that exists along the sequence Sa to Sb
to Sc to Sd galaxies.

RUBIN : We do not yet have H magnitudes for any of our galaxies but
Aaronson has been observing them and we hope his results will be
available soon. However I agree with your comment. To make the Hubble
type dependence that we observe in the blue disappear at the H magnitudes
requires that differential B-H colors be larger than two magnitudes between
types Sa and Sc, independent of luminosity. This seems like an excessive
amount but I am willing to wait for the observations.

AOKI : There will be a difference in the mass estimate depending on
whether you assume spherical symmetry or allow for flattened shapes.

How does this influence your estimated M/L ratios ? I think your data are good enough to permit a fine analysis to bring out the difference.

RUBIN : Because we lack information concerning the distribution of the non-luminous matter, we have chosen to model all of the galaxies in the most simple fashion, as spheres. The M/L ratios we derive then follow. If instead the mass distribution is in a flattened ellipsoid, then the masses will be decreased by a small factor ($\sim$ 20 or 30 %), but the relative M/L ratio will stay the same. However, if the mass distribution differs in galaxies of different Hubble types, or in galaxies of different luminosities, then the relative M/L values will change. We presently have no way to get the detailed distribution of the non-luminous matter, so a more detailed analysis is not yet warranted.

WIELEN : Some of the correlations you find could be caused by systematic errors in the galaxy inclination as a function of Hubble type or absolute magnitude. Can you exclude significant systematic errors of this kind in the inclinations you use ? What is the typical inclination of a galaxy in your sample ?

RUBIN : With only very few exceptions, inclinations for program galaxies range roughly from 50° to 80°. Hence errors as large as 10° in the adopted inclination will produce velocity errors no larger than 10 %, and generally smaller. I do not think this is a problem for this set of galaxies.

TOHLINE : I would like to emphasize at the opening of this symposium that the often quoted ratio M/L is in fact the ratio $V^2 r/L$ of the directly observable quantities V, r and L. This ratio $V^2 r/L$ can only be interpreted as an indicator of mass to light ratio if we assume that Newton's law of gravitational attraction is correct on the scale of galaxies. Since Keplerian behavior is essentially never seen in extra-galactic systems, I might be so bold as to suggest that the validity of Newton's law should now be seriously questioned. I hope that observers who have definitive evidence that Keplerian behavior has been observed in any system will emphasize that evidence at this meeting.

RUBIN : While we have observed that most Sa, Sb and Sc, galaxies have flat or slightly rising rotation curves, a few have slightly falling curves. However, I know of no isolated galaxy with rotation velocities decreasing as rapidly as $V \propto r^{-1/2}$. The point you raise is worth keeping in mind although I believe most of us would rather alter Newtonian gravitational theory only as a last resort.

HI VELOCITY FIELDS AND ROTATION CURVES

A. Bosma
Sterrewacht, Leiden

ABSTRACT

Recent results on 21-cm line velocity fields of spiral galaxies are
reviewed. Attention is drawn to the sometimes discrepant results on
the spatial orientation of galaxies inferred from various methods.
Major classes of deviations from circular, planar, motion of the
neutral gas are discussed. Some comments are made about HI rotation
curves.

I. INTRODUCTION

HI velocity fields are now available for several tens of spiral
galaxies. Most of these data come from aperture synthesis studies in
the 21-cm line, and, because of the two dimensional mapping involved,
much more information becomes available about a galaxy than just its
axisymmetric rotation curve. Moreover, the HI usually extends to the
outer parts, and often beyond the optical image so that rotation data
are known at larger distances from the centre than be obtained by
optical spectroscopy. However, the relative angular resolution (ratio
of beamsize to diameter of the galaxy) is in most cases rather low
(~0.02 to 0.2) so that resolution effects play a significant role in
the interpretation of the raw data.

As a starting point for the discussion let me review the results
of a comparative study of about 20 galaxies with detailed HI data
obtained in the 1970's and published recently (Bosma, 1981b). Based
on the velocity fields from a variety of sources the following
conclusions were arrived at:
1. Many spiral galaxies have large scale deviations from axial
symmetry. Four categories of distinctive signatures in the velocity
fields can be recognized: a) spiral arm motions, b) bars and oval
distortions, c) warps, and d) asymmetries.
2. Rotation curves are fairly flat out to large radii. For those
galaxies with both photometry and kinematical data available one can

11

E. Athanassoula (ed.), Internal Kinematics and Dynamics of Galaxies, 11–22.
Copyright © 1983 by the IAU.

make estimates of the local mass-to-light ratio, which is found to
increase with increasing radius (cf. Bosma and Van der Kruit, 1979).
The HI extent of a galaxy is not very predictable: in some cases
large HI envelopes are found, while in other cases the HI is confined
to the optical image.
3. From HI data alone it is difficult to find a dynamical basis for
Hubble types. This is partly due to lack of angular resolution in the
central parts of galaxies and partly because HI has not been detected
in the central parts of some early type galaxies with large bulges.

II. ORIENTATION PARAMETERS

Before discussing some of the intrinsic properties of spiral galaxies
one has to know the orientation of the galaxy in space. There are
several ways of determining the orientation parameters, but all
assume that the disk of a spiral galaxy is circular and planar at a
certain range of radii so that we need to know only the position
angle of the line of nodes, pa, and the inclination angle, i. These
methods are:
- photometrically, from the orientation of the major axis and the
ratio of major to minor axis diameter at a given isophote, e.g. the
25th magn arsec^{-2}, or from isophotal maps obtained in detailed
surface photometry studies.
- kinematically, by assuming circular motion and solving for pa and i
on the basis of observed radial velocities from either long slit
spectra or from two dimensional mapping with a Fabry-Perot interferometer or with HI line synthesis.
- spiral rectification, assuming usually that spiral arms have a
logarithmic form. This method goes back to Von der Pahlen (1911) and
was extensively used by Danver (1942). Recently, Fourier analysis of
the spiral pattern in e.g. HII regions has attracted some interest
and also yields orientation parameters (e.g. Considère and
Athanassoula, 1982). This method is qualitatively related to Danver's
method.

    If all is well with a spiral galaxy the various methods will
yield the same orientation parameters. If this is not the case it
might be due to several causes like the presence of a bar or oval
distortion in the disk, a strong asymmetry, or the spiral pattern
might not be in the form assumed to do the rectification.

    I have compared the results for the orientation parameters
obtained from HI data with those from spiral rectification. Though
there is general agreement I will call attention to two galaxies for
which there is clear disagreement between the various methods.
a. NGC 4321. The outer spiral arms of this bright Virgo cluster
spiral outline a relatively face-on disk in position angle ~15° (see
e.g. Sandage, 1972). Kinematics of the nuclear region by Van der
Kruit (1973) shows a pa of ~117°, while Rubin et al. (1980) found for
the main body pa ~155° ± 20°, a value confirmed by a recent HI study

using Westerbork by Warmels (1982). Anderson, Hodge and Kennicutt (1982) find from rectification of the principal spiral arms pa ~108°. This galaxy most likely has a fat bar which might account for some of the misalignments indicated by the disagreements between the various estimates of the position angle of the major axis.

b. M33. The photometric study by De Vaucouleurs (1959) gives pa = 23° ± 2 and i = 55° ± 2°. These orientation parameters are corroborated by the HI synthesis studies of Warner et al. (1973), Rogstad et al. (1976) and Newton (1980a). However, Danver (1942) found for the two principal arms pa = 49° and i = 40°. This discrepancy prompted Sandage and Humphries (1980) to reexamine the spiral structure of this galaxy. They delineate 5 pairs of arms; the inner two at the orientation found by Danver and the others with pa of order 15° and i of order 65°. They then suggest that the optical disk of M33 must be severely warped in order to produce such an arrangement. This warping continues smoothly outside the optical disk in the HI plane, where it is inferred from the kinematics as discussed by Rogstad et al. (1976) and Reakes and Newton (1978). Yet it is difficult to see why there is no kinematic signature of the warp in the optical part of M33. None of the above mentioned HI studies have found substantial deviations from axial symmetry there even at the level of 5-10 km/sec. However, the warp proposed by Sandage and Humphries should show clearly its signature in a kinematical major axis which changes its position angle as function of radius in the optical parts of the disk. Since this is not seen I think that the assumption underlying the rectification, i.e. that the arms are described by a logarithmic spiral,

Table 1. Orientation parameters, comparison between Danver & HI results.

Galaxy NGC	Danver's results pa	i	HI results pa	i	(ref)	Δpa (corrected for winding)
300	111.5	39.8	108	55	1	3.5
598	48.9	40.4	23	55	2	-25.9
628	0.4	34.9	15	--	3	14.6
2403	125.0	54.5	125.5	60	4	0.5
2841	147.8	64.0	148	68	5	- 0.2
3031	153.6	59.2	149	59	6	4.6
3198	41.8	72.7	36	70	5	5.8
5033	166.6	58.5	172	63	5	- 5.4
5055	98.9	58.6	98	55	5	0.9
5194	41.7	35.2	- 8	20	7	49.7
6946	52.2	31.1	62	30	8	9.8
7331	165.4	69.1	167	75	5	1.6

references for HI data. 1. Rogstad et al. (1979), 2. Rogstad et al. (1976), 3. Briggs (1982), 4. Shostak (1973), 5. Bosma (1978, 1981a), 6. Visser (1978), 7. Tully (1974), 8. Rogstad et al. (1973).

is incorrect.

In most cases though there is relatively good agreement between the orientation parameters determined by Danver (1942) and those from HI velocity fields, as is shown in table 1 for a dozen galaxies. Most of that table is self explanatory, only in the last column it should be noted that the difference is always taken in the same sense with respect to the winding pattern of the spiral arms. The outstanding deviators are M33, NGC 628, and M51, all very interesting galaxies in their own right.

III. SPIRAL STRUCTURE, ASYMMETRIES

Only for a few galaxies the angular resolution at 21 cm is good enough to enable a detailed study of the motions associated with the spiral arms. The best known case is M81, where the observations by Rots and Shane (1975) show a regular pattern of wiggles, which has been reproduced by Visser (1978) in the context of the Lin-Shu-Roberts density wave theory. Such wiggles are indeed associated with the change in velocity of the gas as it passes through the spiral arm shock region and can be reproduced also by other density wave theories (cf. Toomre, 1981).

There are several hints of these wiggles in other galaxies, but in the best studied cases, M31 and M33, there are no clear patterns as in M81, either because it is not there (in M33, cf. Newton, 1980a) or because the pattern is not due to a simple 2-arm spiral (in M31, see contribution by Brinks in this volume).

It may be expected that the new generation of Fabry-Perot interferometers like TAURUS or the French system will generate a number of pictures in which these motions can be recognized. The preliminary results from TAURUS on NGC 2997 and M83 presented by Atherton and Allen at this meeting are very suggestive in this respect. Note that the axisymmetric rotation curve cannot be derived directly from the mean circular velocity in annuli if strong spiral arm perturbations are present, but that a correction is necessary (cf. Visser, 1978).

Asymmetries occur in nearly every galaxy and it is useful to distinguish several categories:
- HI intensity asymmetries. It occurs very often that one side of a galaxy is stronger in HI intensity (sometimes up to 50% in the arms) than the other side, e.g. M81 inside 12 kpc (Rots and Shane, 1975) and NGC 3198 (Bosma, 1981a).
- Optical asymmetries in the inner parts. Galaxies like M33, NGC 1313 and NGC 4027 are asymmetrical in the central parts in that the nucleus, or the bar in the magellanic barred spirals, is displaced from the centre of the disk. Dynamical models have been constructed for these situations (e.g. Colin and Athanassoula, this volume);

alternatively it has been suggested that stochastic self propagating star formation can also account for the appearance of asymmetries in magellanic systems (cf. Feitzinger et al., 1981).
- Optical asymmetries in the outer parts. Some giant galaxies like M101 show a distinct asymmetry, predominantly in the outer parts. In both M101 and NGC 2805 the HII regions are more dominant on one side of the galaxy and the HI on the other side (cf. Bosma et al., 1980). The velocity field of M101 is very distorted in an asymmetric manner (Rogstad and Shostak, 1971; Bosma et al., 1981).
- HI distribution asymmetries in the outer parts. Sometimes the detectable HI extends very far out at only one side of a galaxy, e.g. in IC342 (Newton, 1980b) or NGC 891 (Sancisi and Allen, 1979); sometimes there are only moderate differences in HI extent and a single extended cloud seems to be responsible for the asymmetry, e.g. NGC 2841 and NGC 5033 (Bosma, 1981a).

Baldwin et al. (1980) reviewed the last two asymmetries and argue that these m = 1 distortions can be relatively long lived since $\kappa - \Omega$ does not vary all that much in the outer parts of the disk. Note that asymmetries do hamper the determination of accurate rotation curves in the asymetric parts of a galaxy, and it is not at all obvious to include the outer points in e.g. NGC 891 in a rotation curve analysis since a large amount of non-circular motion could be present there.

IV. OVAL DISTORTIONS AND WARPS

In a substantial number of spiral galaxies the velocity field deviates from axial symmetry in a systematic bisymmetric pattern. Two kinds of patterns are seen, and in both cases the kinematical major axis (line of extreme radial velocities) changes its position angle as function of radius. If this change is in the inner parts usually there is misalignment with the major axis of some of the structures seen on optical photographs, the HI distribution in the outer parts is confined to the optically visible arm or ring structures and the kinematical major axis is not perpendicular to the kinematical minor axis. If this is the case we suspect an <u>oval distortion</u> to be the principal cause of the bisymmetric deviation from axial symmetry. If the change in the position angle of the kinematical major axis is in the outer parts the major axis of the inner regions is usually well aligned, the HI continues smoothly outwards beyond the optically visible material on sky limited plates and the kinematical minor axis is perpendicular to the kinematical major axis. If this is the case we suspect a <u>warp</u> in the outer disk, and since it is inferred from the kinematics we speak about a kinematical warp. Obviously the distinction is not watertight since both phenomena can occur in the same galaxy, and in a couple of cases this actually happens.

There is now more data around than at the time the above distinction was formulated. It appears necessary to consider various

subdivisions in both oval distortions and warps, basically to isolate
properly the astrophysical problems on which the observations seem to
have a bearing.

   Oval distortions can be distinguished as follows:
- True bars like NGC 5383, NGC 3359 and the large southern ones. In
  these cases the HI data clearly lack angular resolution compared to
  the optical observations. The theoretically expected signature of
  the bar distortion in the velocity field is much more pronounced
  than the observed one. A promising way to get around this is to use
  surface photometry and optical spectroscopy to obtain parameters
  like bar strength etc., and to use the HI data to get a good
  estimate of the axisymmetric mass distribution in the outer parts.
  This approach has been tried for the best studied barred spiral,
  NGC 5383, by Duval and Athanassoula (1982), and can now also be
  done for the other barred spirals which have been observed in the
  HI with moderate angular resolution, such as NGC 3359 (Gottesman,
  1982), NGC 3992 and NGC 4731 (Gottesman and Hunter, this volume),
  NGC 1097 and NGC 1365 (Van der Hulst et al, this volume) and NGC
  1398 (Bosma, in preparation).
- Ovals (i.e. no classical bar visible in a blue photograph but a fat
  bar (oval) instead). Some of the kinematical properties of these
  galaxies are reminiscent of those of barred spirals. Clear examples
  are NGC 4151 (Bosma et al., 1977a), NGC 4258 (Van Albada, 1980) and
  NGC 4736 (Bosma et al., 1977b). Yet no bar but a strong oval can be
  seen in blue photographs. However, it may well be that in the red
  or near infrared one can detect such a bar. From plates of M83
  taken by Ken Freeman one can easily see a steady progression in the
  dominance of the bar when one goes from the ultra-violet via blue
  and visual to the near infrared. Some of the pictures published by
  Elmegreen (1981) also show this.
- Suspected bar/oval systems. In these cases the velocity field has
  only a weak signature of an oval distortion i.e. small residual
  velocities when one subtracts the axisymmetric field based on the
  rotation curve from the observed field. The characteristic pattern
  for a bar distortion can still be noticed, as for example in NGC
  3198 (Bosma, 1981a), but the bar is not very impressive and one has
  a hard time to convince skeptics about its existence. Often I
  suspect that galaxies where the axial ratio changes as function of
  radius, e.g. NGC 300 (De Vaucouleurs and Page, 1962) or IC 342
  (Ables, 1971) have a weak oval in the central parts, but again the
  evidence in velocity data is not clearcut. Nevertheless these
  galaxies have been classified as SAB or SB galaxies by De
  Vaucouleurs (1963).

   Warps can either be seen directly in some edge-on galaxies or
argued on the basis of the kinematics. Kinematical warps are usually
decribed in terms of tilted ring models, which were first introduced
by Rogstad et al. (1974) for the galaxy M 83. Nowadays there are many
more examples of warped HI disks being reported (cf. Bosma, 1981b),
but it comes always as a bit of a surprise when one finds the next

one, like e.g. NGC 628 (Briggs, 1982). Several forms can now be
distinguished:
- the simple integral sign form. Here the tilted ring models show a
  monotonic change in the position angle and inclination of the rings
  as function of radius. Good examples are M 83 (Rogstad et al.,
  1974), M 33 (Rogstad et al., 1976), NGC 300 (Rogstad et al., 1979)
  and NGC 2841 (Bosma, 1981a).
- a more complex shape with the plane crossed twice. The type example
  is the optical warp in NGC 4762 (cf. Sandage, 1961). Other examples
  argued on the basis of the kinematics are NGC 5055 and NGC 7331
  (Bosma, 1981a).
- somewhere in between are the cases where the warp turns back a bit
  but does not cross the plane twice. New data by Sancisi (1982) show
  this to be the case in the direct warps of NGC 5907 and NGC 4565,
  although the effect is predominantly seen at one side.

It should be noted that although the direct warps of NGC 5907
and NGC 4565 do start approximately at the cutoff radius of the
stellar disk this is not a universal phenomenon. Certainly in all the
cases with complex shapes the warp starts well within the Holmberg
radius. Also in the case of IC 342 (Newton 1980 b, c) the spiral
structure extends into the warped region.

An interesting case of a galaxy with oval distortion and a warp
has now been found by Wevers (1982) in his study of NGC 2903. There
most of the HI strongly coincides with the outer arms, leaving gaps
between these arms and the main optical disk quite similar to an oval
galaxy like NGC 4151. But outside the strong outer arms there is
faint HI emission with velocities clearly indicating the pattern of a
kinematical warp. There is structure in this emission in the sense
that one of the tilted annuli is rather pronounced in HI, and it will
be interesting to find out whether this has something to do with the
1:1 resonance as e.g. discussed by Binney (1981).

V. ROTATION CURVES.

A compilation of HI rotation curves has been given in Bosma (1978,
1981 b). These curves go usually much farther out than curves
determined from optical spectroscopy. However the sample of galaxies
with good HI curves is much more heterogeneous than the samples
discussed by Rubin and her collaborators. The curves are usually
flat, also in the outer parts beyond the optical image, but effects
of oval distortions, warping and asymmetries have to be taken into
account. In the absence of succesful dynamical models one has to be
cautious in interpreting the results from e.g. tilted ring models.
Moreover, in the central region of galaxies beamsmearing effects are
important and therefore additional optical spectroscopy is necessary
to obtain the correct rotation curve. For early type disk galaxies,
like Sa's, no HI is detected in the inner parts, and sometimes there
are no HII regions either, hence determination of the circular

velocity there might be very difficult. Absorption line data measure
the mean velocity of the stars, and a full theoretical treatment
involving also the velocity dispersion and the volume density of
matter is necessary to obtain the circular speed. Thus a systematic
study of HI rotation curves and the relationship with morphological
types is rather difficult.

Comparison of HI and Hα rotation curves is very useful since
there is now much more overlap than 10 years ago. A technical problem
is presented by galaxies having inclinations around 65°-80°. These
are probably too inclined in order to use an intensity weighted mean
HI velocity, but not inclined enough to justify use of the terminal
velocity as is done for edge-on galaxies (cf. Sancisi and Allen,
1979). Resolution effects in the HI data play certainly a big role
here. For UGC 2885 the HI data by Roelfsema and Allen (1983) do agree
with the optical data by Rubin et al. (1980) if one ignores the inner
parts and takes the HI peak velocity. For NGC 6503 Shostak et al.
(1979) took the terminal velocity and obtained higher rotational
values than those found later by De Vaucouleurs and Caulet (1982),
although here the overlap in radius range is not that large.
Certainly care has to be taken here, and it is desirable to construct
geometrical models taking beamsmearing into account, which then
should be compared with the observations. The necessity for correcting the rotation curve becomes then obvious, as was the case for
NGC 2841 and NGC 7331 (cf. Bosma, 1981 a).

There is some evidence for falling rotation curves in the far
outer parts of some galaxies, although careful consideration
indicates that this evidence is not as solid as one would wish.
Already the data on NGC 891 presented by Sancisi and Allen (1979)
indicated that the rotation curve in the outer parts beyond the
optical image might drop. However this occurs only in the southern
tail, which has no counter part at the other side. Similar asymmetries occur also in NGC 5907 and NGC 4565, though there the HI does
extend beyond the optical image at both sides of the galaxy. Considerations about fitting a tilted ring model to these warped edge-on
galaxies suggest that their warp is quite complicated and that if a
drop is observed in the rotational velocity it is probably real.
Another case where a drop-off has been found is in the galaxy NGC
5908 (Van Moorsel, 1982); here the drop is symmetric and might well
be real. Since all these galaxies are seen edge-on, however, a couple
of objections remain: 1) it is not clear whether the gas we see is
actually on the line of nodes and 2) it is not clear whether the gas
is in circular orbits. Therefore one has yet to be cautious in
treating these observations as evidence for falling rotation curves.

Finally, a brief comment on the determination of the mass
distribution from a rotation curve. There is much room for ambiguity
here. Even if one takes e.g. a constant M/L for the disk, one still
has the freedom to choose specific disk and halo models which all fit
the handful of independent data points well. As an example for NGC

2841 one can calculate a halo+disk model in which M_{halo} (out to 40 kpc)/M_{disk} ~ 2.8 with a $(M/L)_{disk}$ of 5.8 which fits the observed rotation curve reasonably well. So if one wants to have M_{halo}/M_{disk} = 10 one has to go out to ~ 150 kpc or to reduce $(M/L)_{disk}$. Clearly we need other ways to determine the properties of the massive halo, if it exists, than rotation curves alone.

REFERENCES:

Ables, H.D., 1971, Publ. U.S. Naval Obs. XX, part. IV
Anderson, S., Hodge, P., and Kennicutt, R.C., 1982, Astrophys. J.,
    (in press)
Baldwin, J.E., Lynden-Bell, D., and Sancisi, R., 1980, M.N.R.A.S.,
    193, 313
Binney, J.J., 1981, M.N.R.A.S. 196, 455
Bosma, A., 1978, Ph. D. Thesis, University of Groningen
Bosma, A., 1981a, Astron. J. 86, 1791
Bosma, A., 1981b, Astron. J. 86, 1825
Bosma, A., Ekers, R.D., and Lequeux, J., 1977a, Astron. Astrophys.
    57, 97
Bosma, A., Van der Hulst, J.M., and Sullivan, W.T., 1977b, Astron.
    Astrophys. 57, 373
Bosma, A., and Van der Kruit, P.C., 1979, Astron. Astrophys. 79, 281
Bosma, A., Casini, C., Heidmann, J., Van der Hulst, J.M., and Van
    Woerden, H., 1980, Astron. Astrophys. 89, 345
Bosma, A., Goss, W.M., and Allen, R.J., 1981, Astron. Astrophys. 93,
    106
Briggs, F.H, 1982, Astrophys. J. 259, 544
Considère, S., and Athanassoula, E., 1982, Astron. Astrophys. 111, 28
Danver, C.G., 1942, Ann. Obs. Lund No. 10
De Vaucouleurs, G., 1959, Astrophys. J. 130, 728
De Vaucouleurs, G., 1963, Astrophys. J. Suppl. 8, 31
De Vaucouleurs, G., and Page, J., Astrophys. J. 136, 107
De Vaucouleurs, G., and Caulet, A., 1982, Astrophys. J. Suppl. 49,
    515
Duval, M.F., and Athanassoula, E., 1982, Astron. Astrophys., (in
    press)
Elmegreen, D.M., 1981, Astrophys. J. Suppl. 47, 229
Feitzinger, J.V., Glassgold, A.E., Gerola, H., and Seiden, P.E.,
    1981, Astron. Astrophys. 98, 371
Gottesman, S.T., 1982, Astron. J. 87, 751
Newton, K., 1980a, M.N.R.A.S. 190, 689
Newton, K., 1980b, M.N.R.A.S. 191, 169
Newton, K., 1980c, M.N.R.A.S. 191, 615
Reakes, M.L., and Newton, K., 1978, M.N.R.A.S. 185, 277
Roelfsema, P., and Allen, R.J., 1982, (in preparation)
Rogstad, D.H., and Shostak, G.S., 1971, Astron. Astrophys. 13, 99
Rogstad, D.H., Shostak, G.S., and Rots, A.H., 1973, Astron.
    Astrophys. 22, 111
Rogstad, D.H., Lockhart, I.A., and Wright, M.C.H., 1974, Astrophys.

J. 193, 309

Rogstad, D.H., Wright, M.C.H., and Lockhart, I.A., 1976, Astrophys.
    J. 204, 703

Rogstad, D.H., Crutcher, R.M., and Chu, K., 1979, Astrophys. J. 229,
    509

Rots, A.H., and Shane, W.W., 1975, Astron. Astrophys. 45, 25

Rubin, V.C., Ford, W.K., and Thonnard, N., 1980, Astrophys. J. 238,
    471

Sancisi, R., and Allen, R.J., 1979, Astron. Astrophys. 74, 73

Sancisi, R., 1982, (in preparation)

Sandage, A., 1961, The Hubble Atlas of Galaxies, Carnegie Institution
    of Washington

Sandage, A., 1972, Astrophys. J. 176, 21

Sandage, A., and Humphries, R.M., 1980, Astrophys. J. 236, L1

Shostak, G.S., 1973, Astron. Astrophys. 24, 411

Shostak, G.S., Willis, A.G., and Crane, P.C., 1981, Astron.
    Astrophys. 96, 393

Toomre, A., 1981, in "The structure and evolution of normal
    galaxies", ed. S.M. Fall and D. Lynden-Bell, Cambridge, p. 111

Tully, R.B., 1974, Astrophys. J. Suppl. 27, 415

Van Albada, G.D., 1980, Astron. Astrophys. 90, 123

Van der Kruit, P.C., 1973, Astrophys. J. 186, 807

Van Moorsel, G., 1982, Astron. Astrophys. 107, 66

Visser, H.C.D., 1978, Ph.D. Thesis, University of Groningen

Von der Pahlen, E., 1911, Astron. Nachr. 188, 249

Warmels, R.H., 1982, (in preparation)

Warner, P.J., Wright, M.C.H., and Baldwin, J.E., 1973, M.N.R.A.S.
    163, 163

Wevers, B.M.H.R., 1982, (in preparation)

DISCUSSION:

LINDBLAD: Would Dr. Sandage like to comment on the spiral rectification problem in M33?

SANDAGE : Humphries and I were impressed by the change in position
angle of the ten optical arms as one goes outward. Your HI kinematic
position angles seem to agree with this optical change if you add the
far outer HI positional data of the Cambridge deep map (Newton et.
al.) which continues the change of position angle beyond the outer
optical arms. These, together with the double peaked HI profiles in
the outer HI disk still suggest to us that a severe warp exists in
the plane of the arm system as well as the HI disk. But the position
angle of the old disk remains the one found by de Vaucouleurs, also
according to the new work of the Lyon group.

BOSMA: I do not agree that the HI kinematical major axis follows the
warp proposed on the basis of the rectification of the arms. There is
a difference of at least 20° in position angle between the major axis

you propose and the one found in the Cambridge and Owens Valley 21-cm velocity fields in the inner parts of the optical disk. Such a situation seems to me irreconcilable, and therefore I'm inclined to abandon the underlying assumption you make i.e. that all the spiral arms in M33 are logarithmic. Of course I'm well aware of the warp in the HI layer farther out in M33 - beyond the optical image mainly - but the issue here is whether there is a severe warp in the inner 2 kpc where the restoring forces of the disk are quite strong.

BINNEY: Is it not true that in the standard Rogstad et al. ring models of kinematic warps, each ring rotates in its own plane? In reality the rings will wobble or precess and this contributes substantially to a kinematic warp. Thus we probably should not take particular ring models very seriously until they include more dynamics.

BOSMA: Yes, in principle you are correct. In practice the tilted ring models serve as a convenient geometrical description of the observations. Sometimes rotation curves based on tilted ring models are used e.g. to construct mass models, but one always has to be aware of the fact that non-circular motions are present in a warped disk.

TOHLINE: Would you comment on what fraction of HI disks show strong $m=1$ asymmetries, that is, how many disks show substantially more HI gas on one side of the galaxy's center than on the other?

BOSMA: There are not too many galaxies with strong $m=1$ asymmetries in the HI distribution. The paper of Baldwin et al. (see ref. in text) mentions 4 cases. There are several more with minor $m=1$ asymmetries, but I think the total fraction of strong asymmetries does not exceed 10-20% in the case of big spirals.

FEITZINGER: A displacement is often observed in late type galaxies between bar center and rotation center or between optical center and bar center (Feitzinger 1980, Space Science Rev. 27, 35). This may be also a reason for distortions. Do you have some detailed information on this type of distortions in your galaxy sample?

BOSMA: No, not much has been done on these, although there is some unpublished Westerbork work. The good examples are, as always, in the south (NGC 1313, NGC 4027, Magellanic Clouds).

RUBIN: A few years ago, I noticed that all galaxies with dynamical warps, those in your thesis and those published elsewhere were warped in the sense that with increasing radius the inclination became more edge-on. Does this curious circumstance still hold? Is it an observational artifact?

BOSMA: Yes, most of them become indeed more edge-on, but Van der Kruit tells me that NGC 1058 (pure face-on) is warped and there may now be more cases yet to be published. A given ring with a certain

column density in the main plane is more easily detected when it's
more edge-on but I think we'll sooner or later find the ones where
the outer rings become less inclined. If the edge-on's follow the
tilted ring model they become more face-on at increasing radius.

KENNICUTT: Do you ever find galaxies that do not possess warps or
distortions in HI?

BOSMA: There are still some regular galaxies. Roelfsema and Allen (in
prep.) mapped the large Sc UGC 2885 and find it very regular. Also
NGC 3198 (Bosma 1981a) is regular, with deviations from axial symmetry at the 5-10% level.

BAJAJA: The results of high resolution HI observation of M31 clearly
show the presence of warps in this galaxy in the form of morphological distortions and overlapping of features along the line of
sight. The consequence of these projection effects is the presence of
more than one velocity component in almost every point of the galaxy
One has to be careful then in deriving the rotation curves and the
velocity field from the mean velocities in the cases of galaxies with
high inclinations and probable warping. In the particular case of M31
we noticed (Bajaja and Shane, Astron. Astrophys. Suppl. 49, 745) that
there is large difference between the published rotation curves as
derived from optical and from radio observations, the latter being
much flatter than the optical. I would like to ask V. Rubin whether
this situation has been modified specially in view of the results for
Sb galaxies mentioned in her talk this morning?

RUBIN: There are several points to note. In the case of M31, our
optical observations are of HII regions very near the major axis,
with a very large range of nuclear distances. Thus the outer parts of
the optical rotation curve come from HII regions well beyond those
observed at 21-cm by you and Shane. For your 21-cm rotation curve,
velocities for regions well off the major axis, closer to the nucleus
on the plane of the sky, are deprojected to give the rotational
velocities for large R. This deprojection can be treacherous, if non-
circular velocities, warps, or other complex phenomenon are
operating. The most meaningful comparison of optical and 21-cm
observations will come when velocities can be compared region by
region in the observed domain rather than in the projected domain.
    For galaxies farther away, and hence of smaller angular size
than M31, a similar problem can arise if the galaxy is viewed at
relatively high inclination. If the angular size of the galaxy is on
the order of 10 times the size of the 21-cm beam, then observations
along the major axis may not suffer too severely from the effects of
beam smearing. But for regions well off the major axis, beam smearing
becomes a problem, for the velocity field is changing rapidly within
a single beam. Hence if velocities from these fields are included in
the derivation of the rotation curve, I would expect significant
differences from those observed optically along the major axis of the
galaxy.

A HIGH RESOLUTION HI SURVEY OF M31

E. Brinks
Leiden Observatory, Leiden, The Netherlands

1. INTRODUCTION

The first results of a new high resolution 21-cm HI line survey of M31 made with the Westerbork Synthesis Radio Telescope are presented. Five areas were mapped, covering the galaxy except for the extreme northern and southern parts, at a resolution of $\Delta\alpha \times \Delta\delta \times \Delta V = 24" \times 36" \times 8.2$ km s^{-1}. The spatial resolution corresponds to 30 x 120 pc at the distance of M31. This is of the same order as the resolution at the distance of the center or our own galaxy given by a 25-m dish. Consequently the M31 survey is comparable to surveys of the Milky Way galaxy in wealth of detail as well as in amount of data ($\sim$ 1 Gigabyte).

2. RESULTS

The presence of local galactic foreground HI in the observations and calibrations was accounted for during the data reduction. The continuum background was removed by subtracting a separate 21-cm broadband continuum survey from the line survey (Walterbos, Brinks, and Shane, in preparation). The interferometer spacings of 0-m and 18-m were Fourier filtered from the survey of M31 made by Cram, Roberts, and Whitehurst (1980) using the 100-m Effelsberg telescope. Finally, the five survey areas were combined to give in total 147 channel maps each covering M31 at 4.1 km s^{-1} velocity intervals. Figure 1 shows the total HI surface density map at full resolution. The noise varies across the map and increases rapidly at the extreme northern and southern ends. The HI emission is concentrated in a ring at about 10 kpc. This main ring coincides with the ring of radio continuum radiation, of HII-regions, OB-associations, etc. The HI is also strongly correlated with dust. Due to its unfavourable inclination, no firm statements can as yet be made about the spiral structure of M31, although it is possible to trace over several kpc arm segments.

Figure 2 shows isovelocity contours based on the *mean* velocities superimposed on the total surface density map. Because of the presence of two or more velocity components in the line of sight, this mean velocity map should be interpreted cautiously. This map shows simply the HI-intensity

23

E. Athanassoula (ed.), Internal Kinematics and Dynamics of Galaxies, 23–26.

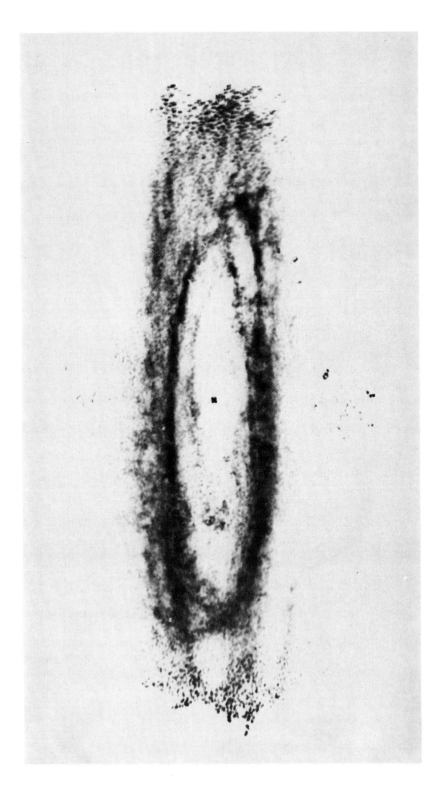

Figure 1: Total HI surface density map at full resolution. The cross indicates the nucleus. North is to the left.

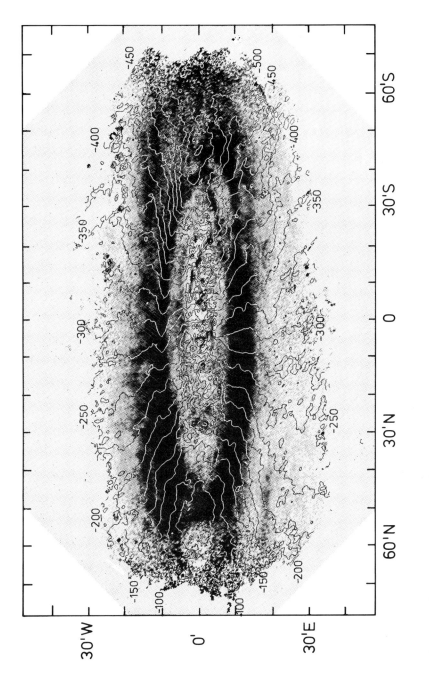

Figure 2: Total HI surface density map at full resolution with the mean velocity field, obtained after smoothing the data to twice the beam, superimposed. Iso-velocity contours are drawn at 25 km s^{-1} intervals.

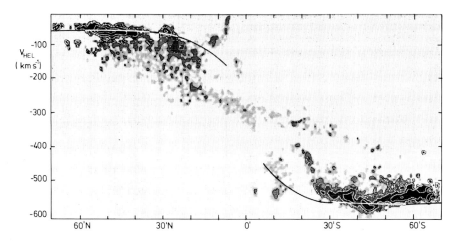

Figure 3: Position-velocity map along the major axis. Contour levels are
2.5, 5, 10 and 25 K. The thick line is based on the mean rotation curve
used by Bajaja and Shane (1982).

weighted mean velocity at each position of M31, after smoothing to twice
the original beamsize. The isovelocity contours show the signature of a
differentially rotating disk, although there are marked deviations from
pure circular rotation. Indications of streaming motion in the gas are
present on both the near and the far, i.e. South East, side of the galaxy.
The strong deviations on the North-West side, corresponding to the dust arm
at a distance of 5 kpc from the nucleus, agree with the findings of Bajaja
and Shane (1982) based on previous, rather incomplete, observations of the
central area.

     Figure 3 shows a position velocity map made along the major axis of
M31. The map gives a first order approximation of the rotation curve. The
thick line is a rotation curve based on previous HI data. The map shows
that the kinematics of M31 is far from simple. At least three subsystems
can be recognized in this and similar plots. In addition to the differentially rotating disk, which follows roughly the drawn line, there is a
second component present across the whole galaxy. This component shows up
as an almost straight line running diagonally across the map. It is attribu
ted to warping of the outer part of M31 into the line of sight. The third
subsystem is restricted to the central 4-5 kpc, where velocities as high as
200 km s^{-1} with respect to systemic velocity are found. This inner part of
M31 is discussed in a separate contribution in this volume.

REFERENCES

Bajaja, E., Shane, W.W.: 1982, *Astron. Astrophys. Suppl. Ser. 49*, 745
Cram, T.R., Roberts, M.S., Whitehurst, R.N.: 1980, *Astron. Astrophys.
     Suppl. Ser. 40*, 215
Walterbos, R.A.M., Brinks, E., Shane, W.W.: in preparation

HIGH VELOCITY HI IN THE INNER 5 KPC OF M31

E. Brinks
Leiden Observatory, Leiden, The Netherlands.

1. INTRODUCTION

New observations made with the Westerbork Synthesis Radio Telescope of M31 in the line of neutral hydrogen (Brinks, this volume) show the presence of high-velocity HI in the inner 4-5 kpc of the galaxy. There were indications from previous 21-cm line studies that HI gas is moving with peculiar velocities (Whitehurst and Roberts, 1972; Unwin, 1979) but the new maps, made with a resolution of $\Delta\alpha \times \Delta\delta \times \Delta V = 24" \times 36" \times 8.2$ km s^{-1} and sufficient sensitivity show for the first time the distribution of this material.

2. RESULTS

Figure 1 shows three position-velocity maps made parallel to the minor axis at +6', 0' and -6' distance from the nucleus. These maps were constructed after smoothing to twice the original beamsize. For all three cross-cuts, we would expect the HI to be distributed more or less in a narrow band around systemic velocity if the gas clouds were moving around the nucleus in circular orbits in the plane of the galaxy. Figure 1 shows that this is clearly not the case. At 1.2 kpc North of the nucleus, roughly on the major axis, HI gas with a radial velocity exceeding 250 km s^{-1} with respect to systemic velocity is found (see figure 1a). Figure 1c shows that at the same distance South of the nucleus the velocity is about 200 km s^{-1} lower than expected from the circular rotation situation. On the minor axis (figure 1b) there is a steep velocity gradient on both sides of the nucleus. The feature labeled "a" lies outside the central region at about 5 kpc and coincides with the prominent dust arm North-East of the nucleus. This arm has already been discussed by Bajaja and Shane (1982) and is also detected in CO by Stark (1979).

3. DISCUSSION

The high-velocity HI features inside 4-5 kpc show the same signature as the high velocities measured in the optical by Rubin and Ford (1970,

E. Athanassoula (ed.), Internal Kinematics and Dynamics of Galaxies, 27–28.

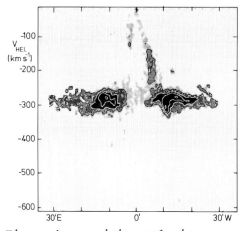

Figure 1a: position-velocity map, 6'N of the minor axis. Contour levels are 2.5, 5, 10 and 25 K.

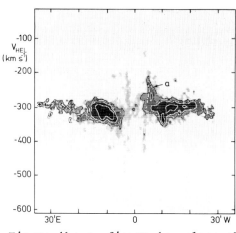

Figure 1b: as figure 1a, along the minor axis.

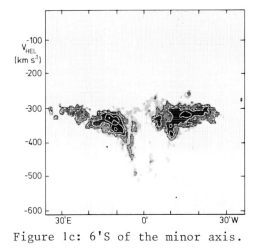

Figure 1c: 6'S of the minor axis.

1971). We have not detected any HI inside 500 pc corresponding with their measurements, perhaps because gas inside 500 pc is mostly in ionised form. The position-velocity maps support the idea that the high velocities are due to rotation rather than produced by infall of material or ejection from the nucleus. This idea is strengthened by the fact that the central radio source in M31 is intrinsically weak and shows no evidence for recent activity. From figure 1 it is clear that the gas does not move in circular orbits in the plane of the galaxy. It has been shown that the isophotes of the bulge show a gradual change in position angle and ellipticity (Matsumoto et al, 1977; Lindblad, 1956), so it is quite possible that the gas moves along elliptical streamlines, perhaps even tilted out of the plane, much like the distribution proposed for the central region of our own galaxy.

REFERENCES

Bajaja, E., Shane, W.W.: 1982, *Astron. Astrophys. Suppl. Ser. 49*, 745
Lindblad, B.: 1956, *Stockholms Obs. Ann. 19*, No. 2
Matsumoto, T., Murakami, H., Hamajima, K.: 1977, *Publ. Astron. Soc. Japan, 29*, 583
Rubin, V.C., Ford, W.K.: 1970, *Astrophys. J. 159*, 379
Rubin, V.C., Ford, W.K.: 1971, *Astrophys. J. 170*, 25
Stark, A.A.: 1979, Ph. D. Thesis, Princeton University
Unwin, S.C.: 1979, Ph. D. Thesis, Cambridge University
Whitehurst, R.N., Roberts, M.S.: 1972, *Astrophys. J. 175*, 347

COMPARISON OF GLOBAL 21 cm VELOCITY PROFILES WITH Hα ROTATION CURVES

Norbert Thonnard
Department of Terrestrial Magnetism
Carnegie Institution of Washington

ABSTRACT. 21 cm profiles are compared with high resolution long slit Hα spectra. Asymmetries in intensity and velocity gradient in the spectra are apparent in the global profiles. When measured at 0.5 of peak intensity, the 21cm profile width and optical rotation curve match best.

High resolution long slit spectra in the Hα region are available for 60 field spiral galaxies of Hubble types Sc, Sb, and Sa (Rubin, Ford and Thonnard, 1980; Rubin et al., 1982, 1983). We also have global 21 cm neutral hydrogen profiles for approximately 3/4 of these galaxies (Thonnard et al., 1983).

Ever since the discovery of the correlation between luminosity and 21 cm profile width, ΔV_{21} (Tully and Fisher, 1977), ΔV_{21} has become an important parameter in characterizing galaxy properties. Unfortunately, observers use various methods to determine ΔV_{21}. Since the global HI profile is a convolution of the HI distribution and the velocity field over the entire galaxy, while the optical rotation curve is derived from a narrow ($\sim 1\overset{''}{.}5$) sample along the major axis, an analysis procedure giving the same dynamical information would be very useful.

Observations of face-on galaxies with very steep profile edges, ones most affected by instrumental resolution, show that for resolutions ranging from 20 to 2.5 km s^{-1}, ΔV_{21} is independent of resolution if measured at the 0.50 to 0.75 peak intensity level. Computer models indicate that gently rising rotation curves (typical of low luminosity galaxies) generate 21 cm profiles whose shape is quite sensitive to the HI distribution, whereas rotation curves that are essentially flat over most of the radius (typical of high luminosity galaxies) generate profiles that are independent of HI distribution. For a wide range of rotation curve shapes, inclinations, resolution (or velocity dispersion) and HI distribution, the measurement level at which the model profile width matches the input maximum rotational velocity ranged between 0.43 and 0.58, except for perfectly flat rotation curves, where the level was 0.8 (which drops to 0.5 with only a 5% positive velocity gradient).

29

E. Athanassoula (ed.), Internal Kinematics and Dynamics of Galaxies, 29–30.

The examples shown in Fig. 1 illustrate some of the points mentioned above. NGC 4605, lowest luminosity Sc studied, has a shallow velocity gradient across its disk and a centrally peaked profile. UGC 2885, highest luminosity Sc, having nearly constant rotational velocity over a large radius range, has sharp horns and a deep central minimum in its profile. Note that in both cases, the side in which Hα emission extends to larger radii is the one with more HI. NGC 1087 and 1620 are average luminosity galaxies with similar rotation curves but radically different profiles. A strong central concentration of HI in N 1087 and a lack of central HI in N 1620 could account for this. Also, in N 1087, the steeper velocity gradient seen at large radii on the high velocity side manifests itself as a shallower gradient on the corresponding HI profile edge. NGC 7606 is one of the few galaxies with decreasing rotational velocities in the outer regions. We note the excellent agreement between the maximum optical rotational velocities and velocities at the half power points of the HI profiles in these five extreme cases. For 42 galaxies, the difference in systemic velocity, $V_{opt} - V_{21} = -1.5 \pm 3.9$ km s^{-1} (m.e.).

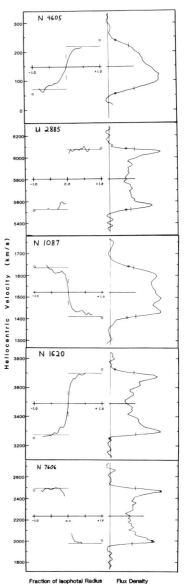

Fig. 1. Comparison of major axis Hα velocities (left) with the HI profile (right). Horizontal lines ranging from −1.0 to 0.0 and 0.0 to +1.0 indicate velocities of the half power points (short vertical lines) on the HI profile. Open circles indicate the extrapolated rotational velocity at R_{25}. The horizontal line going through center of optical and radio data indicates the systemic velocity.

REFERENCES

Rubin, V.C., Ford, W.K., Jr., and Thonnard, N.: 1980, Ap. J., 283, 471.
Rubin, V.C., Ford, W.K., Jr., and Thonnard, N.: 1983 (in preparation).
Rubin, V.C., Ford, W.K., Jr., Thonnard, N., and Burstein, D.: 1982,
        Ap. J., p.261 (in press).
Thonnard, N., Roberts, M.S., Ford, W.K., Jr., and Rubin, V.C.: 1983 (in
        preparation).
Tully, R.B. and Fisher, J.R.: 1977, Astron. Astrophys. 54, p.661.

TAURUS - A Wide Field Imaging Fabry-Perot Spectrometer

P. D. Atherton* and K. Taylor†
*Imperial College, London
†AAO, Sydney and RGO, Herstmonceux, Sussex.

TAURUS, an imaging Fabry-Perot system, was developed as a collaborative project by the Royal Greenwich Observatory and Imperial College London and is capable of obtaining seeing-limited velocity field information over a 9 arcminute field on a 4 m telescope. At the detector (the IPCS) the image of the source is modified by the fringe pattern of the capacitatively stabilized servo-controlled Fabry-Perot. As the Fabry-Perot is scanned, this fringe pattern tracks radially across the field and each pixel of the detector maps out a spectral line profile within the bandpass of the "blocking" interference filter. At each F-P spacing a picture (Typically 256 x 256 pixels) is recorded on computer disk. 100 pictures make up a complete scan, covering for example 1,200 km sec^{-1} free spectral range (this range in practice depends on the particular etalon). In this manner a 3 D data cube of the field is built up where (X, Y, Z) are typically (256 pixels, 256 pixels, 100 steps).

If we plot the intensities recorded by 1 pixel through the scan range, we see the spectral line profile for that part of the object being observed. Because the light from different parts of the field passes through the etalon at different angles, there is a shift in the wavelength zero-point as a function of field angle. Because of this the (X, Y) pictures of raw data are not monochromatic. This positional dependence of wavelength is calibrated and then corrected for by shifting each profile in the computer so as to line up the zero point at all (X, Y) positions - a process known as phase correction. The phase corrected data cube has X, Y as spatial dimensions and Z as a wavelength dimension. These data are then analysed by viewing X, λ or Y, λ plots or as a "movie" of a rapid sequence of (X, Y) pictures as a function of λ, enabling us to view different parts of the objects at different velocities. By summing all the pictures in λ we may produce the equivalent narrow band inter-ference filter pictures, as if the Fabry-Perot were not in the system. By summing in X or Y we may produce an X, λ or Y, λ plot

E. Athanassoula (ed.), Internal Kinematics and Dynamics of Galaxies, 31–32.

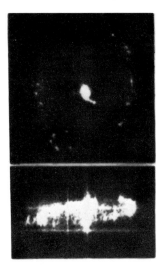

Fig. 1 (upper) is an image of the barred spiral galaxy NGC 1365 in Hα, resulting from summing the data cube in the z or λ direction. Fig. 1 (lower) is a position velocity plot obtained by summing all the phase corrected data in the Y direction and shows velocity as a function of the X coordinate (in this size R.A.) over the whole galaxy. The vertical straight lines represent the continua from stars in the field whilst horizontal streaks are night sky features. The ellipsoidal structure traces the regular large scale motion of the HII regions in this galaxy.

similar to the display of 2D long slit spectra. In these, night sky lines should appear straight across the field, indicating the accuracy of the phase-correction process.

The data may be analyzed to produce velocity maps and line width maps at the seeing-limit using techniques similar to those employed when reducing channel maps obtained through HI radio synthesis observations.

This technique is superior to grating spectrographs for obtaining velocity field information from extended emission line sources (such as galaxies, supernova remnants, planetary nebulae, HII regions etc.) because the information is concentrated in only a small part of the spectrum but over a wide field. Whilst a grating spectrograph looks at all the spectral elements all the time across the field defined by a narrow slit most of these elements contain no information concerning the velocity field. In TAURUS we have traded the spectral multiplicity for a spatial multiplicity and obtain spectral elements sequentially over a small wavelength region centred on the emission line of interest. This method is typically 10 to 20 times faster than grating spectroscopy when studying large objects (> 1' diameter).

References

Taylor K., Atherton P.D., MNRAS 191, 675, 1980.
Atherton P.D., Taylor K., Pike C.D., Harmer C., Parker N.M.,
     Hook R.N., MNRAS, 200, 1982.

HI OBSERVATIONS OF THE IRREGULAR GALAXY IC 10

G.S. Shostak and H. van Woerden
Kapteyn Astronomical Institute
University of Groningen, The Netherlands

IC 10 is seen optically as a patchy nebulosity, approximately 3 x 4 arcmin in size, possibly heavily obscured due to its low (-3°) Galactic latitude. De Vaucouleurs and Freeman (1972) have classified it as a Magellanic barred irregular. With a heliocentric radial velocity of -350 km/sec, its distance is uncertain, and various investigators have placed it as far away as 3, and as near as 1 Mpc (Bottinelli et al., 1972; Roberts, 1962). For the latter value, IC 10 would be a Local Group member.

Recent single-dish HI studies (Cohen, 1979; Huchtmeier, 1979) have shown IC 10 to have a hydrogen envelope ∿70 arcmin in size, or more than 20 times the optical diameter. The linear size of the envelope is 20 kpc for a distance of 1 Mpc. The envelope has an overall velocity gradient from -300 to -360 km/sec, south to north, although substantial irregularities are present.

We have made radio synthesis observations of IC 10 with resolutions of 30 arcsec and 8 km/sec in the neutral hydrogen line using the Westerbork telescope. These confirm Shostak's (1974) result that, in the central region of IC 10, the velocity gradient is opposite to that later measured by single-dish in the outer regions. The suggestion by Cohen (1979) that the velocity gradient reversal is due to IC 10 being nearly face-on and warped is consistent with the new data. Note, however, the following:

- The reversal in velocity gradient, and therefore the presumed warp, begins to the south within 5 arcmin (1.5 Δ kpc, where Δ is the distance to IC 10 in Mpc) of the dynamic center, or at only 15% of the total radius measured in HI.

- The central velocity field is consistent with a rotation having a maximum amplitude 20/sin i km/sec. For i = 15°, this is 77 km/sec, or typical of the rotation speed measured for irregulars of moderate luminosity.

- For reasonable assumptions about the shape of the rotation curve in the outer regions and the galaxy's inclination, the required warp

33

E. Athanassoula (ed.), Internal Kinematics and Dynamics of Galaxies, 33–34.
Copyright © 1983 by the IAU.

along the major axis is generally mild, with a maximum of $<35°$ from
the central plane. However, there are clearly strong azimuthal
asymmetries, and in the extreme south a warp angle of $\sim90°$ is indicated.

- The velocity dispersion in the central regions is 8 - 10 km/sec,
justifying the assumption of a thin HI disk.

Because of the improved resolution and sensitivity of the present
observations, they have revealed considerable detail in the central,
patchy, high surface brightness HI. This consists primarily of

- a roughly elliptical region with column density $N_H > 6 \ 10^{20}$ cm^{-2},
with a position angle of $\sim50°$ and dimensions 7 x 10 arcmin.

- a "tail" extending ~10 arcmin south, with $N_H \sim 1.5 \ 10^{20}$ cm^{-2}.

- a "spur" to the northwest, about 5 arcmin long, and similar in
brightness to the tail.

The main body of the bright HI is rent with clearly defined holes.
The most conspicuous of these have half-intensity diameters ~1 arcmin
(0.3 Δ kpc), and imply, on the basis of the observed HI deficit, the
removal of $\sim5 \ 10^5 \ \Delta^2$ M$_\odot$ of HI. At least one shows a double profile
suggesting radial motions of ~20 km/sec. Assuming this to be a simple
expansion, the energy required to remove HI from the hole is 2 $10^{51} \ \Delta^2$
ergs, or similar to or larger than that in massive supernovae. The time
to evacuate the hole is $\sim10^7 \ \Delta$ yr which, because of the small rotation
velocity of IC 10, is less than,or comparable to,the time scale for
smearing of such features by differential rotation, thus making such
features visible.

Acknowledgement

The Westerbork Observatory is operated by the Netherlands Foundation for Radio Astronomy, with financial support from ZWO.

References

Bottinelli, L., Gouguenheim, L., Heidmann, J. 1972, Astron.Astrophys.
    18, 121
Cohen, R.J. 1979, Mon. Not. R.A.S. 187, 839
Huchtmeier, W.K. 1979, Astron. Astrophys. 75, 170
Roberts, M.S. 1962, Ap.J. 67, 431
Shostak, G.S. 1974, Astron. Astrophys. 31, 97
De Vaucouleurs, G., Freeman, K.C. 1972, Vistas in Astronomy, Vol. 14,
    Pergamon Press, Oxford, p. 163

THE DISTRIBUTION OF MOLECULAR CLOUDS IN SPIRAL GALAXIES

P. M. Solomon,
Astronomy Program, State University of New York
Stony Brook, N.Y. 11794, USA

The use of millimeter wave CO emission as a tracer of molecular hydrogen in the Galaxy (Scoville and Solomon 1975) showed that most of the H_2 unlike HI is concentrated in the inner part of the Galaxy in a "ring" between 4-8 kpc and in the inner 1 kpc. Subsequent surveys (Gordon and Burton 1976, Cohen and Thaddeus 1977, Solomon et al. 1979) confirmed this picture with more extensive data. The molecular interstellar medium was shown to be dominated by Giant Molecular Clouds with individual masses between 10^5 and $3 \cdot 10^6 M_\odot$ (Solomon et al. 1979, Solomon and Sanders 1980). The GMC's confined to a layer with a half thickness of only 60 pc are an important component of the galactic disk, and the most massive objects in the galaxy. They affect the dynamics of the disk by contributing significantly to the surface density and through their individual gravitational interactions with stars.

During the past two years significant progress has been made in measuring the CO emission from the disks of external galaxies. In this review I will concentrate on those galaxies which have been mapped sufficiently to determine the radial distribution of CO emission. For several of these systems, Sc and Scd galaxies, it has been shown that the average CO emission is a monotonically decreasing function of radius with a strong correlation between the CO integrated intensity $I = \int T_A dv$ and the blue surface brightness.

M101 (N5457)

M101 has the largest angular extent of any face on Sc I galaxy. We use this as an example. Solomon et al. (1982 a) obtained observations of CO emission at $\lambda = 2.6$ mm, from 40 locations in M101 and compared the radial dependence of CO surface emission with the optical emission and the HI distribution (Schweizer 1976, Allen and Goss, 1979). Figure 1 shows a spatial velocity map of CO emission near the major axis. The strong emission 2' south of the center is at a spiral arm crossing. Figure 2 shows the CO integrated intensity I, as a function of galactocentric radius in M101. All of the strongest emission was from the central 6 kpc although weak emission was found in one of the points at

E. Athanassoula (ed.), Internal Kinematics and Dynamics of Galaxies, 35–42.

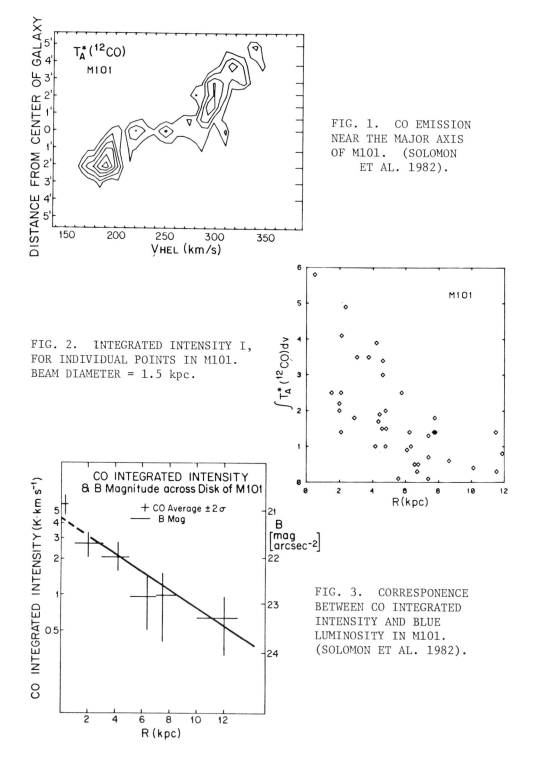

FIG. 1.   CO EMISSION
NEAR THE MAJOR AXIS
OF M101. (SOLOMON
    ET AL. 1982).

FIG. 2.   INTEGRATED INTENSITY I,
FOR INDIVIDUAL POINTS IN M101.
BEAM DIAMETER = 1.5 kpc.

FIG. 3.   CORRESPONENCE
BETWEEN CO INTEGRATED
INTENSITY AND BLUE
LUMINOSITY IN M101.
(SOLOMON ET AL. 1982).

12 kpc. The general trend of increasing I with decreasing R is obvious.
There is, however, a great deal of real scatter which shows a contrast
as high as a factor of four between points at the same R. Of the 40
positions observed, 25 show positive CO detections, 21 of these are
stronger than CO observed at an H II region in the outer part of M101
by Blitz et al. (1982). Clearly the outer regions of M101 are weak in
CO emission.

Figure 3 shows the average CO integrated intensity as a function
of radius on a semi-log plot employing all 40 observations. The mean
integrated intensity shows a clear decrease with radius by about a factor of eight from the center to 12 kiloparsecs radius. Since the central
point may be affected by peculiar activity associated with the nuclear
region, the distribution of CO emission may be better indicated by the
points between 2 and 12 kpc; over this range the mean CO integrated intensity declines by a factor of four. This systematic decline is completely unlike the 21 cm emission. The solid line in Figure 3 shows the
blue surface brightness normalized to the CO integrated intensity at R
= 4 kpc. There is a strong correlation between the CO emission and the
exponential falloff of the optical disk indicated by the blue surface
brightness first demonstrated for spirals by Freeman.   The scale
length of the CO emission is the same as the optical emission to within
the errors of the determination. From Figure 4 which shows HI and H_2
surface densities, it is clear that in the inner part of M101 (R<6 kpc)
the interstellar medium is predominantly in the form of molecular hydrogen. The surface density of molecular hydrogen has been determined from
the CO surface brightness using the expression

$$N_{H_2} = 3.6 \times 10^{20} \int T_A^* \, dv \quad (cm^{-2})$$

The predominance of H_2 will remain true even if the conversion
factor were decreased by a factor of 4 or more. Thus these CO observations show that there is no deficiency of interstellar matter in the inner disk as suggested by HI measurements.

IC 342 and NGC 6946

Young and Scoville (1982a) have convincingly demonstrated the correspondence between blue luminosity and CO integrated intensity for the
two Scd galaxies IC 342 and NGC 6946. They show a monotonic decrease of
CO integrated intensity across the disks of these galaxies along both
the major and minor axis as can be seen for IC 342 in Fig. 5. A different interpretation based on independent data has been given for these
galaxies by Rickard and Palmer (1981) who state that a nuclear source
plus a flat disk will fit the radial distribution equally well. Actually a close inspection of the IC 342 data does show a flat stretch between 3 and 6 kpc but the CO data of Young and Scoville which goes out
to 9 kpc correlates very well with the optical emission. Young and Scoville find a factor of 20 times more surface density in H_2 than HI in
the center of N 6946 (see Fig. 6) with H_2 dominating the interstellar
medium inside of 10 kpc.

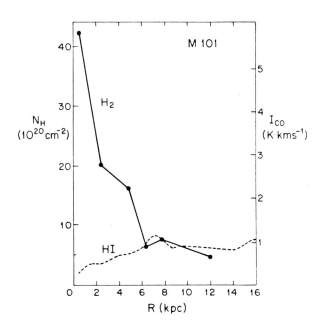

FIG. 4.   COMPARISON OF H_2 AND HI SURFACE DENSITIES
IN M101 (SOLOMON ET AL. 1982)

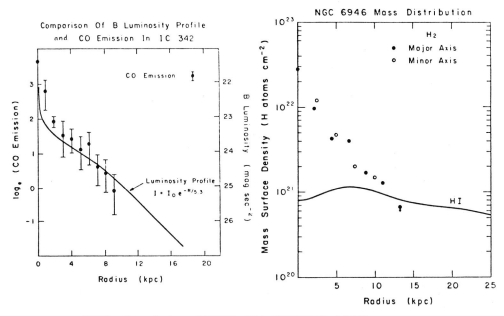

FIGS. 5 and 6.   (YOUNG AND SCOVILLE 1982)

M51

M51 has been observed by Rickard and Palmer (1981) and Scoville and Young (1982). Scoville and Young have shown that the CO emission, Far I.R. and radio continuum all correlate strongly, with a scale length of 3.9 kpc. This galaxy has 3/4 of all interstellar matter in H_2. From Fig. 7 the HI appears to be a very minor constituent in the inner few kpc.

Sc and Scd systems

The four galaxies discussed above are all Sc I or Scd I systems. They have in common a linear relationship between CO emission and optical surface brightness. This has been interpreted by Young and Scoville (1982) and Solomon et al. (1982) as indicating that the star formation rate is proportional to the first, rather than the second, power of the average molecular density. This generalization must be regarded as preliminary in view of the small sample. In N628 (Solomon 1983) there is some indication that CO falls off more slowly than the optical light.

NGC 891

N891 is an edge on (i = 88°) Sb I galaxy which is often compared to the Milky Way. The CO emission across the disk has been measured by Solomon et al. (1982 The large line of sight through the disk enables the observer to determine the radial distribution of emission from relatively strong signals by assuming an axisymmetrick disk. The average CO emission as a function of radius from the galaxy, NGC 891, can be determined out to 15 kpc by 11 observations spaced every 0.75 (3 kpc if D = 14 Mpc) along the major axis. These CO observations are approximately analgous to observing our Galaxy with a very small (~ 1 cm) antenna or just a feed horn. The difference between the HI and H_2 distribution is given in Fig. 8 which depicts the integrated intensity of CO, HI (Sancisi and Allen 1979) and nonthermal radio continuum as a function of distance along the major axis. The half power projected radius of the CO emission is only 2.5 corresponding to less than half of the size of the optical disk while the HI extends out to 4.5 to its half power points. The CO emission has a half power size close to that of the nonthermal radio continuum, and both are concentrated in the inner galaxy. The central dip in Fig. 8 in HI and possibly CO is effected by overlapping clouds at the same velocity along the line of site.

The total mass of molecular hydrogen, H_2, in N891 inferred from the total integrated intensity across the disk assuming D = 14 Mpc is $M(H_2) = 7 \times 10^9$ $M_{\odot}$ Sancisi and Allen (1979) find for total atomic H mass, $M(HI) = 8 \cdot 10^9 M_{\odot}$. Thus approximately one half of all ISM in N891 is in the form of H_2 but the two components have completely different radial distributions. The inner half of N891 has an ISM dominated by H_2 and the outer half dominated by HI.

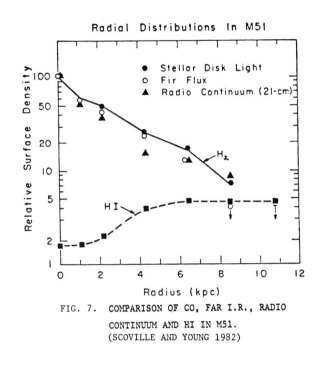

FIG. 7.   COMPARISON OF CO, FAR I.R., RADIO

CONTINUUM AND HI IN M51.

(SCOVILLE AND YOUNG 1982)

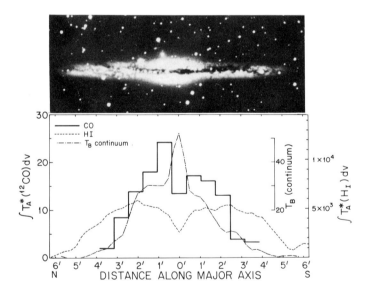

FIG 8.   COMPARISON OF CO, HI AND RADIO CONTINUUM
IN N891 (SOLOMON ET AL. 1982).

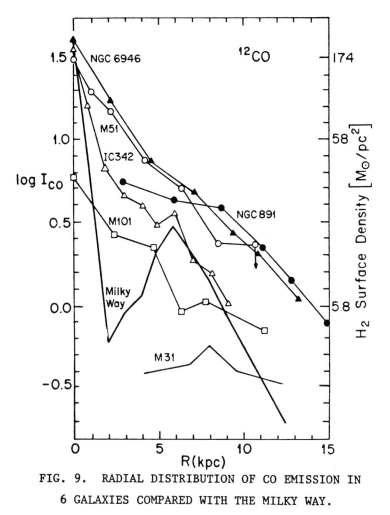

FIG. 9.   RADIAL DISTRIBUTION OF CO EMISSION IN
6 GALAXIES COMPARED WITH THE MILKY WAY.

M31

M31 has been mapped in CO emission by Stark (1980) and by Linke
(1982) who reports a very close correspondence between the HI "arm" and
CO emission. However in sharp contrast to all of the other galaxies
discussed here, M31 is a very weak CO emitter, almost an order of magnitude lower in I(CO) than the Galaxy. The CO emission appears to be confined to the region of optical obscuration where HI emission peaks with
extremely weak CO emission in the central 6 kpc. Thus there is no
region in M31 corresponding to the molecular ring of the Galaxy or to
the inner 1 kpc of the Galaxy. The interstellar medium in M31 unlike
all of the others discussed here is mainly atomic hydrogen throughout.

Summary of Radial Distribution and Comparison with the Galaxy

Figure 9 shows the radial dependence of CO integrated intensity I,

for the 6 galaxies discussed above, compared with that of our Galaxy
(determined from the Stony Brook-Massachusetts survey of the Galaxy by
Sanders, Solomon and Scoville 1982). In all cases I(CO) is the equivalent face on integrated intensity. The determination of H_2 surface
density assumes a constant (with radius) conversion factor given above.
The magnitude of I(CO) in the Galaxy is similar to that of M101 and IC
342 at R ~ 6 kpc but the shape of the distribution is clearly different.
There is no inner gap in the molecular cloud distribution at 3 or 4 kpc
in the disk of the ScI galaxies. The deficiency of molecular clouds at
R = 1-4 kpc may be a characteristic of Sb galaxies; N891, which is classified SbI, does show some indication of a turnover in I and there is
some evidence of a hole in two Sb galaxies observed by Young and Scoville
(1982b).

Another interesting consequence of these observations, which
remains to be tested with many more galaxies, is that although the HI
surface density of galaxies seldom is more than about 5 $M_\odot/pc^2$ and has a
fairly flat distribution (with R) with a central dip, the H_2 varies by
more than an order of magnitude. Whenever the total surface density of
interstellar matter exceeds about 5 $M_\odot/pc^2$, the excess appears to be in
the form of H_2 in molecular clouds (Solomon et al. 1982). From this
point of view the most important property of a galactic disk is the total
surface density of interstellar matter which determines the star formation rate and luminosity of the galaxy.

References

Allen, R.J. and Goss, W.M., 1979 A. and A. Suppl. 36, 136.
Blitz, L., Israel, F.F., Neugebauer, G., Gatley, I., Lee, T.J., Beattie,
    D. H., 1981, Ap.J. 249, 76.
Cohen, R.S. and Thaddeus, P., 1977, Ap.J.Letters 217, 155.
Gordon, M.A. and Burton, W.B. 1976, Ap.J. 208, 346.
Linke, R. 1982, NRAO "Symposium on Extragalactic Molecules" ed. M.Kutner.
Rickard, L.J. and Palmer, P. 1981, Astron. Astro. 102, L13.
Sancisi, R. and Allen, R.J., 1979, A.and A.74, 73.
Sanders, D.B., Solomon, P.M. and Scoville, N.Z. 1982 (submitted Ap.J.)
Schweizer, F., 1976, Ap.J. Suppl. 31, 313.
Scoville, N.Z. and Solomon, P.M., 1975, Ap.J. 199, L105.
Scoville, N.Z. and Young J. 1982 (Preprint).
Solomon, P.M., Sanders, D.B., Scoville, N.Z. 1979, IAU Symposium No. 84
    ed. B. Burton, p. 35.
Solomon, P.M. and Sanders, D.B., 1980, "Giant Molecular Clouds in the
    Galaxy" eds. P. Solomon and M. Edmunds, Pergamon Press.
Solomon, P.M., Barrett, J., Sanders D.B. and deZafra, R. 1982a (to be
    published Ap.J. Letters).
Solomon, P.M., Barrett, J. and deZafra R. 1982b, submitted to Ap.J.
Solomon, P.M. 1983 (to be published).
Stark, A. 1980, Ph.D. Thesis, Princeton University.
Young, J. and Scoville N.Z. 1982a, Ap.J. July 15.
Young, J. and Scoville, N.Z. 1982b (to be published Ap.J. Letters).

THE CO ROTATION CURVE OF THE MILKY WAY :
ACCURACY AND IMPLICATIONS

Leo Blitz
University of Maryland

The CO rotation curve exhibits an increase in circular velocity
beyond R = 12 kpc (Blitz, Fich, and Stark 1980) which may be as much as
50 km s^{-1} at R = 18 kpc. Recently, Blitz and Fich (1983) have examined
the uncertainties in the rotation curve at large R and have concluded
that : 1) Of all the uncertainties, small changes in R_o and θ_o have the
most serious effect on the rise at large R. However, unless ω_o < 20 km
s^{-1} kpc^{-1}, a value smaller than that accepted by all observers, the
rotation curve rises beyond R = 12 kpc. 2) Systematic errors in stellar
distances and non-circular motions might have an effect on the magnitude
of the rise. Both effects are thought to be small, but probably work to
make the rotation curve even steeper. 3) The global value of the Oort
A constant is < 12.5 km s^{-1} kpc^{-1}, and the local value of 15 km s^{-1}
kpc^{-1} is most likely due to a local velocity perturbation.

One of the implications of a rising rotation curve is that most of
the mass in the outer Milky Way can be shown observationally to reside
outside the disk. This may be done as follows : Spitzer (1942) has shown
how the scale height of a gas layer in a flattened system is related to
the velocity dispersion of the gas and the density. This can be generalized to include all sources of gas pressure by substituting the ratio
of the total gas pressure to gas density for the square of the velocity
dispersion (Kellman 1972). If one assumes that all of the mass implied
by the rotation curve is in the disk, and if the stellar scale height is
much larger than the gas scale height, it can be shown (Kulkarni, Blitz
and Heiles 1982) that

$$\frac{z_2^2}{z_1^2} \leqslant \left(\frac{R_2}{R_1}\right)^\alpha \frac{H_{*2}}{H_{*1}} .$$ (1)

where Z is the rms scale height of the gas, R is galactic radius, H_* is
the stellar scale height, and the subscripts 1 and 2 refer to different
radii. The equality holds if the gas pressure is independent of radius.
The exponent α = 1 for a flat rotation curve, and is < 1 for a rising
curve. If $H_* \ll Z$, equation (1) becomes

43

E. Athanassoula (ed.), Internal Kinematics and Dynamics of Galaxies, 43–44.
Copyright © 1983 by the IAU.

$$\frac{Z_2^2}{Z_1^2} \leqslant \left(\frac{R_2}{R_1}\right)^\alpha \tag{2}$$

with the same conditions on α and the inequality.

Locally $H_* \gg Z$. Either this condition continues to be satisfied at $R > R_o$ or there is a distance R_c beyond which $H_* \ll Z$. In the former case, equation 1 is valid at all $R > R_o$. In the latter case, equation 2 is valid at $R > R_c$. Kulkarni, Blitz and Heiles (1982) have shown that the HI scale height increases almost linearly from ~ 200 pc at $R = 10$ kpc to ~ 2 kpc at $R = 30$ kpc. Thus if equation 1 is appropriate, the left hand side is ~ 100 for these two values of R and $H_{*2}/H_{*1} > 30$. This implies that the stellar scale height at $R = 30$ kpc is comparable to the radius of the Galaxy, a notion inconsistent with the concept of a thin disk. If equation (2) is valid, there is no location beyond R_c at which equation (2) can be satisfied. This is due to the linear increase of Z with R_2 which makes the left hand side of equation (2) proportional to R^2. Since the right hand side is proportional to R, the inequality cannot be satisfied for $R_2 > R_1$. Thus, it is not possible to satisfy either equation (1) or (2) with all of the galactic mass at large R in a thin disk.

By relaxing the condition that all of the mass implied by the rotation curve resides in the disk, it is possible to avoid these difficulties. Thus, without any assumptions about the mass-to-light ratio or the form of the disk mass distribution, the CO rotation curve and the run of HI scale height imply that a significant fraction of the mass of the Milky Way lies outside the disk.

References

Blitz, L., Fich, M., and Stark, A.A., 1980, in Interstellar Molecules,
      B.H. Andrew, ed. Reidel : Dordrecht, p; 213.
Blitz, L., and Fich, M., 1983, in Structure, Dynamics and Kinematics of
      the Milky Way, W. L. H. Shuter, ed., D. Reidel: Dordrecht, p. 143.
Kellman, S. A., 1972, Ap. J. 175, 363.
Kulkarni, S. R., Blitz, L., and Heiles, C., 1982, Ap. J. (Letters),
      259, L63.
Spitzer, L., 1942, Ap. J. 95, 329.

COLD MOLECULAR MATERIAL IN THE GALAXY

Neal J. Evans II and S. R. Federman
Department of Astronomy and Electrical Engineering Research
Laboratory, The University of Texas at Austin

F. Combes and E. Falgarone
Observatoire de Meudon

The kinetic temperatures in molecular clouds are usually considered to range upward from about 10 K (e.g., Dickman 1975). These temperatures are generally measured by observing the CO $J = 1 \rightarrow 0$ transition and assuming that this line is optically thick and thermalized. This assumption also underlies estimates of the total mass and distribution of molecular material in our galaxy based on CO surveys. Because a significant amount of molecular material could in principle be missed by galactic CO surveys, a search was undertaken for "ultra-cold" molecular gas, by which is meant an excitation temperature, $T_{ex} < 5$ K. No evidence was found for a large amount of such material (Evans, Rubin, and Zuckerman 1980), but many clouds with T_{ex} between 5 and 10 K were found. To determine if this low T_{ex} is due to low kinetic temperature, low density, or low CO abundance, we have undertaken observations of a large number of clouds in the $J = 2 \rightarrow 1$ CO line and the $J = 1 \rightarrow 0$ ^{13}CO and CO lines. These observations will be analyzed to determine the properties of these clouds.

The clouds for our analysis were obtained from two sources. First, the survey of Downes et al. (1980) was used to select clouds with a small equivalent width for the 6-cm line of formaldehyde, W (6-cm) $\lesssim$ 0.150 km s^{-1}. A preliminary version of the formaldehyde survey was used by Evans et al.; here, we observed all directions between $\ell = 15°$ and 49° with W (6-cm) $\lesssim$ 0.150 km s^{-1}. A total of 33 directions with 100 velocity components were observed in the $J = 1 \rightarrow 0$ and $J = 2 \rightarrow 1$ lines of ^{12}CO. Second, the study of carbon monoxide toward extragalactic radio sources showing 21-cm absorption (Combes et al. 1980) was used. Using the $J = 1 \rightarrow 0$ and $J = 2 \rightarrow 1$ lines of ^{12}CO and the $J = 1 \rightarrow 0$ line of ^{13}CO, the clouds in the directions of 3C27, 3C111, 3C154, and 3C353 were studied in detail. The ^{12}CO data were obtained on the 4.9m telescope at the Millimeter Wave Observatory at Fort Davis, Texas; the ^{13}CO data were obtained on the 2.5m telescope at the Observatoire de Bordeaux in Bordeaux, France.

For some of the components, the excitation temperatures of the $J = 1 \rightarrow 0$ and $J = 2 \rightarrow 1$ CO lines agree, suggesting that the

E. Athanassoula (ed.), Internal Kinematics and Dynamics of Galaxies, 45–46.
Copyright © 1983 by the IAU.

transitions are thermalized. In addition ^{13}CO has been detected towards 29 of the 44 velocity components in which it has been searched for, indicating that the CO lines are often optically thick. We find that $\sim$ 80% of the excitation temperatures are less than 10 K, with a peak at 6 K. Thus there seems to be a substantial population of cold clouds.

The most interesting result of our analysis to date is the extremely low ratio, $R = \dfrac{T_A^* (2\rightarrow1)}{T_A^* (1\rightarrow0)}$, seen toward several clouds. At least 10 percent of the clouds from the H_2CO survey have R sufficiently low that no fit to an LVG model for the line transfer is possible. The uncertainty in the number of clouds arises from the uncertainty in the beam efficiency caused by uncertainties in source size. Of the clouds toward extragalactic radio sources, maps were used to assess this effect and we conclude that all of these sources show low values for R, with the possible exception of 3C353.

We are completing maps in all the CO transitions toward the extra-galactic radio sources. The bulk of the material toward 3C111 and 3C123 lies $\sim$ 1/2° from the direction of the radio source. These two clouds appear to be similar to the dark, molecular clouds studied by Snell (1981). Although the ^{12}CO lines toward 3C111, 3C123, and 3C353 are similar, the ^{13}CO line is $\sim$ 3-5 times weaker in 3C353 relative to the lines in 3C111 and 3C123. Understanding the cause of the weaker ^{13}CO will help in unraveling the physical conditions in the three clouds.

References

Combes, F., Falgarone, E., Guibert, J., and Rieu, N. Q. 1980:
    Astr. and Ap., 90, 88.
Dickman, R. L. 1975: Astrophys. J., 202, 50.
Downes, D., Wilson, T. L., Beiging, J., and Wink, J. 1980:
    Astr. and Ap. Suppl., 40, 379.
Evans, N. J., II, Rubin, R. H., and Zuckerman, B. 1980:
    Astrophys. J., 239, 939.
Snell, R. 1981: Astrophys. J., 45, 121.

MOLECULAR CLOUDS IN THE GALACTIC NUCLEUS

R.J. Cohen and W.R.F. Dent
University of Manchester,
Nuffield Radio Astronomy Laboratories, Jodrell Bank,
Macclesfield, Cheshire, SK11 9DL England

The galactic nucleus contains the largest concentration of dense molecular clouds in the Galaxy. A new survey of the region has been carried out recently at Jodrell Bank in the 1667 and 1665 MHz lines of OH (Cohen 1981). The OH lines appear primarily in absorption against the galactic continuum background. The survey reveals many new clouds within the previously known nuclear concentration, and also a more extensive outer distribution of clouds extending to projected distances of ±1 kpc from the centre. The outer part of the distribution is tilted with respect to the galactic plane, reaching z-distances of -200 pc at positive longitudes and +200 pc at negative longitudes. The same tilt or warp has been seen in the high-velocity HI "nuclear disk".

The molecular clouds have complex noncircular motions. We have used the axisymmetric velocity model by Burton and Liszt (1978) to locate the clouds in the nucleus according to their radial velocities. Figure 1 shows the results. Molecular clouds at all z-distances have been projected onto the galactic plane and summed. Thus Figure 1 shows the nucleus as it would appear to an observer located outside the Galaxy, above the North Galactic Pole. The molecular clouds are strongly concentrated into a bar-like structure. The side of the bar furthest from the Sun shows up weaker in OH, but this may be partly a selection effect due to the location of the clouds relative to the continuum sources against which the OH absorption lines are seen, as discussed by Cohen and Few (1976). A bar in this particular orientation has been suggested to explain the non circular motions of many HI features out to and including the 3 kpc arm (Peters 1975; Liszt and Burton 1980; and references therein). It is remarkably that such a bar can be deduced starting from an axisymmetric velocity field.

We are currently extending the analysis to other velocity fields.

47

E. Athanassoula (ed.), Internal Kinematics and Dynamics of Galaxies, 47–48.

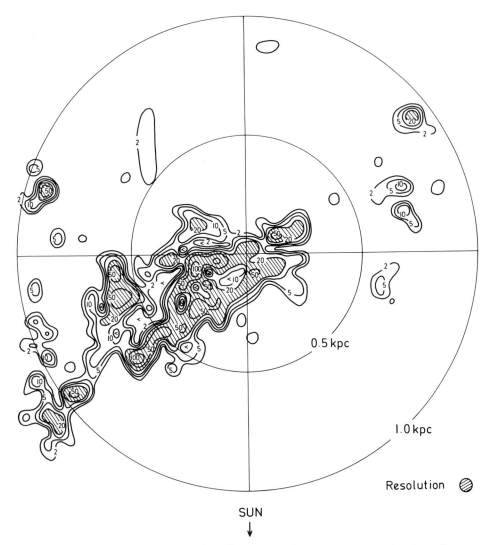

Fig. 1 Distribution of OH in the galactic nucleus, calculated
assuming the velocity model by Burton and Liszt (1978). Contours
show the apparent surface density of OH, in arbitrary units.

REFERENCES

Burton, W.B. and Liszt, H.S., 1978. Astrophys.J., 225, pp815-842.
Cohen. R.J., 1981. QMC Conference on "Submillimetre Wave Astronomy",
    eds. J. Beckman and J. Phillips, Cambridge University Press,
    in press.
Cohen, R.J. and Few, R.W., 1976. Mon.Not.R.astr.Soc., 176, pp.495-523.
Liszt, H.S. and Burton, W.B., 1980. Astrophys.J., 236, pp.779-797.
Peters, W.L., 1975. Astrophys.J., 195, pp.617-629.

MOLECULAR CLOUDS IN SPIRAL GALAXIES

Judith S. Young
Five College Radio Astronomy Observatory, U. of Massachusetts

ABSTRACT

A large observational program investigating the 2.6 mm CO line in
spiral galaxies is being conducted by myself and Nick Scoville using the
14 m telescope of the Five College Radio Astronomy Observatory
(HPBW = 50"). Thus far we have observed 46 galaxies of types Sa, Sb, Sc
and Irr, detected 31, and mapped 16. Our major findings are:
    (1) In several late type spiral galaxies (IC 342, NGC 6946 and M51)
the radial distribution of molecular gas out to 10 kpc follows the exponential blue luminosity profile of the disk within each galaxy (Young
and Scoville 1982a, Scoville and Young 1983).
    (2) From a comparison of the CO and B luminosities of the central
5 kpc in a sample of Sc galaxies, we find that the blue luminosity is
proportional to the first power of the CO content (Young and Scoville
1982b). We interpret this to mean that the star formation rate per
H_2 in Sc galaxies (indicated by the B luminosity) is constant.
    (3) No molecular rings like the one in the Milky Way at radii 4 to
8 kpc were seen in the Sc galaxies.
    (4) We have found molecular rings in two Sb galaxies, NGC 7331 and
NGC 2841, with peaks at radii of 4-5 kpc (Young and Scoville 1982c).
The central holes in the CO distributions are possibly related to the
presence of large nuclear bulges in these galaxies.

1. INTRODUCTION

In order to investigate the relationship of the dense molecular
clouds to star forming activity in other galaxies, we have been
observing the 2.6 mm CO line using the 14 m telescope of the Five
College Radio Astronomy Observatory (HPBW = 50"). The aims of these
observations are to determine (1) the radial distributions of molecular
gas in spiral galaxies, (2) the dependence of the CO distributions on
morphological type and galaxy luminosity, (3) the relative amounts of
star-forming material in the nuclei of normal and active galaxies, and
(4) the relative confinement of molecular clouds to spiral arms.

E. Athanassoula (ed.), Internal Kinematics and Dynamics of Galaxies, 49–52.

2. SC RADIAL DISTRIBUTIONS

In IC 342, NGC 6946 and M51 (types Sc and Scd) we have mapped the CO radial distributions out to 10 kpc at 1-2 kpc resolution (Young and Scoville 1982a, Paper I; Scoville and Young 1982, Paper II). In each of these galaxies we have compared the CO distributions with the optical light profiles and find that the radial distribution of molecular gas out to 10 kpc follows the exponential blue luminosity profile of the disk within each galaxy, as shown in Figure 1 for NGC 6946. Although the B luminosity may not itself be primarily from the youngest stars, it is an indicator of the recent star formation rate for several reasons. First, the B-V colors are relatively constant within these Sc galaxies. Second, as we have shown for M51 (Paper II), in galaxies for which Hα light distribution has been measured the Hα and B luminosity profiles have similiar exponential scale lengths. Assuming the abundance of molecular gas is proportional to the CO emission, the close correspondence between the CO emission and blue light indicates that the star formation rate per H_2 is constant within a particular galaxy.

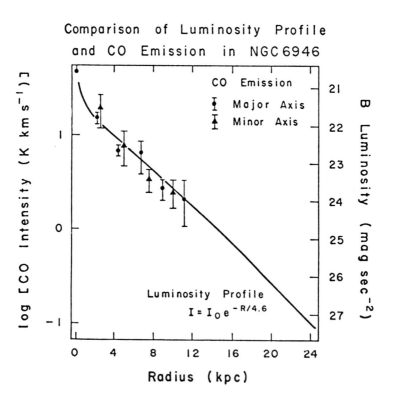

Figure 1. Comparison of CO emission with the exponential B luminosity profile of Ables (1971) for NGC 6946. Points plotted are the mean CO intensities at each radius, with bars indicating the spread in observed intensities.

We have derived an empirical relationship for determining H_2 column densities (N_{H_2}) from CO intensities (I_{CO}) based on visual extinction and CO observations in our own galaxy. In the Appendix of Paper I we showed that both dark cloud and giant cloud samples are consistent with $N_{H_2}/I_{CO} = 4 \pm 2 \times 10^{20}$ H_2 cm^{-2}/(K km s^{-1}). Using this conversion in the external galaxies we find that H_2 dominates HI by as much as a factor of 100 in the interiors of the high luminosity Sc galaxies. The H_2 and HI distributions diverge at the centers of these galaxies; the H_2 exhibits an exponential increase while the HI profiles are flat with central holes. In the Sc galaxies the H_2 masses within ~ 10 kpc are comparable to those of HI interior to 25 kpc. Global spiral structure is not evident in the molecular data at the present resolution.

3. COMPARISON OF SC GALAXIES

The correlation of CO intensity with blue luminosity within a particular Sc galaxy led us to investigate a larger sample of Sc'c covering a wide range of size, mass, and total luminosity. We have compared the CO and blue luminosities of the central 5 kpc for each of nine galaxies and an approximately <u>linear</u> correlation is revealed as shown in Figure 2. Assuming the CO emission is proportional to the abundance of molecular gas, this correlation implies that low luminosity regions have little H_2 while high luminosity regions have large amounts. If the B luminosity is an indicator of the recent star formation rate, these results suggest that <u>the star formation rate per H_2 in Sc galaxies is constant</u>.

Within this sample the molecular masses out to a fixed radius, R < 2.5 kpc, range from < 6×10^7 M_0 for M33 to ~ 2×10^9 M_0 for NGC 6946. However, the H_2 mass to blue luminosity ratio is relatively constant, with $M_{H_2}/L_B = 0.17 \pm 0.08$ M_0/L_0 over two orders of magnitude in L_B. In contrast, these galaxies all have similar amounts of HI in the central R < 2.5 kpc, so that H_2/HI ratio varies from < 0.4 in M33 to ~ 32 in M51. However, the <u>total</u> ISM mass (H_2 + HI) to $\overline{B}$ luminosity ratio is also relatively constant in these galaxies.

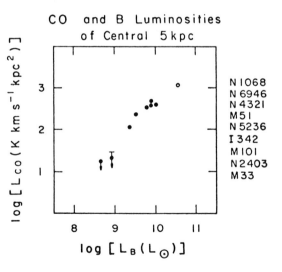

CO and B Luminosities of Central 5 kpc

N 1068
N 6946
N 4321
M 51
N 5236
I 342
M 101
N 2403
M 33

Figure 2. Comparison of CO and blue luminosities in regions 5 kpc in diameter in 8 Sc galaxies and NGC 1068. Over 2 orders of magnitude in luminosity a linear correlation is evident: $L_B = 4 \times 10^7$ $L_{CO}^{1.0}$.

4. MOLECULAR RINGS

    No molecular rings like the one in the Milky Way at radii 4 to 8 kpc
were seen in the Sc galaxies as shown in Figure 3a. In Paper I we
suggested that the difference in the radial distributions is the absence
of gas at R ~ 1 to 4 kpc in the Milky Way. Recently, however, we
discovered molecular rings in two Sb galaxies, NGC 7331 and NGC 2841,
with peaks at radii of ~ 4 kpc (Young and Scoville 1982c) as shown in
Figure 3b. In these galaxies, the central CO holes are coincident with
the size of the bulge components measured by Boroson (1981). The CO
distributions in NGC 7331, NGC 2841 and possibly the Milky Way appear to
be related to the nuclear bulges. Rather than being in molecular clouds
the gas which was present during the formation of the galaxy may have
been depleted in forming stars in the nuclear bulge.

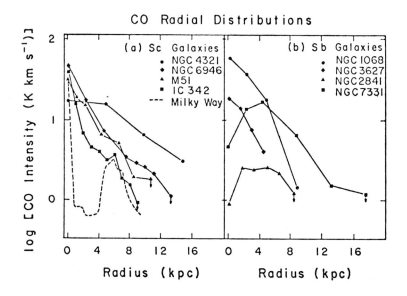

Figure 3. (a) Observed CO radial distributions in 4 relatively face-on
Sc galaxies do not show a molecular "hole" like that in the Milky Way at
R ~ 1 to 4 kpc. (b) Central CO holes are observed in several Sb galaxies
with large nuclear bulges.

REFERENCES

Ables, H.D.: 1971, Publ. U.S. Naval Obs. Sec. Ser., Vol XX, Part IV,
    Washington, D.C.
Boroson, T.: 1981, Astrophys. J. Supp., 46, 177.
Scoville, N.Z. and Young, J.S.: 1983, Astrophys.J., in press.
Young, J.S. and Scoville, N.Z.: 1982a, Astrophys. J., 258, 467.
_____ 1982b, Astrophys.J.(Letters), Sept. 1.
_____ 1982c, Astrophys.J.(Letters), Sept. 15.

CO OBSERVATIONS OF M51

G. Rydbeck, G. Pilbratt, Å. Hjalmarson, H. Olofsson,
and O.E.H. Rydbeck

Onsala Space Observatory
S-439 00 Onsala, Sweden

We present observations, which are part of an ongoing investigation, of the CO (J=1-0) emission in the spiral galaxy M51. The spectra were obtained in a beamswitched on-on mode with the Onsala 20 m antenna (beam size ∿33"), equipped with a cooled mixer and a 512×1 MHz multichannel receiver, and are shown in Figure 1. The inset diagram shows the observed positions superposed on the optical outline of the galaxy. With the present signal-to-noise ratio there is no evidence for an arm-interarm intensity contrast. This is even more apparent in integrated intensity. This result agrees with the lower resolution findings of Rickard et al. (1981). We have observed ^{13}CO in one position (22" south of the center). The ^{13}CO to ^{12}CO ratio, ∿0.1, agrees with Bell results from observations with a 1!7 beam (Encrenaz et al. 1979).

In the central region of the galaxy we note differencies in velocity dispersion along the east-west and north-south axes. The spectra along the east-west axis through the center are all broad (∿150 km s^{-1}), while the spectra along the north-south axis are narrow (∿40 km s^{-1}), with a sharp velocity shift (∿125 km s^{-1}). The average of the four spectra offset 22" from the central position is an almost exact synthesis of the observed spectrum here, which may indicate a central neutral gas depression. It seems impossible to interpret these data in terms of galactic rotation only. An elliptical velocity pattern over an extended region (∿20"×60") seems to be the "simplest" explanation and might point at a bar driven spiral density wave picture (cf. Lin and Roberts 1981). The observed ionization enhancement and broad emission lines in the M51 nucleus, indicating a nonstellar source of radiation (Rose and Searle 1982), may be of interest in this context. Goad et al. (1979) have observed more complex, small-scale, structure in the ionized gas.

In Figure 2 we compare the CO velocity structure along the north-south axis with ionized gas velocities derived by Goad et al. (1979) and the HI velocity structure of Shane (1975). There appear to be some systematic kinematic differences between CO and the ionized gas. The HI gas shows the largest dispersion, and appears to cover the velocity pattern of both the ionized gas and the CO molecular cloud ensemble.

53

E. Athanassoula (ed.), Internal Kinematics and Dynamics of Galaxies, 53–54.
Copyright © 1983 by the IAU.

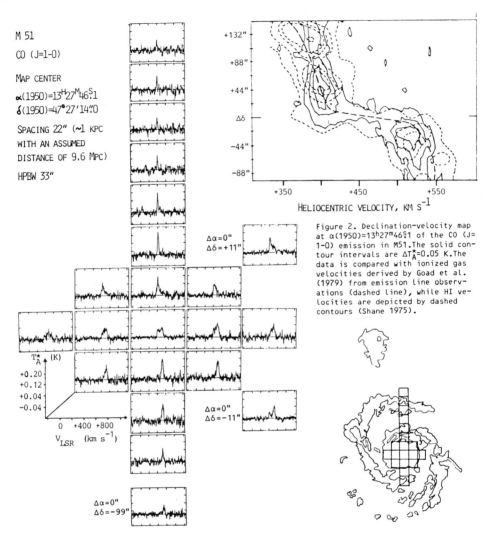

Figure 2. Declination-velocity map at $\alpha(1950)=13^h27^m46^s1$ of the CO (J=1-0) emission in M51. The solid contour intervals are $\Delta T_A^\star=0.05$ K. The data is compared with ionized gas velocities derived by Goad et al. (1979) from emission line observations (dashed line), while HI velocities are depicted by dashed contours (Shane 1975).

Figure 1. CO emission in M51. The inset diagram shows the position of the spectra on a tracing of the optical outline of the galaxy. The intensity scale is given in antenna temperature corrected for radome and atmospheric attenuation. The main beam efficiency is 0.3 and the moon efficiency is 0.6. To get from V_{LSR} to heliocentric velocity, subtract 11.7 kms^{-1}.

REFERENCES

Encrenaz, P.J., Stark, A.A., Combes, F., Wilson, R.W. (1979) Astron. Astrophys. 78, L1
Goad, J.W., De Veny, J.B., Goad, L.E. (1979) Astrophys. J. Suppl. 39, 439
Lin, C.C., Roberts, Jr, W.W. (1981) Ann. Rev. Fluid. Mech. 13, 33
Rickard, L.J., Palmer, P. (1981) Astron. Astrophys. 102, L13
Rose, J.A., Searle, L. (1982) Astrophys. J. 253, 556
Shane, W.W. (1975) in La Dynamique des Galaxies Spirales, ed. Weliachew, L., CNRS

GAS AT LARGE RADII

R. Sancisi
Kapteyn Astronomical Institute
University of Groningen,
Groningen, the Netherlands

A considerable fraction of the neutral hydrogen in spiral galaxies is generally found at large radii and beyond the optical image. This outlying gas either forms an extension of the stellar disk and of the inner HI layer, or, is concentrated in a ring. It shows well ordered motion around the galaxy, although the plane of the orbits is generally inclined with respect to the inner disk. The study of this gas is important for at least three reasons: i) although it represents only a very small fraction of the total mass it can be used as a tracer of the kinematics to determine the total mass and the mass distribution of the system, ii) it contributes to the galactic cross-sections needed to explain QSO absorption lines, iii) its physical and chemical properties may throw some light on the formation of galactic disks and their evolution.

The general kinematical properties of the HI have already been reviewed by Bosma (this symposium). I will therefore discuss the main properties of the density distribution and add only a few comments on the rotation curves in the outermost regions of galaxies.

1.   THE TOTAL RADIAL EXTENT

In the majority of isolated spiral galaxies, the neutral hydrogen is found to extend beyond the bright optical image, generally out to one or two Holmberg radii, at column densities of about 1×10^{20} atoms cm^{-2} (Bosma, 1981b). In a number of galaxies, such as Mkn 348, HI emission has been detected out to several Holmberg radii (cf. Sancisi, 1981), but these must probably be regarded as exceptional cases. In fact, observations of a dozen isolated galaxies down to column densities of about 2×10^{18} cm^{-2} (Briggs et al., 1980), indicate that the HI layer at such low density levels generally does not extend beyond 2-3 Holmberg radii. It is unlikely that in the near future significantly lower detection limits will be reached and that HI emission will be traced at levels much fainter than 1×10^{18} cm^{-2}. But, absorption spectra of faint QSO and background galaxies may provide a more sensitive test to the presence and extent of gas around galaxies.

E. Athanassoula (ed.), Internal Kinematics and Dynamics of Galaxies, 55–62.

2. Z-DISTRIBUTION

In the outer parts of galaxies, the HI layer generally exhibits large departures from the galactic plane. The bending is clearly visible in HI maps of galaxies seen edge-on or inferred from the kinematics of galaxies viewed more face-on. A comparison of HI and surface photometry maps shows that the bending of the gas layer generally begins almost exactly at the optical edge. This is clearly demonstrated in Fig. 1 where channel maps of HI are reproduced on the same scale as the "optical" map of NGC 5907. In this galaxy the stellar disk does not show any appreciable bending, whereas in NGC 4565 a slight warp is also noticeable in the optical disk (see Van der Kruit and Searle, 1981).

All galaxies viewed more face-on and with HI extending beyond the optical image show large-scale disturbances in their velocity fields which are usually considered to be the "signature" of a warp (cf. Rogstad et al., 1974). This effect is particularly strong in galaxies of very low inclination, as in NGC 628, recently studied by Briggs (1982) and by Van der Kruit and Shostak (1982, private communication).

All observations of warps show or suggest the integral sign shape seen in NGC 5907 (Fig. 1), but the two sides are generally not completely symmetrical in their radial extent, degree of warping, and position angle. NGC 4565 and our Galaxy are good examples of such asymmetries. In NGC 5907 and 4565 and in the southern side of our Galaxy, the warping does not continue to increase at large radii: the HI layer seems to settle to a new plane or perhaps turn back toward the principal plane. The outer HI layer of our Galaxy exhibits in addition to the warping, a "scalloping" up and down with respect to the galactic plane (Henderson et al., 1982; Blitz et al., 1982). Finally, in addition to these "smooth" warps, there are also the more discontinuous distributions found, for example, in NGC 4736 (Bosma et al., 1977), Mkn 348 (Heckman et al., 1982) and in some S0 galaxies (Van Woerden, this symposium), where the outlying HI is concentrated in a ring tilted with respect to the inner disk.

The thickness of the HI layer is known to increase in the outer parts of our Galaxy (cf. Henderson et al., 1982) and of NGC 891 (Sancisi and Allen, 1979). But the actual z-structure is not known in detail, and may be less simple than the picture suggested by a one-component layer of increasing thickness.

3. RADIAL DISTRIBUTION

The detailed distribution of HI surface density is known for a number of spirals. For some, photometric data also exist and therefore can be compared with the luminosity profile. The systems are often lopsided (Athanassoula, 1979; Baldwin et al., 1980; Bosma, 1981b), however, and the radial profiles obtained by averaging over all azimuthal angles are unfortunately a poor representation of the actual distributions. The

Figure 1. HI and optical pictures of NGC 5907 (on same scale, north at top). Left: map of HI emission at radial velocities 882 km/s (top) and 452 km/s (bottom). The contours are 2, 4, 8, 12, 16, 20, 28, 36 and 44 K. The FWHP beam is 25"× 30" (Sancisi, 1982; in preparation). Right: Isophote map of the surface brightness distribution (Van der Kruit, 1979). The contour interval is $0^m.5$ with the faintest contour at 27.5 J-mag $arcsec^{-2}$.

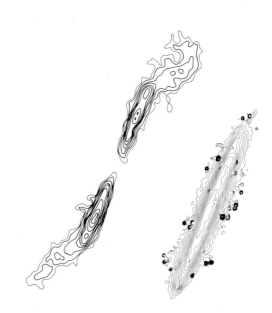

lopsidedness is not only present in the outer parts, but generally continues in the inner parts, where it is observed also in the distribution of the light and of the radio continuum emission. In the inner parts the displacement is often opposite to that of the outer parts, suggesting perhaps a one-arm spiral structure. Some substantial asymmetry is also present in the rotation curves.

Fig. 2 shows separately the projected HI density distribution along the northern and southern sides of NGC 891 major axis (Sancisi and Allen, 1982; in preparation). The flatness of the distribution in the inner region (R < 4') is partly due to the effect of large optical depth (cf. Sancisi and Allen, 1979). Also, a large fraction of the gas in this region may be in molecular form and therefore missing from the diagram. At the edge of the disk, the density has a steep drop-off followed at larger radii by a "shoulder" or "tail", which is more pronounced and extended on the southern than on the northern side. This shape, and in particular the "shoulder", seems to be a general characteristic of the radial density profiles of spiral galaxies. The shoulder may be more or less pronounced, but is found in the majority of galaxies which have extended HI layers, and usually starts near the Holmberg radius. Some striking examples of such distributions are shown in Fig. 3a. Other cases are those of NGC 2841 and 3198 (see Bosma, 1981a) and NGC 4565 (Sancisi, in preparation). The gas forming the shoulder is the same component which is generally warped (as in NGC 628), and is often asymmetrically distributed around the galaxy (as in NGC 891). The asymmetry suggests that this gas may be moving in off-set eccentric orbits instead of circular orbits. Also the HI rings found around some galaxies (e.g. Mkn 348) may represent extreme examples of such "shoulders". The radial and z-distributions observed in NGC 891 suggest that the shoulder gas may form an envelope (in all directions) surrounding the entire galaxy. The face-on column density in these shoulders is typically a factor five

Figure 2: Northern and
southern major axis pro-
files of the projected HI
density in NGC 891, obtained
by integrating the 21-cm
line emission in velocity
and in the direction perpendicular to the major axis
(Sancisi and Allen, 1982;
in preparation). The beam-
width along the major axis
is 30". Note "shoulder" at
6-10 arcmin.

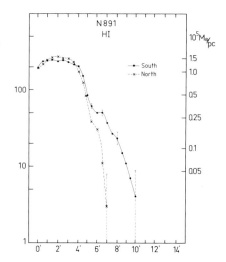

lower than in the peak of the inner disk. The amount of gas in the
shoulder is of the order of 10 to 50% of the total galactic HI content.

The shape of the HI distribution and the origin of the shoulder may
be related to the process of star formation, which causes depletion of
the gas at smaller radii (cf. e.g. Fall and Efstathiou, 1980). Alternatively, there may be dynamical effects in the region where the
shoulder begins, which cause the gas to move inward and/or outward.

In general the HI-to-luminosity ratio increases with radius, but
the relative amount of HI and stellar mass at the edge of the disk seems
to vary greatly from galaxy to galaxy. At the optical edge of NGC 891
less than 20% of the disk mass is in HI and more than 80% in stars. This
is based on the luminosity model of Van der Kruit and Searle (1982),
their assumption of $(M/L)_{old\ disk} = 7$, and the HI observations of
Sancisi and Allen (1979). The opposite situation is found in NGC 5907,
where at the optical edge most of the disk mass is in HI. The latter, in
fact, seems to extend the exponential stellar disk beyond the optical
cut-off radius. It is interesting to note in this regard that the
stellar disk of this galaxy appears to have a large scalelength (5.7
kpc), but to cut-off at a remarkably small radius of only 3.4 scale-
lengths. This is about one scalelength less than found in most spirals
of the sample of Van der Kruit and Searle (1982). It is conceivable that
the unusually large warping of the outer HI layer observed in this
galaxy (Fig. 1) may be related to these properties of the stellar and HI
disk (i.e. large scalelength, small truncation radius, extended HI
layer).

The amount and distribution of total ("dark") mass, as inferred
from rotation curves, are so uncertain that a comparison with the HI is
of little use. There is a suggestion, however, that in the outer parts
of galaxies the HI and the "dark mass" may decrease similarly (see
Bosma, 1981b) and may therefore be coextensive.

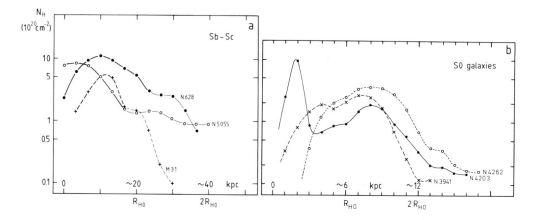

Figure 3. Radial distributions of averaged HI surface densities. a) Spiral galaxies NGC 628 (Van der Kruit and Shostak, 1982; private communication), NGC 5055 (Bosma, 1981a) and M 31 (cf. Sofue and Kato, 1981). b) S0 galaxies observed at Westerbork by Van Driel, Van Woerden, Schwarz et al. (1982, private communication; Van Woerden, this symposium). R_{HO} is the Holmberg radius. Note the much larger approximate size (kpc) of the three spirals as compared to that of the three S0's.

4. DEPENDENCE ON MORPHOLOGICAL TYPE AND GROUP OR CLUSTER LOCATION

    A clear and simple dependence of the total HI extent and of the shape of the HI distribution on the morphological type of the systems does not seem to exist. Extended HI disks are found around late-type spirals and irregular galaxies as well as around early type spirals and S0's. But it is possible now to compare the HI density distribution of spirals described above with that recently found in S0 galaxies. For a dozen S0's recent Westerbork observations have shown that the gas is concentrated in rings surrounding the optical image of the galaxy (Van Woerden, this symposium). Fig. 3b shows the radial distributions of HI surface density for three S0 galaxies normalized to the Holmberg size. The HI is concentrated in a broad component peaking slightly outside the Holmberg radius.

    A comparison with the diagram of the Sb-Sc galaxies (Fig. 3a) shows that inside the Holmberg radius the HI in the S0's is deficient by a large factor, whereas the HI rings of the S0's have similar HI densities and similar locations with respect to the Holmberg radius as the "shoulder" of the Sb-Sc systems. If instead of referring to the photometric radius, the comparison is made between the HI densities at a given distance from the center, the HI appears to be deficient in both types of galaxies inside 6 kpc (i.e. in the bulge region) and the peak of the radial density distribution is between 5 and 12 kpc, with the density in the S0's down by a factor 4. The galaxies compared here, especially the

SO's, are too few and are not representative of their type. Therefore, no conclusions seem justifiable at this stage.

The HI distribution in the outer parts of a galaxy depends on its location in a group or cluster. HI tails and bridges can be traced to large distances from a galaxy when there is a strong tidal interaction with another galaxy or a group. The dependence of the HI distribution on the location in a cluster has been studied recently by Giovanardi et al. (1982) for the galaxies of the Virgo Cluster. These authors find that the hydrogen diameters, defined in terms of scalelength, are significantly smaller for galaxies in the core of the cluster than in the outer parts, and suggest that rampressure stripping of the outer galactic disks by intracluster gas and/or tidal disturbances may be the explanation. An obvious question, soon to be answered by higher angular resolution observations, is whether and how the shape of the radial density profile is affected, and whether for instance the "shoulder", described above, has vanished in the galaxies of the cluster core.

5. ROTATION CURVES

In order to obtain the mass distribution of galaxies and to establish or rule out the presence of massive halos, the kinematical properties of the gas at large radii must be accurately known. It has already been emphasized that the derivation of rotation curves in these outer regions is often complicated by geometrical effects, such as warping, and by large-scale non-circular motions, whose existence is suggested by the asymmetries in the HI density and velocity distributions. Yet the improved sensitivity and techniques of data analysis, and a better understanding of the physical properties of the outlying gas, should lead for at least a number of carefully selected objects to accurate values of the rotational velocities out to 1-2 Holmberg radii.

In all three edge-on galaxies studied in detail and reported above (NGC 891, 4565 and 5907), the rotation curves are approximately flat out to the optical edge found by Van der Kruit and Searle (1981). Beyond this radius, the maximum deviations of observed radial velocities from the systemic velocity, which usually give the rotational velocity, decrease by about 5 to 20%. At approximately the same radius, the properties of the gas change: the gas layer is warped and probably thicker, the density drops off and the "shoulder" begins. In NGC 891 there is no warp but the distribution becomes lopsided. It is therefore not clear that the deviation observed in the radial velocities means a declining rotation curve. The consequences of declining rotation curves on the disk halo masses of these galaxies have been explored in recent work by Casertano (1982, and this symposium) in a model for NGC 5907 and by Bahcall (1982), who estimates for NGC 891 in the case of a declining curve, a halo mass as low as 1-2 disk masses.

Also in other galaxies (e.g. M 31 and NGC 3198), a smooth decrease in the rotation curves takes place inside or near the edge of the optical disk (cf. Sofue and Kato, 1981; Bosma, 1981a), and in some systems (e.g. NGC 2841) a similar effect may be present but masked in the published curves by beam-smearing effects, which tend to cause an underestimate of the rotational velocities in the region of the optical disk. It should be emphasized, however, that this decrease is not in itself evidence against the hypothesis of massive halos. This will depend on whether the rotation curves continue to drop, as suggested by NGC 891, or remain approximately constant as they seem to be doing in M 31 and NGC 3198.

SUMMARY

The outlying gas of spiral galaxies seems to undergo a more or less abrupt change of its properties at some radius, a radius which is approximately at the Van der Kruit and Searle (1981) edge of the optical disk and generally close to the Holmberg radius. The main effects observed at approximately this same radius are the following:
1) The gas layer is warped and probably thicker.
2) The density profile shows a drop-off by a factor of five and a "shoulder" appears in the outer parts.
3) A decrease of radial velocity takes place, which may imply a drop-off in the rotation curve.

ACKNOWLEDGEMENTS

I am indebted to G.D. van Albada, J. N. Bahcall, S.M. Fall, P.C. van der Kruit, N. Thonnard and S. Tremaine for valuable discussions. It is a pleasure to acknowledge the hospitality of Prof. J. Bahcall and the support enjoyed during my visit at the Institute for Advanced Study in Princeton, where most of the work for this manuscript was done.

REFERENCES

Athanassoula, E.: 1979, Ann. Phys. 4, 115
Bahcall, J.N.: 1982, preprint
Baldwin, J.E. Lynden-Bell, D. and Sancisi, R.: 1980, Mon. Not. R. Astr. Soc. 193, 313
Blitz, L., Fich, M., and Kulkarni, S.: 1982, preprint
Bosma, A.: 1981a, Astron. J. 86, 1791
Bosma, A.: 1981b, Astron. J. 86, 1825
Bosma, A., Hulst, J.M. van der, Sullivan III, W.T.: 1977, Astron. Astrophys. 57, 37
Briggs, F.H.: 1982, preprint
Briggs, F.H., Wolfe, A.M., Krumm, N., and Salpeter, E.E.: 1980, Astrophys. J. 238, 510
Casertano, S.: 1982, Mon. Not. R. Astr. Soc., in press
Fall, S.M., and Efstathiou, G.: 1980, Mon. Not. R. Astr. Soc. 193, 189
Giovanardi, C., Helou, G., Salpeter, E.E., and Krumm, N.: 1982, preprint
Heckman, T.M., Sancisi, R., Sullivan III, W.T., and Balick, B.: 1982,

Mon. Not. R. Astr. Soc. 199, 425
Henderson, A.P., Jackson, P.D., and Kerr, F.J.: 1982, preprint
Kruit, P.C. van der: 1979, Astron. Astrophys. Suppl. 38, 15
Kruit, P.C. van der, and Searle, L.: 1981, Astron. Astrophys. 95, 105
Kruit, P.C. van der, and Searle, L.: 1982, Astron. Astrophys. 110, 61
Rogstad, D.H., Lockhart, I.A., Wright, M.C.H.: 1974, Astrophys. J. 193, 309
Sancisi, R.: 1981, in "The Structure and Evolution of Normal Galaxies", pp. 149, Ed. S.M. Fall and D. Lynden-Bell, Cambridge Univ. Press
Sancisi, R., and Allen, R.J.: 1979, Astron. Astrophys. 74, 73
Sofue, Y., and Kato, T.: 1981, Publ. Astron. Soc. Japan 33, 449

DISCUSSION

INAGAKI: You mentioned that in some galaxies there is a bending in the optical picture. Do you think that the stellar disk of old stars is also bending? If only the stellar disk of young stars is bending, then this might be due to the bending of the HI.

SANCISI: The bending seen in the optical picture of NGC 4565 (Van der Kruit and Searle, 1981) and of other galaxies, such as NGC 3187, is most probably a bending of the old stellar disk.

BLITZ: Is it possible that the turnover in the warp that you see in some galaxies is related to the "scalloping" which has recently been detected in the Milky Way?

SANCISI: The turnover in the warp of NGC 5907 and NGC 4565, if it is indeed present, is of a larger scale than the "scalloping", and more similar to the turnover in the southern warp of our Galaxy. Effects such as corrugations or scalloping of the HI layer would be difficult to observe in edge-on galaxies because of the line-of-sight integration.

WAMSTEKER: Referring to your remark that these extended HI warps and envelopes could represent a phenomenon which only takes place in the gas, I would like to point out my results on NGC 5236. For this galaxy we find ordered optical features corresponding to the warped HI at distances at least out to radii of 40 kpc (r > 2R Holmberg). Some of this seems to be in the form of what is apparently diffuse stellar material. Possible HII regions seem to be indicated in other places. This appears to rule out any process that involves only the gas. The presence of HII regions – still to be confirmed by spectra – would also indicate recent star formation in the outermost regions of the galaxy.

SANCISI: There are well-known cases, as already mentioned, of warped "old" stellar disks, and there may also be, as you point out, some recent star formation in the region of the HI warp. But in isolated systems, like NGC 5907, the bending seems to affect primarily the HI layer.

ROTATION CURVE AND MASS MODEL FOR THE EDGE-ON GALAXY NGC 5907

S. Casertano
Scuola Normale Superiore, I-56100 Pisa, Italy

The rotation curve of the edge-on disk galaxy NGC 5907 can be interpreted
in terms of a two-component model, consisting of a truncated exponential
disk and of a spherical halo. About 60% of the total mass inside the edge
of the disk turns out to be in the halo. Indicative masses for disk and
halo are 9 and $13.5 \cdot 10^{10}$ solar masses respectively.

The photometric study by van der Kruit and Searle (1981 a, b) of
some edge-on galaxies has indicated that their exponential disks are
probably truncated. The rotation curve of a galaxy with a truncated exponential disk shows (Casertano 1982) a characteristic shape, the "signature" of the truncation being a step-like decrease of velocity at the
cut-off radius (see Fig. 1).

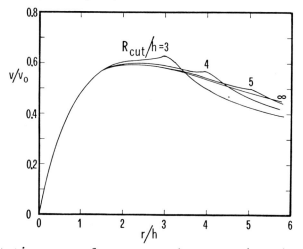

Fig. 1: Rotation curve for truncated exponential disks with different
values of the cut-off radius R.

The rotation curves available for two of these galaxies, NGC 4565
and NGC 5907, extend beyond the optical cut-off radius. They both show
the expected velocity decrease at the optical truncation. Moreover, at

E. Athanassoula (ed.), Internal Kinematics and Dynamics of Galaxies, 63–64.
Copyright © 1983 by the IAU.

the same radius the warping of the galactic plane begins. The two facts indicate that the mass in the disk should also decrease, following the behaviour of the light emission, and therefore support the view that the mass-to-light ratio in the disk is approximatively constant in the outer part of the disk itself.

On the other hand, the velocity decrease observed at the cut-off radius is in both cases smaller than it would be if all the mass were in the disk. Therefore it is natural to assume that part of the mass is smoothly distributed in a different component, such as a spherical halo. The combination of a truncated exponential disk and of a spherical halo with mass density proportional to r^{-2} gives a satisfactory reproduction of the observed rotation curve of NGC 5907 (see Fig. 2). The model has two free parameters, the total mass $M_h + M_d$ and the ratio M_h/M_d (here M_d and M_h are the disk and halo masses inside the disk cut-off radius). The total mass is related to the total rotation velocity, the ratio of the masses to the shape of the rotation curve. The resulting best-fitting values of M_d and M_h are 9 and $13.5 \cdot 10^{10}$ solar masses respectively. The corresponding value of M/L (light in the J band) for the disk is 11, in solar units. The uncertainty on the above figures, in the framework of the present model, is about 20%.

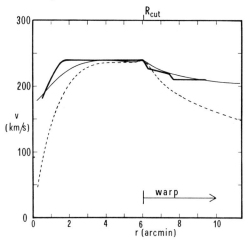

Fig. 2: Observed and model rotation curves for NGC 5907. The observed curve is the thick solid line, the model curve for the disk is dashed and the thin solid line represents the best-fitting disk-halo model.

A similar modelling has not been attempted for NGC 4565, since the uncertainties in the rotation curve are comparatively too large. The best value for the parameter M_h/M_d should lie between 2 and 5.

REFERENCES

Casertano, S.: 1982, M.N.R.A.S. (to be published)
Kruit, P.C. van der, Searle, L.: 1981a, Astr. Astrophys. 95, pp. 105-115
Kruit, P.C. van der, Searle, L.: 1981b, Astr. Astrophys. 95, pp. 116-126

OPTICAL FEATURES ASSOCIATED WITH THE EXTENDED HI ENVELOPE OF M83.

W.Wamsteker   J.J.Lorre   H.E.Schuster
Astronomy Division   NASA-JPL   E.S.O.
ESA-ESTEC   Pasadena   La Silla Observatory
c/o ESA-VILSPA   Calif.,U.S.A.   Casilla 16317
P.O.Box 54065   Santiago,Chile
Madrid,Spain

Abstract.
Deep red Schmidt plates (127-04 + RG630) of the galaxy M83 show the
presence of extensive faint optical structure well beyond the Holmberg
radius of the galaxy.Digital processing has been used to bring these
features out in detail.They consist of arc-like structures and clumps
of H II regions,which seem to follow the H I distribution and thus are
likely to participate in the warping of the H I disk.

Observations and results

The galaxy M 83(=NGC 5236) is a bright Suothern Sc/SBb I-II galaxy
seen nearly face-on ($i=24°$).It shows a very large H I envelope as was
found by Rogstad et al.(1974).Huchtmeier and Bohnenstengel (1981) found
that this envelope extended at the   sensitivity of their observations,
out to about 4-5 Holmberg radii of M 83.The neutral hydrogen velocity
field can be quite well represented by a warped disk (Rogstad et al.1974).
We have obtained two deep red exposures of the galaxy to search for faint
optical extensions.The plates were taken with the ESO-Schmidt telescope
at the La Silla Observatory.
     Inspection of the plates showed the presence of very faint features
fairly far away from the main body of the galaxy.To bring these details
out more clearly,a sizable portion (10*10 cm) of the plates was digitized
at the JPL Image Processing Laboratory.Various forms of processing were
applied to the data (see also Lorre,1981).No calibration was available
at the telescope when the plates were taken,so the actual brightness of
the features is undetermined.

     Qualitatively , the following was found:
1).When considering progressively fainter levels,the shape of the main
   disk grows from nearly circular to more irregular.The asymmetry of
   the disk changes orientation with fainter brightness levels.
2).Far to the north of the galaxy ($r \approx 40$ kpc),a faint partial circle or
   spiral arm can be traced over a distance of $\sim$ 35 kpc. The width of
   this feature is $\sim$ 5 kpc (Feature A).

65

E. Athanassoula (ed.), Internal Kinematics and Dynamics of Galaxies, 65–66.
Copyright © 1983 by the IAU.

3).To the NW,closer in (r$\approx$30 kpc),a somewhat longer ($\sim$50 kpc) and
   thinner ($\lesssim$3 kpc) arc is easily distinguished (Feature B).
4).At variuos distances from the main body clumps of what appear to be
   H II regions are present.These are confined to the H I envelope
   (especially the so-called H I finger towards the South of the galaxy).
5).A rather drastic difference in the background level between the SW
   and the SE side of the galaxy seems present.The reality of this and
   its association with M 83 can however not be unambiguously established.

Some of the above,especially Feature A,appear to represent a diffuse
stellar population,Others are most likely H II regions.Both feature A
and feature B occur at the edges of the H I distribution as maesured by
Rogstad et al.($N_{H\,I} > 1*10^{20}$ atoms cm^{-2}).Feature A is a direct
extension of the large H I counterarm,while feature B delineates the
outer edge of the steep H I gradient at the NW side of the galaxy.
Although no velocities have been determined for the optical features,
the close correspondence between the optical and the H I distribution
makes it very likely that the optical material participates in the
warping shown by the neutral hydrogen.

These results indicate that at least in the case of M 83 the warping
process is not restricted to the cold gas solely.The presence of H II
regions and a diffuse stellar population would be in conflict with the
concept that star formation stops rather abrupt when the H I column
drops below a certain minimum value.Of course,it is very well cocievable
that external influences have caused the formation of both the outer
neutral hydrogen and the stellar material discussed here.

References:

Huchtmeier,W.K.,Bohnenstengel,H.D.,1981,Astron.Astrophys.,100,72.
Lorre,J.J.,JPL Publication 81-8,2-1.
Rogstad,D.H.,Lockhardt,I.A.,Wright,M.C.H.,1974,Astrophys.J.,193,309.

M82 - TILT AND WARP OF ITS PRINCIPAL PLANE

P. Notni, W. Bronkalla
Zentralinstitut für Astrophysik der AdW der DDR
Potsdam - Babelsberg

The main plane of a late type galaxy is best determined by its population
I components : young stars, gas and dust. In the case of M82 the determination of its principal plane is complicated by its very irregular appearance and the presence of scattered light. We have used three-colour UBV-
photometry of this galaxy to determine the stellar population, using the
extinction- free parameter $Q = (U-B) - 0.65(B-V)$, and the extinction $A_B =
4 ((B-V) - (B-V)_{oo}$; $(B-V)_{oo}$ being the intrinsic colour of the population,
taken, e.g. from population synthesis calculations. Corrections for scattered light were made using the degree of polarization as well as the excess ultraviolet light (see Notni and Bronkalla, 1983 ; Notni et al., 1981
for details).

The southern half of the galaxy differs markedly from the northern one.
Firstly, the bulk of the extinction is located south of the major axis
(cf Fig. 1). The major axis has been drawn through the dynamical centre as
given by Beck et al. (1978). Secondly, the stellar population of the smoother northern region differs from that of the more irregular southern
region, as shown in Fig. 2. In the south, where the degree of polarization
and hence the amount of scattered light is small, the $Q = -0.3$ line defines
the beginning of the blue population I belt. In the north, this line signi-
fies only the stronger contribution of scattered light and does not imply
a population change.

We conclude from both figures : 1) The southern side of the principal plane
is the near one (Lynds and Sandage, 1963, found the opposite, using the
minimum in the H - distribution near the minor axis). The principal tilt,
ϕ, can be crudely estimated from the axial ratio of an ellipse in the central regions : $10 < \phi \sim 12° < 20°$. 2) The plane is warped to the north,near
the centre and west of it. The direction of this disturbance is in accord
with the reaction of the dust to a mass of gas rotating about the major
axis as deduced by Cottrell(1977). The existence of a simular disturbance
of the plane defined by the stellar population I suggests, however, a more
direct tidal influence as the main cause of the warp.

Beck,S.C., Lacy,J.H., Baas,F., and Townes,C.H.: 1978, Astrophys.J. 226,545.
Bingham,R.G., McMullan,D., Pallister,W.S., White,C., Axon,D.J., and Scar-

E. Athanassoula (ed.), Internal Kinematics and Dynamics of Galaxies, 67–68.

rott,S.M.: 1976, Nature 259, 463.
Chesterman,J.F. and Pallister,W.S.: 1980, Mon.Not.R. Astr. Soc. 191, 349.
Cottrel,G.A.:1977, Mon.Not.R. Astr. Soc. 178, 577.
Lynds,C.R. and Sandage,A.R.:1963, Astrophys. J. 217, 425.
Notni,P., Tiersch,H. and Bronkalla,W.:1981, Astron. Nachr. 302, 259.
Notni,P., Bronkalla,W.: 1983, Astron. Nachr., in preparation.
O'Connell,R.W. and Mangano,J.J.:1978, Astrophys. J. 221, 62.

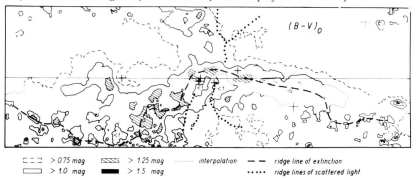

Fig.1. Plot of the colour index, $(B-V)_0$, in M82, corrected for scattered
light. The size of the figure is 315" x "110", north is 25° left of vertical. Large crosses mark reference stars, small crosses the positions of
features B,A (double), C and F (O'Connell and Mangano, 1978) and the dynamical centre (above A). To derive the colour excess, subtract 0.6 in
regions with Q = 0 (north of Q =-0.05 line in Fig.2.), and subtract 0.2
in regions of pop. I (south of lower Q +-0.30 line in Fig.2.). The ridge
line of the colour excess, defining the plane of the dust, lies mainly
south of the major axis, and is warped to the north near the centre and
west of it. Lines of maximum scattered light are also indicated ; some
parts of the -0.75 isoline in the north have been omitted for clarity.

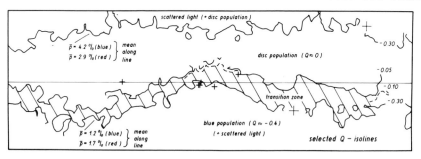

Fig.2. Plots of selectec isolines of the population index, Q, in M82, on
the same scale as Fig.1. No corrections for scattered light have been made
(increasing scattered light decreases Q). The transition zone between old
disk population and young population can be seen, as well as the similarity in position between the blue population I belt (Q<-0.30) and the ridge
line in extinction south of the major axis (compare Fig.1.). The southern
extent of the pop. I belt cannot be determined because of the increasing
scattered light contribution. Mean polarization values, approximately corrected for measuring errors, have been indicated (Bingham et al. 1976,
Chesterman and Pallister, 1980).

VERTICAL MOTION AND THE THICKNESS OF HI DISKS:
IMPLICATIONS FOR GALACTIC MASS MODELS

P.C. van der Kruit and G.S. Shostak
Kapteyn Astronomical Institute
University of Groningen, the Netherlands

1. INTRODUCTION

Most studies of the mass distribution in spiral galaxies have been based on the observed rotation curves. A serious ambiguity in this approach has always been that the rotation curve contains in itself no information on the mass distribution in the direction perpendicular to the galactic plane. The usual assumption has been that the mass in late type galaxies is distributed as the light, namely outside the central bulge in a highly flattened disk. In recent years it has been found that the rotation curves decline little or not at all, indicating large increases in the local value of M/L with increasing galactocentric radius (e.g. Bosma and van der Kruit, 1979). On the basis of dynamical arguments involving stability it has been suspected that the material giving rise to the large values of M/L - the "dark matter" - is distributed in the halos of these galaxies, so that the assumption of a flat mass distribution would have to be wrong.

This question can be elucidated by an independent measurement of the mass distribution near the plane of the disk. One does this by comparing the z-distribution of a disk component with its vertical velocity dispersion $\langle V_z^2 \rangle^{\frac{1}{2}}$. A component that is observationally accessible in external galaxies is the neutral hydrogen.

Assume that the total mass distribution in a galactic disk can be approximated by that of a locally isothermal sheet:

$$\rho(R,z) = \rho(R,o) \operatorname{sech}^2 (z/z_0) , \qquad (1)$$

which corresponds to a surface density $\sigma(R) = 2z_0\rho(R,o)$. The velocity dispersion $\langle V_z^2 \rangle^{\frac{1}{2}}$ corresponding to this mass distribution is

$$\langle V_z^2 \rangle^{\frac{1}{2}} = \left[\pi G \sigma(R) z_0 \right]^{\frac{1}{2}} \qquad (2)$$

and the force field in the z-direction is

69

$$K_z = -2 \pi \, G \sigma(R) \, \tanh \, (z/z_o) \; . \tag{3}$$

In a series of papers on surface photometry of edge-on spiral galaxies, van der Kruit and Searle (1981a, b; 1982a, b; hereafter KSI-IV) have shown that the <u>light</u> distribution of the old disk population in spiral galaxies has the distribution of (1) with z_o independent of R and luminosity density $L(o,R) = L(o,o) \exp \, (-R/h)$ out to a relatively sharp edge at $R_{max} \approx 4\text{-}5 \; h$. In the appendix to KSIII they showed that (3) is indeed a reasonable approximation to K_z for $z \lesssim z_o$ when the mass is distributed as the light in this model: i.e. $\rho(R,z) = (M/L) \, L(R,z)$.

A second isothermal component with velocity dispersion $\langle V_z^2 \rangle_{II}^{\frac{1}{2}}$ that does not add significantly to the total force field will be distributed in the field (3) as

$$\rho_{II}(R,z) = \rho_{II}(R,o) \; \mathrm{sech}^{2p} \, (z/z_o) \tag{4}$$

with $p = \langle V_z^2 \rangle / \langle V_z^2 \rangle_{II}$. For the case of HI we have $p \gtrsim 1$; the FWHM of the HI-layer can then be approximated by (see van der Kruit, 1981)

$$(z_{\frac{1}{2}})_g \sim 1.7 \; \langle V_z^2 \rangle_g^{\frac{1}{2}} \left[\pi \, G \sigma(R)/z_o \right]^{-\frac{1}{2}} , \tag{5}$$

where $\langle V_z^2 \rangle_g^{\frac{1}{2}}$ is the velocity dispersion of the HI gas.

The most recent work regarding the vertical HI distribution in our own Galaxy is that of Celnick et al. (1979) who also give references to earlier work. Their conclusions can be summarized as follows: (1) The z-distribution of HI can be understood as the equilibrium distribution of an isothermal gas in the Galactic force field. (2) A model with a mixture of two isothermal gas distributions each with its own velocity dispersion is <u>not</u> supported by the data. (3) The derived HI parameters can be made consistent with other dynamical constraints if a consider- able fraction of the Galactic mass is distributed outside the disk.

We can test equation (5) for the solar neighbourhood. Oort's (1965) curve for K_z can be fitted to (3) with $z_o = 0.6\text{-}0.7$ kpc and $\sigma(R_o) = 80\text{-}100 \; M_\odot \; pc^{-2}$ (see KSI and KSIII). Stars of spectral type later than early G show a z-distribution corresponding to a z_o of $0.6\text{-}0.8$ kpc in the solar neighbourhood (see also KSI). From the data in Jackson and Kellman (1974) and their own, Celnick et al. imply local values of $(z_{\frac{1}{2}})_g$ around 300 pc with an uncertainty of about 25%. The expected values for $\langle V_z^2 \rangle^{\frac{1}{2}}$ from (5) then are $7\text{-}9$ km s^{-1}. HI observations in our Galaxy and in external galaxies summarized by Baldwin (1981), van der Kruit (1981) and below suggest that values in the range $7\text{-}10$ km s^{-1} are typical for the velocity dispersion in the HI gas in galactic disks. Evidently eq. (5) applies to the solar neighbourhood, and can be expected to be also applicable to external galaxies. The result also suggests that magnetic fields and cosmic ray pressure play no important role in establishing the z-distribution of the HI.

2. HI OBSERVATIONS OF FACE-ON GALAXIES

The determination of the velocity dispersion in the gas is only possible in those systems that show only a small gradient in their line-of-sight velocity field across the beam of the radio telescope. Clearly the safest approach is to observe spirals that are very close to face-on with a synthesis instrument. We have selected three such galaxies on the basis of the small width of their integrated HI profiles to observe with the Westerbork synthesis radio telescope, and these three systems are now briefly discussed in turn.

<u>NGC 3938</u> (van der Kruit and Shostak, 1982) is an ScI with an optical radius determined from our surface photometry of about 3 arcmin. The velocity field is consistent with pure rotation and the corresponding rotation curve is constant from within 1' out to 4' at a velocity of about 38 km s^{-1} in the line-of-sight. Profiles are always fitted well by Gaussians and the median velocity dispersion in rings of constant galactocentric radius is 10±1 km s^{-1} at all radii. For a discussion of the reduction techniques we refer to our paper.

<u>NGC 628</u> (M 74) (in preparation) is also an ScI, and our deep surface photometry traces the galaxy out to about 5 arcmin radius. At 21 cm we detect HI out to at least 13 arcmin and Briggs (1982) has found HI-emission out to 20 arcmin at Arecibo. Beyond the optical radius the velocity field indicates that the HI layer is warped with respect to the inner plane, but the area within the optical boundary has a well organized velocity field that is to within an r.m.s. residual of 3 km s^{-1} consistent with pure rotation. The rotation velocity in the line-of-sight is about 25 km s^{-1} over this area at all radii larger than 1'. The HI distribution follows the optical spiral structure and in these arms the 21-cm emission is so strong that our profiles have peaks in excess of 10 times the r.m.s. noise in our channel maps. Even at this high S/N Gaussians provide excellent fits to the profiles and we find no evidence for extended wings corresponding to a significant second component with a higher velocity dispersion.

The ring averaged velocity dispersion has a median value of about 9-10 km s^{-1} in the central area declining to about 7 km s^{-1} at the optical boundary and remains at this level to the edge of the observed HI distribution. Over the optical extent we find that in the arms the mean dispersion is about 10 km s^{-1} compared to about 7 km s^{-1} in the interarm regions. The radial decline probably reflects the fact that further from the centre a larger fraction of the solid angle of a ring is occupied by interarm regions. The higher velocity dispersion in the arms may result from recent energy input from stellar winds and supernovae in regions of vigorous star formation. In NGC 3938 we do not separate arm and interarm regions and if most HI is concentrated in arms, the arm emission will dominate the emission from any one beam area. This may explain the observed constancy of the velocity dispersion in that galaxy.

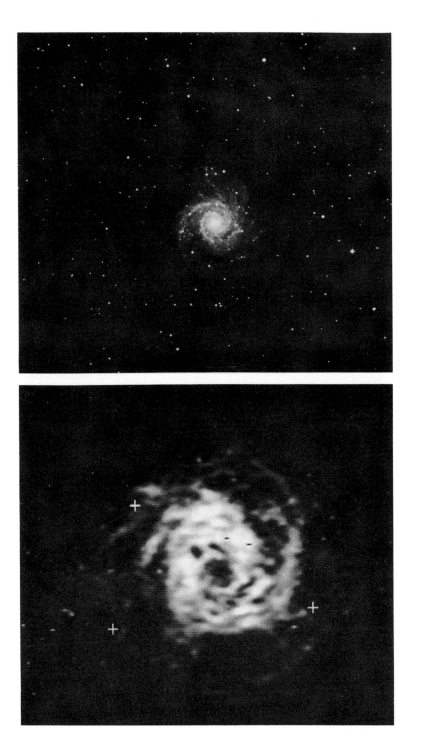

Figure 1 Optical and HI distributions in the spiral galaxy NGC 628. The two sharp dark features in the HI map are a processing artifact. Both distributions are shown to the same scale.

NGC 1058 (in preparation) is an ScII-III galaxy and has the narrowest integrated HI profile known to us. Our integrated Westerbork profile is fitted very accurately with a Gaussian of dispersion 11.1 km s^{-1}. In spite of the fact that Boroson (1981) traced the galaxy optically out to only 2' radius we find HI out to almost 7 arcmin. Again we find beyond the optical radius evidence from the velocity field of a warped HI layer. The inner velocity field is consistent to within a 4 km s^{-1} r.m.s. residual with pure rotation. The rotation velocity in the line-of-sight is about 12 km s^{-1}.

The median velocity dispersion is 7-8 km s^{-1} at all radii. The HI distribution shows distinct spiral structure out to the maximum extent of the gas, but the velocity dispersions in the arms and the interarm regions are indistinguishable. Since NGC 1058 shows less evidence for vigorous star formation and most of this spiral structure occurs beyond the optical extent this is not in contradiction to our result on NGC 628.

In view of the results presented we conclude that: (1) There is no evidence for patterns of systematic z-motions across the optical extents of the disks. (2) There is no evidence for two distinct gas components with different velocity dispersions. (3) The HI in galactic disks has velocity dispersions in the range 7-10 km s^{-1}, where the higher values may correspond to areas of vigorous star formation.

3. THE MASS DISTRIBUTION IN SPIRAL GALAXIES

The edge-on spiral galaxy NGC 891 can be used to apply the method described in the Introduction. Sancisi and Allen (1979) have measured the HI distribution, while in KSII the light distribution of the old disk population has been derived. Van der Kruit (1981) showed from this that the observations of Sancisi and Allen can be represented very well by a model in which M/L is constant throughout the disk, while a model in which all the mass indicated by the rotation curve is concentrated in the disk is in disagreement with these data. The best fitting model had $(z_{\frac{1}{2}})_g = 0.22$ exp (R/2h) kpc with h = 4.9 kpc at the assumed distance of 9.5 Mpc (model I in table 1 of van der Kruit, 1981). The distribution of light from the old disk population with the revised luminosity scale of KSIII has L(o,R) = 2.4 × 10^{-2} exp (-R/h) L$_{\odot,J}$ pc^{-3}. Using (5) with $\sigma(R) = 2z_0\rho(o,R)$ and $z_0 = 0.99$ kpc this gives $(M/L)_{old\ disk} = 9.2 \times 10^{-2} \langle V_z^2\rangle_g$. Our range above for $\langle V_z^2\rangle_g^{\frac{1}{2}}$ of 7-10 km s^{-1} then implies $(M/L)_{old\ disk} = 4.5-9$ M$_{\odot}$/L$_{\odot,J}$. In the solar neighbourhood the old disk population contributes about half of the total surface brightness of the disk, so that M/L for the entire disk population would be expected to be a factor two or so lower.

The constancy of $(M/L)_{old\ disk}$ implies that the dark matter resides in the halo. For a value of 7 for $(M/L)_{old\ disk}$ it follows that of the total mass within the optical radius of NGC 891 only about one-third

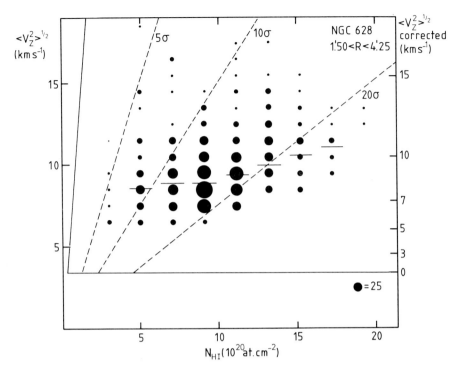

Figure 2

The rms z-velocity dispersion as a function of HI surface density over the optical extent of NGC 628. Area of circles is proportional to the number of pixels having the given values. Note the increase in the median value of $\langle v_z \rangle^{\frac{1}{2}}$ (indicated by horizontal lines) with increasing surface brightness. Dotted lines give signal-to-noise values.

is found in the disk. Van der Kruit and Searle use photometry of 6 other edge-on galaxies and estimated parameters for our Galaxy to show that in general $(M/L)_{\text{old disk}} \approx 7$ implies that only one-third to half of the total mass within the optical radius resides in the disks (see KSIII).

With the mass distribution obtained above for NGC 891 we can make some estimates of properties relevant to our assumptions. At 10 kpc from the centre for example we find that the mass density at z = 0 for the disk exceeds that of the (spherical) halo by a factor of about 4, while this ratio drops to about 2 at the edge of the disk at 21 kpc. This confirms that the disk can in first approximation be described as self-gravitating over all of its radial extent. Everywhere the surface density of HI is much smaller than that of the disk. At R_{max} the modelling shows that of the total surface density of the disk 10-20% is in the form of HI so that the assumption necessary for eq. (4) also is confirmed.

In KSIV a study is made of the edge-on Sab galaxy NGC 7814 whose light distribution is dominated by that of the spheroid. They estimate that the disk contains only 2-3% of the total mass within the optical radius of 22 kpc, so that the rotation curve can be used directly to infer the local M/L in the bulge/halo. It is then found to increase from about 15 in the inner regions to about 160 at 22 kpc. If the dark matter is in the form of faint stars of low mass and black dwarfs that formed along with the luminous stars in the spheroid, the local M/L and the mean stellar metal abundance should be inversely proportional. The colour gradient in the spheroid of NGC 7814 is consistent with this (see KSIV for a more detailed discussion).

Our final conclusion, then, is that the halos of spiral galaxies contain more than half the mass within the optical boundaries. The local M/L values in the halos go up to at least 10^2 at these maximum galactocentric distances, but are close to constant throughout the disks.

REFERENCES

Baldwin, J.E. 1980, The Structure and Evolution of Normal Galaxies (ed. D. Lynden-Bell and S.M. Fall), Cambridge Univ. Press, p. 137.
Boroson, T. 1981, Ap. J. Suppl. 46, 177.
Bosma, A., Kruit, P.C. van der 1979, Astron. Astrophys. 79, 281.
Briggs, F.H. 1982, preprint.
Celnick, W., Rohlfs, K. Braunsfurth, E. 1979, Astron. Astrophys. 76, 24.
Jackson, P.D., Kellman, S.A. 1974, Ap. J. 190, 53.
Kruit, P.C. van der 1981, Astron. Astrophys. 99, 298.
Kruit, P.C. van der, Searle, L. 1981a, Astron. Astrophys. 95, 105 (KSI).
Kruit, P.C. van der, Searle, L. 1981b, Astron. Astrophys. 95, 116 (KSII)
Kruit, P.C. van der, Searle, L. 1982a, Astron. Astrophys. 110, 61 (KSIII).
Kruit, P.C. van der, Searle, L. 1982b, Astron. Astrophys. 110, 79 (KSIV).
Kruit, P.C. van der, Shostak, G.S. 1982 Astron. Astrophys. 105, 359.
Oort, J.H. 1965, Stars and Stellar Systems V: Galactic Structure (ed. A. Blaauw and M. Schmidt), Univ. of Chicago Press, p. 455.
Sancisi, R., Allen, R.J. 1979, Astron. Astrophys. 74, 73.

DISCUSSION

WHITE : The inverse relation between M/L ratio and metallicity which
you suggest, requires each radial shell of stars to evolve like a
"closed box" enrichment model over several stellar generations. Such a
requirement puts some constraints on the formation of your systems, and
suggests that they may have collapsed slowly like LARSON's early models
for elliptical galaxies.

VAN DER KRUIT : One can infer from the absence of populations intermediate between disk and spheroid in NGC7814,that the abundance gradient
in the spheroid (and on the hypothesis above also the M/L gradient)
existed before virialisation and collapse (see Van der Kruit and Searl,
Astron Astrophys 110, 79). If so, the abundance and M/L are correlated
with binding energy in the protogalaxy and this correlation can be
expected to survive the collapse and violent relaxation (see Van Albada,
1982, Mon. Not. R. Astr. Soc., in press).

DRESSLER : You found a local M/L of $\sim$ 160 for the outermost point in
NGC7814. What is the global M/L implied within that radius, i.e. how
much has the global M/L risen from the center in your model ?

VAN DER KRUIT : Our plates are over-exposed in the central region so
that we have difficulty estimating the total light of the galaxy.
Reasonable extrapolations of the radial light profile to the center
give integrated M/L out to 22 kpc of the order of 10 $M_\odot/L_\odot$.

WIELEN : I would like to draw attention to a paper by B. FUCHS and
myself on the results of the workshop on "The Milky Way", held in
Vancouver in May 1982 (Astrophysics and Space Science Library, Vol. 100,
D. Reidel Publishing Company). In this paper we also discuss the
thickness of stellar disks in which orbital diffusion operates.

DO SPIRAL GALAXIES HAVE A VARIABLE DISK THICKNESS?

Kristen Rohlfs
Lehrstuhl für Astrophysik der Ruhr-Universität Bochum

Estimating the forces in the z-direction that affect the disks of spiral galaxies in reasonable galactic mass models we find (Rohlfs and Kreitschmann 1981) that all external forces are small compared to the self-gravity of the disk so that Spitzer's (1942) self-consistent sheet model should give a good description for the z-distribution of the disk where it is well visible. But then the three parameters describing this shape are connected by the formula

$$\Delta(r) = \frac{<v_z^2>}{\pi G \cdot \mu(r)} \tag{1}$$

Now $\mu(r)$ varies strongly with r according to Freeman (197o)

$$\mu(r) = \mu_o \cdot \exp(-ar) \tag{2}$$

and therefore either $\Delta(r)$ or $<v_z^2>$ or both should vary with r too.

NGC 4244 and NGC 59o7 are spiral galaxies seen edge-on that seem to be pure disk systems, and thus should be well suited to test the question whether Δ is variable or not. Such a test was made by van der Kruit and Searle (1981), and they obtained a satisfactory fit to their photometry using a model with a constant disk thickness.

Some systematic deviations between model and observation remained, however, and therefore this fit was repeated by Rohlfs and Wiemer (1982) using models with variable disk thickness

$$\Delta(r) = b + c \exp(ar). \tag{3}$$

Both galaxies resulted in a fit with c > o thus giving an increase of Δ with r, but the mean error of c is so large that only a 2.4 σ (NGC 4244) or 1.8 σ (NGC 59o7) effect was found.

Such effects are usually not considered to be statistically significant, but here some theoretical arguments will be given that causes to accept c > o as a viable result. For this let us expound the consequences of (3) on $<v_z^2(r)>$.

77

E. Athanassoula (ed.), Internal Kinematics and Dynamics of Galaxies, 77–79.
Copyright © 1983 by the IAU.

$\langle v_z(r)\rangle$ could vary in almost any way if its shape resulted from the processes leading to the formation of the disk. But as Wielen (1977) showed this is highly improbable, it would require that the gaseous disk keeps a highly supersonic turbulence over timescales of several 10^9 a.

We are thus left with a velocity dispersion $\langle v_{zo}^2\rangle$ = const. which the stars attain at the moment of their birth, and the magnitude of which perhaps would be determined by the mechanism of formation. As Wielen (1977) has shown for stars of the solar neighbourhood their velocity dispersion slowly increases with the time t according to the relation

$$\langle v_z^2(t)\rangle = \langle v_o^2\rangle (1 + \frac{t}{\tau})$$ (4)

here t is the age and τ a relaxation time, which is of the order of $\tau \simeq 2 \cdot 10^8$ a in the solar neighbourhood.

If we now assume that (4) describes the general behaviour of the stellar velocity dispersion in galactic disks, and if the relaxation time depends on the position in the disk, then we would observe

$$\langle v_z^2(r)\rangle = \langle v_o^2\rangle (1 + \frac{\bar{t}}{\tau(r)})$$ (5)

where $\langle v_o^2\rangle$ is the (constant) velocity dispersion of the stars at birth, and $\bar{t}$ their mean age. Combining (1), (2), (3) and (5) we then find

$$c = \frac{\langle v_o^2\rangle}{\pi G \cdot \mu_o}$$ (6)

$$b = \frac{c}{\tau_o} \cdot \bar{t} = \frac{\langle v_o^2\rangle}{\pi G \cdot \mu_o} \cdot \frac{\bar{t}}{\tau_o}$$ (7)

provided

$$\mu(r) \tau(r) = \mu_o \tau_o$$ (8)

Thus c will remain constant with time, while b will grow linearly. The observable ratio $b/c = \bar{t}/\tau_o$ gives the dynamical age of the disk in units of the relaxation time at the centre. NGC 4244 and NGC 5907 are thus fairly old system with $\bar{t}/\tau_o$ = 316 and 163 resp. It is remarkable that these results can be obtained without specifying which relaxation processes are responsible for (4).

If such processes are specified, formulae relating τ to other disk parameters can be given. As Wielen (1977) and van der Kruit and Searle (1982) showed, the Spitzer-Schwarzschild (1953) mechanism is a definite possibility. The resulting formula for τ is derived by Rohlfs and Wiemer (1982) and is given here in Table 1, but the required abundance of giant clouds is far greater than observed. Therefore contributions of all kinds of density perturbation from spiral arms to interstellar clouds were included in in an expression derived by Rohlfs (1982) given here in Table 1 too. They are compared in the following synopsis

Table 1

Spitzer-Schwarzschild mechanism	general perturbation of surface mass density
$$\frac{\tau}{a} = 8.27 \cdot 10^6 \frac{\left\langle \left(\frac{v_z}{km\ s^{-1}}\right)^2 \right\rangle^{3/2}}{\gamma \cdot \left(\frac{m_c}{10^6 M_\odot}\right)\left(\frac{\mu(r)}{M_\odot pc^{-2}}\right)}$$	$$\frac{\tau}{a} = 3.26 \cdot 10^7 \frac{\left\langle \left(\frac{v_z}{km\ s^{-1}}\right)^2 \right\rangle^{1/2}}{\eta^2 \cdot \left(\frac{\mu(r)}{M_\odot pc^{-2}}\right)}$$
$\gamma \cdot \mu(r)$: total mass density con- tained in massive clouds	$\eta = \dfrac{\mu_s}{\mu(r)}$ average rms perturbation of surface mass density
$\gamma \cdot m_c = \begin{cases} 6.2 \cdot 10^4\ M_\odot \\ 3.2 \cdot 10^4\ M_\odot \\ 5.0 \cdot 10^4\ M_\odot \end{cases}$	$\eta = \begin{cases} 9.9\ \% & \text{Solar neighbourhood} \\ 7.1\ \% & \text{NGC 4244} \\ 8.9\ \% & \text{NGC 5902} \end{cases}$

Both mechanisms obey the relation (8) and therefore cause a run of Δ as given by (3). But both mechanisms have difficulties too in explaining the short required relaxation times of $\tau \sim 10^8$ a. To do this either a large abundance of massive interstellar clouds with $\gamma \cdot m_c \simeq 5 \cdot 10^4\ M_\odot$ is required, or a rms variation of the surface mass density of 8 - 10 % is needed. It should, however, be noted, that adopting $\Delta \sim$ const. makes this situation even worse, because then $\bar{t}/\tau_o$ should be even larger.

References

Freeman, K.C. 1970, Astrophys. J. 160, 811
Rohlfs, K. 1982, in prep.
Rohlfs, K. and Kreitschmann, J. 1981, Astrophys. Space Sci. 79, 289
Rohlfs, K. and Wiemer, H.J. 1982, Astron. Astrophys. 112, 116
Spitzer, L. 1942, Astrophys. J. 95, 329
Spitzer, L. and Schwarzschild, M. 1953, Astrophys. J. 118, 106
van der Kruit, P.C. and Searle, L. 1981, Astron. Astrophys. 95, 105
van der Kruit, P.C. and Searle, L. 1982, Astron. Astrophys. 110, 61
Wielen, R. 1977, Astron. Astrophys. 60, 263

MASS DISTRIBUTION AND DARK HALOS

U. Haud and J. Einasto
Tartu Astrophysical Observatory

One of the most controversial astronomical paradoxes seems to be the mass paradox : the mass of large astronomical systems depends on the mass determination method. This paradox can be explained by supposing the existence of huge invisible coronas of unknown origin around galaxies. In the present paper we discuss basic empirical evidence of hidden matter in galaxies and examine the total mass distribution.

1. MASSIVE CORONAS AROUND GALAXIES

Photometric observations show that the density of ordinary galactic populations decreases exponentially with increasing distance from the galactic centre (de Vaucouleurs 1953). Due to this fact the mass inside a sphere of a given radius R approaches a certain limit M_{gal} with increasing R and therefore it is expected that the circular velocity should decrease, according to the Keplerian law. The observed rotation curves (Bosma 1978) as well as the dynamics of companion galaxies (Einasto et al. 1974a) show, however, that on the periphery rotation velocity is not decreasing, but is practically constant. This means that the internal mass must linearly grow with radius, indicating the presence of a previously unknown population, a massive corona or halo (Einasto 1972) that dominates in the outer regions of galaxies.

In 1980 Fabricant et al. mapped the X-ray surface brightness of M87. Takahara and Takahara (1980) showed that this X-ray source seems to have two components, and the central component, of which the temperature tallies with the velocity dispersion of M87 companions, can be explained by thermal emission of the gas which is in hydrostatic equilibrium in the gravitational potential of M87. This explanation demands the presence of a very massive halo around M87. Such a picture may also be applied to sources like A1060 and the Centaurus cluster (Mitchell, Mushotzky 1980).

Observations (Bosma 1978) suggest that all giant spirals have flat rotation curves and, consequently, are surrounded by massive coronas. On the other hand, dwarf irregular galaxies do not show this property. Sta-

81

E. Athanassoula (ed.), Internal Kinematics and Dynamics of Galaxies, 81–88.

tistics of double galaxies of different absolute magnitude also indicate that low magnitude (L < 0.1 L_{Galaxy}) galaxies have considerably smaller relative velocities than the brighter ones at the same relative linear distances (Einasto, Kaasik 1982). All this suggests that only giant galaxies have massive coronas.

The presence of coronas is probably connected with the existence of a system of companion galaxies. As noted by Lynden-Bell (1976), all companions of our Galaxy are located in a thin belt of which the inclination to the galactic plane is 70°. Recently (1981) he argued that companions of the Andromeda galaxy also form a flattened disk, highly tilted to the parent galaxy. However, the fact that galactic companions form a flat system does not mean that coronas themselves are also flat. On the contrary, there exists strong theoretical evidence favouring a more or less spherical form of coronas (Peebles 1980). Recent X-ray coronas are fairly spherical. These observations also show that the thermal component of X-radiation in M87 extends to about 320 kpc from the centre of the galaxy. Direct studies of the dynamics of the outer regions of galaxies (Bosma 1978, Ostriker et al. 1974) confirm that coronas extend to at least 100 ÷ 200 kpc.

If the radial mass distribution of the corona is similar to the radial distribution of their visible elements then the density distribution of the corona can be represented by a modified isothermal model (Einasto et al. 1974b)

$$\rho(a) = \begin{cases} \rho_0 \{|1+(a/ka_0)^2|^{-N} - (1+x^2)^{-N}\}^{1/N}, & a \leq a^\circ \\ 0 & , a \geq a^\circ \end{cases} \tag{1}$$

In this formula $\rho_0 = hM/4\pi\epsilon a_0^3$ is the central density of the population, a_0 its harmonic mean radius, ϵ is the axial ratio of the equidensity ellipsoids, M is the mass of the population, N and x are structural parameters, h and k are dimensionless normalizing parameters depending on N and x, and $a^\circ = xka_0$ is the outer limiting radius of the corona. From the distribution of the cumulative number of companions of our Galaxy we can estimate a_0 = 75 kpc, N = 0.5 and log x = 1.4 (Einasto et al. 1976).

A unique possibility to estimate the mass of the corona is given in the Local Group of galaxies, which has two concentration centers, our Galaxy and the Andromeda galaxy. Total mass of the double system can be derived from available kinematical and geometric data. The result is 3 ÷ 6 · 10^{12} $M_\odot$ (Einasto, Lynden-Bell 1982). Gott III and Thuan (1978) have found the value 5.6 · 10^{12} $M_\odot$. Hartwick and Sargent (1978) measured radial velocities for stars in outlying satellites of the Galaxy. They found that the mass inside the 60 kpc radius is 3.4 ÷ 7.6 · 10^{12} $M_\odot$. Combining these mass estimates with earlier results on the mass distribution in the Corona, we can conclude that the total mass of the corona of your Galaxy lies in the range 1.2 ÷ 3.1 · 10^{12} $M_\odot$. For the mass inside the 200 kpc radius Webbink (see Faber and Gallagher 1979) derived the value 1.4. 10^{12}· $M_\odot$, which corresponds to a mass of the corona 1.7·$10^{12}M_\odot$. In the Andromeda galaxy the observed rotation curve goes out to the distances of about 40 kpc (Haud 1981). In these regions most of the inner mass is already due to the corona and therefore we can estimate the parameters of the corona directly from the rotation curve. The results are $a_0 = 90$ kpc, $N = 0.5$, $\log x = 1.5$ and the mass of the corona is $3.3 \cdot 10^{12}$ $M_\odot$. As we can see, the sum of the masses of the coronas of the Galaxy and M31 agrees with the total mass of the Local Group.

2. GALACTIC MODELS

After examination of the overall mass distribution in galaxies we can obtain more detailed information about these systems only by modeling their properties. The most convenient way of determining a model is to use a certain analytic expression for the density of the galactic populations. Our experience has shown that the best representation can be obtained using an exponential function (Einasto 1972)

$$\rho(a) = \rho_0 \exp\left|-(a/ka_0)^{1/N}\right|, \tag{2}$$

where all parameters have the same meaning as in (1). The spatial density of a disk with a central hole can be expressed as the sum of two spheroidal mass distributions, for which $N_- = N_+$, $a_{0-} = a_{0+}/\kappa$, $\varepsilon_- = \kappa\varepsilon_+$ and $M_- = -M_+/\kappa^2$, where $\kappa > 1$ is a parameter determining the amount of the hole in the center of the disk.

Models of M32, M81, M104 and our own Galaxy have been constructed at our institute. A model of the Andromeda galaxy M81 was also finished recently. To find the parameters of this model rotation curve, velocity dispersions of central populations, integral brightness profiles in four colors and distribution of young stars, HII regions and objects of the halo were all taken into account. The results are given in Table 1. As we can see from Figures 1 and 2, the observations have been approximated quite successfully - accuracy is 3.1 % on the average. Most inaccurate are the values of velocity dispersions of the bulge and halo - the deviations are 29 % and 26 %, respectively. These deviations are calculated on the basis of the most recent observations of the dispersions, but recently we found a correlation between the observed velocity dispersion and the resolution of the corresponding observations (Fig. 3). This means that the value for the nucleus - 177 km/s - used in modelling, may be overestimated 1.5 times in comparison to its actual value. If the same happened to the dispersions of the bulge and halo, then the model would represent actual values with the accuracy of 3 %. At the same time, accepting the corrected value for the velocity dispersion of the nucleus would mean that its mass is overestimated 2.25 times in the present model.

There are also rather large mean deviations of the model from the observed rotation curve and from the distribution of surface density in young population - 6.8 % and 13.6 %, respectively, but in this case we have to deal with the effect of spiral structure (Haud 1981). Correcting the observations on account of this perturbation, we find that the model represents the smoothed rotation curve and density profile with the accuracy of 4.4 % and 5.1. %, respectively. The mean deviation of the model falls then to 1.9 %, and considering corrections of dispersions to 1.7 %.

Table 1. M31 model parameters

Population	ε_+	κ	a_{o+} (kpc)	M_+ (10^{10} $M_\odot$)	N	M/L_B	M/L_V	M/L_{379}	M/L_{624}
Nucleus	0.655		0.0021	0.0130	1.394	45.452	45.146		
Core	0.910		0.1347	0.1930	0.323	22.829	12.851		
Bulge	0.724		0.6427	0.3890	1.610	2.011	1.374	3.297	1.294
Halo	0.443		1.3860	0.6064	6.408	1.257	1.199	1.821	3.522
Disk	0.080	1.570	4.9788	28.3042	2.424	18.178	16.178	27.799	15.752
Flat	0.020	1.209	8.8748	2.4516	0.233	1.913	1.318	1.497	2.312
Corona	1		90	330	0.5				

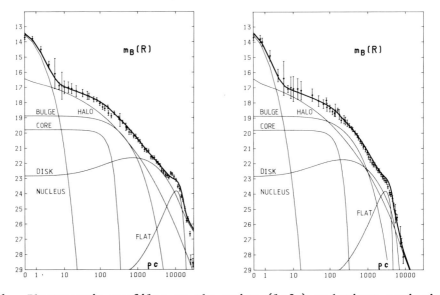

Fig. 1.   Photometric profiles on the major (left) and minor semiaxis.

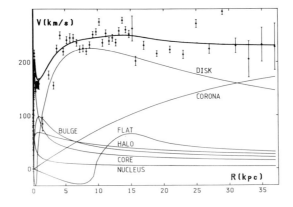

Fig. 2.   Observed (crosses) and model rotation curve of M31.

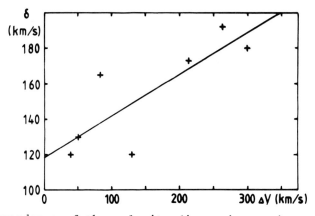

Fig. 3. Dependence of the velocity dispersion estimates of the nucleus of M31 on the resolution of the observations.

3. MASS-TO-LUMINOSITY RATIOS

    Physical properties of galactic populations can be well expressed in mass-to-luminosity ratios, M/L. Very bright hot stars of young populations have small values of M/L. Old populations have higher M/L. M/L also depends on the composition of stars. Metal-rich cores of galaxies have higher M/L than metal-poor halo stars on the galactic outskirts. Thus, going outwards, the mean local value of M/L should decrease. The comparison of photometric and dynamic data indicates that this is indeed the case. However, at the very periphery of the galaxies M/L starts to increase rapidly and at the last measured point has the value M/L = 300 (Fig. 4). From these data a lower limit $(M/L)_{cor} > 10^3$ follows. Other methods give an even higher lower limit $- 10^4 \div 10^5$ (Jaaniste and Saar 1976). At the same time Faber and Gallagher (1979) noted that M/L for spirals within the Holmberg radius R_{HO} is $\simeq 4 \div 6$, not much greater than the local M/L for the solar neighborhood. Our calculations show that the value of M/L within R_{HO} is closely correlated with the mean M/L of the optical populations and is only about 1.13 times higher. This indicates that unseen matter does not strongly dominate the mass within R_{HO} and that the physical properties of the coronal matter are completely different from the properties of conventional stellar populations.

        Table 2. M/L in galaxies and systems of galaxies

Object	M/L	
Galaxies	$(5 \div 10)$ h	Faber, Gallagher 1979
Local Group	160	Peebles 1980
Groups	$(80 \div 300)$ h	Faber, Gallagher 1979
Clusters	$(300 \div 650)$ h	Faber, Gallagher 1979
Virgo supercluster	800 h	Davies et al. 1980
Universe $(\Omega = 1)$	2000 h	Einasto et al. 1980

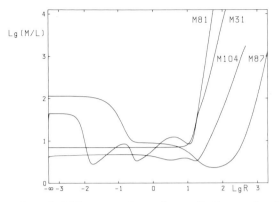

Fig. 4. Mean local M/L as a function of the galactocentric distance.

The comparison of the M/L of ordinary populations with the overall M/L (Table 2) shows that they comprise only 1 % of the whole known mass. Estimates (Bahcall and Sarazin 1977) show that the amount of intergalactic gas is approximately equal to the amount of the matter in galaxies. Taking this into consideration we conclude that about 2 % of the whole matter is in the ordinary form and 98 % is in a hidden form. This ratio is independent of the Hubble constant and we hope that it is correct within a factor of 2. We might also mention that the ordinary matter forms about 1 % of the critical density of the Universe (last entry in Table 2).

REFERENCES

Bahcall,J.N. and Sarazin,C.L. : 1977, Astrophys. J. 213, L99.
Bosma,A. : 1978, Dissertation, University of Groningen.
Davies,M. et al. : 1980, Astrophys. J. 238, L113.
Einasto,J. : 1972, Proc. First European Astr. Meeting 2, 291.
Einasto,J., Kaasik,A. and Saar,E. : 1974a, Nature 250, 309.
Einasto,J., et al. : 1974b, Tartu Astron. Obs. Teated 48,3.
Einasto,J., et al. : 1976, Mon. Not. R. astr. Soc. 177, 357.
Einasto,J., Joeveer,M. and Saar,E. : 1979, Tartu Astron.Obs. Preprint A-2.
Einasto,J. and Kaasik,A. : 1982, Tartu Astron. Obs. Preprint.
Einasto,J. and Lynden-Bell,D. : 1982, Mon. Not. R. astr. Soc. 199, 67.
Faber,S. and Gallagher,J. : 1979, Ann. Rew. Astron. Astrophys. 17, 135.
Fabricant,D., Lecar,J. and Gorenstein,P. : 1980, Astrophys. J. 241, 552.
Gott III,J.R. and Thuan,T.X. : 1978, Astrophys. J. 223, 426.
Hartwick,F.D.A. and Sargent,W.L.W. : 1978, Astrophys. J. 221, 512.
Haud,U. : 1981, Astrophys. Space Sci. 76, 477.
Jaaniste,J. and Saar,E. : 1976, Tartu Astrophys. Obs. Publ. 43, 216.
Lynden-Bell,D. : 1976, Mon. Not. R. astr. Soc. 174, 695.
Lynden-Bell,D. : 1981, Observatory 101, 111.
Mitchell,R. and Mushotzky,R. : 1980, Astrophys. J. 236, 730.
Ostriker,J.P., Peebles,P.J.E. and Yahil,A. : 1974, Astrophys. J. 193, L1.
Peebles,P.J.E. : 1980, The Large-scale Structure of the Universe, Princeton University Press, Princeton, New Jersey.
Takahara,M. and Takahara,F. : 1980, Kyoto Univ. Preprint 405, 1.
Vaucouleurs,G.de : 1953, Mon. Not. R. astr. Soc. 113, 134.

DISCUSSION

KALNAJS :   The customary approach of deducing mass distribution from
rotation curves involves an implicit or explicit extrapolation of the
velocity data, and the often reported rise of M/L usually begins where
the observed information runs out. I would like to show you a slide
depicting four rotation curves computed from photometric data which has
been converted into mass distributions by assuming that M/L is constant
within a galaxy. The photometry extends to faint enough limits to completely determine the rotation curves. For NGC 4378 it was necessary to
decompose the light into a bulge and a disk. For the others the decomposition gave essentially the same curves as would have been obtained
from pure disks.
The rotation curves agree well with the observed velocity points, and
thus demonstrate that the flat rotation curves of NGC 7217 and NGC 4378
need not lead one to conclude that there is dark matter in the outer
parts of these galaxies.

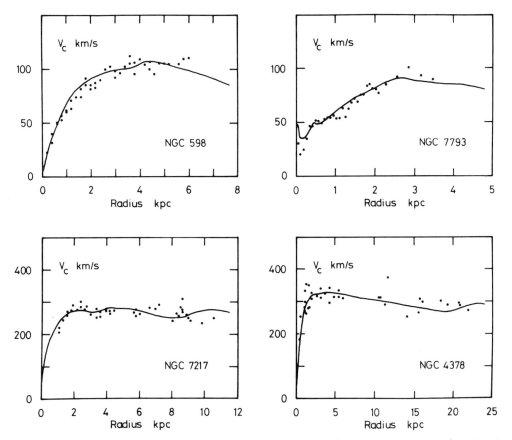

Rotation curves computed from photometry assuming a constant M/L within
each galaxy. The dots are the measured velocities. The values of M/L used
are 5.0, 2.9, 4.2 and 6.5.

```
. . . . . . . . :
. . . . . . . . : ?
. . . . . . . . : !!!!!
somebody   : HA, HA, HA.
. . . . . . . . : ***, ???, !!!
```

(The audience becomes restive and the massive halo enthusiasts slowly
regain their composure).

HAUD : This is very interesting, but note the limited extent of the
rotation curves of these four galaxies. Three of them extend to radii
less than 12 kpc and the fourth one reaches 25 kpc. Usually the M/L
starts to increase rapidly only outside roughly 30 kpc, and only giant
galaxies have coronas.

RUBIN (to Kalnajs) : It is true that the analysis of the rotation curves
presents the mass interior to any R, but not the distribution of the
mass. Thus, while the mass could be in a disk, there are other reasons,
stability especially, that suggest a halo. The velocities you show for
NGC 4378 and 7217 come from our data, and both rotation curves are
fairly exceptional in that the velocities fall slightly with increasing
R. I suspect you would have more difficulty in fitting with constant
M/L a flat or slightly rising rotation curve which extends to very large
radii. In any case, it seems to me, you must be saying that the surface
brightness of these galaxies falls slower than exponentially with
increasing R.

GOTTESMAN : As a contrast to Dr Haud's presentation, I would like to
offer the barred spiral NGC 3992. The HI in this system has been well-
observed at the VLA. This data allows a mass to be determined within a
radial distance of 15 - 20 kpc. There are also three satellites whose
atomic hydrogen emission has been detected. Following the method of
Bahcall and Tremaine one can use the satellites to calculate a mass
within $\sim$ 60 kpc of NGC 3992. Within the errors, the two masses calculated
are the same ($\sim$ 2 10^{11} $M_\odot$).
One can also invert the argument. If NGC 3992 had an isothermal halo, the
expected velocities of the satellites would be 3-4 times greater than
observed values. Jim Hunter and I have therefore concluded that there is
little or no room for a massive halo 10 times greater than the disk mass.
The problem then remains to explain the observed flat rotation curve.

HAUD : I think that a mass calculated from three satellites only may
have large statistical errors.

THE ORIGIN OF DWARF SPHEROIDAL GALAXIES

D. Lynden-Bell
Institute of Astronomy, Cambridge, England.

ABSTRACT

Two dwarf spheroidal galaxies are associated with the Magellanic
Stream with their major axes oriented along it. Evidence suggests that
all the dwarf spheroidal satellites of the Galaxy belong to one or other
of two streams of tidal debris. If this is true the orbits will give
the first reliable determination of the total mass of the Galaxy out to
120 kpc.

In order to measure the mass of our Galaxy's halo we need good data
on well determined orbits far from the galactic centre. Can we get such
information?

After taking account of the parallax due to our offset from the
galactic centre Ursa Minor and Draco lie opposite the SMC and LMC in
the Galacto-centric sky. They therefore lie in the great circle of the
Magellanic stream. When a satellite is tidally torn, the main shearing
is in the plane of the orbit - the part of the satellite closest to the
Galaxy is pulled down towards it and runs forward in the orbital plane
by angular momentum conservation, while the tidal tail is left behind
in the orbit beyond the remnant of the satellite. Thus, from the
Galactic centre a satellite that is being torn will appear as a streak
elongated along the orientation of the orbit. Likewise, the pieces
that are being torn away are likely to be shearing in the plane of the
orbit. If pieces manage to hold themselves together their final spins
are likely to be oriented with axes perpendicular to the orbital plane.
Thus their major axes will be oriented in the orbital plane. In the
figure the tick marks show the orientations of the major axes. Both
Ursa Minor and Draco are oriented along the Magellanic Stream. This
strongly suggests that they were once part of the Greater Magellanic
Galaxy that has been torn up by successive passages. One might suspect
that both objects were torn off long ago when still gaseous, and made

E. Athanassoula (ed.), Internal Kinematics and Dynamics of Galaxies, 89–92.
Copyright © 1983 by the IAU.

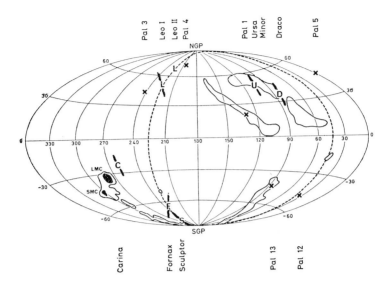

Figure 1. The Fornax-Leo-Sculptor great circle plotted on a map of high-velocity clouds with the dwarf spheroidal galaxies and diffuse globular clusters. The plot is in Galactic coordinates as observed from the Sun. Correction to the Galactocentric coordinates are minor.

stars through the shocks that accompanied the encounter. Hodge & Michie showed that both objects are at or close to their tidal limits even for a Galaxy without a halo.

The fact that two of our dwarf spheroidals were probably made from "tidal debris" suggests that all of them might have that origin. Carina lies near the Magellanic Clouds and although its orientation is 20° out of true, this is no more than is seen in some of the hydrogen clouds that form part of the Magellanic Stream. There may be a real association here but the case is not proven. There remain 4 other dwarf spheroidal satellites, Sculptor, Fornax, Leo I and Leo II. Of these, Sculptor lies very close to the Magellanic Stream but, although close to its tidal limit, it has the wrong orientation some 45° off that of the stream. More interestingly, all four of these systems lie accurately in a great circle in the galactocentric sky and Sculptor, the one closest to its tidal limit, is oriented along that great circle. The largest of these systems is Fornax which has four of its own globular clusters. This suggests the existence of a Fornax-Leo-Sculptor stream caused by the tidal break-up of a greater Fornax.

It will be interesting to search both this great circle and the Magellanic Stream circle for small faint dwarf spheroidals which may have been missed.

If we assume that Leo I, Leo II and Sculptor were torn off Fornax, then each should have approximately the same specific energy and specific

angular momentum in its orbit about the Galaxy. Accurate tests of this
can only be made when velocities for the Leo systems are available, but
guessing that they will be small galactocentrically, one deduces a mass
for the Galaxy out to 120 kpc of $(5 \overset{\times}{\div} 2) \times 10^{11}$ $M_\odot$. Most of the uncertainty comes from our imprecise knowledge of the velocity of Fornax.
More work on the dwarf spheroidals will provide the best data on the
mass of our Galaxy's halo however the orientations of Leo I, Leo II and
Fornax do not lie along the F.L.S. stream. This might be because they
are not close to their current tidal limits, but it could be that the
great positional accuracy (the centres lie off 'the great' circle by
only $0.5°$, $1.4°$, $0.6°$, $0.6°$ while Sculptor agrees in orientation to
$3°±3°$) is after all due to a chance coincidence!

REFERENCES

Lynden-Bell, D., 1976, Mon. Not. R. Astr. Soc. 174, 695.
Hodge, P.W. & Michie, R., 1969, Astronom. J. 74, 597.
Hunter, C. & Tremaine, S., 1977, Astronom. J., 82, 262.
Lynden-Bell, D., 1982, Observatory, 102, 7.
Lynden-Bell, D., 1982, Observatory, 102, 202.

DISCUSSION

    WHITE: Under what conditions do you suspect that debris torn from
a satellite galaxy will be able to overcome the tidal field of the
parent and recollapse to form stellar systems such as Fornax or Sculptor?

    LYNDEN-BELL: I do not believe that the tidal field of the parent
has to be overcome as the debris may be a considerable distance from
its parent before forming a true body. I envisage that the stream of
debris may consist of two parts that try to cross each other and the
resulting shocked region both loses energy and provides the conditions
necessary for star formation. The resulting stellar system would then
be weakly bound and therefore diffuse. This may well be the case for
Ursa Minor and Draco in the Magellanic orbit and for Sculptor and the
Leo systems in the other. Currently I regard Fornax as the most probable
parent of this second suggested stream, in which case Fornax may not
have been made by a much larger parent but merely shaken up as its outer
parts were torn off.

    INAGAKI: Do you think that the stellar debris and the gaseous debris
were torn from the Magellanic Clouds at different epochs? If so have
you a reason why?

    LYNDEN-BELL: Unlike some astronomers, my familiarity with epochs
long ago is much less than it is with the current one. We see gas
currently leaving the surroundings of the Magellanic Clouds and being
drawn out into the Magellanic Stream. It would not amaze me if one of

the larger condensations eventually formed stars and became a dwarf spheroidal galaxy. It would be diffuse for lack of binding and oriented along the orbit as indeed the clouds of the stream are. Since all this may be happening now, it seems natural to suppose that it may have happened before, both to the Magellanic Cloud system and to a probably gaseous proto-Fornax. There are examples in the sky of stellar systems being tidally torn but I get the impression that the smoother streams left behind by them will disperse. Draco and Ursa Minor were probably torn off the very outer parts of the Greater Magellanic Galaxy in gaseous form. I suspect old gaseous debris that does not condense gets too dispersed to be detected. Thus all that is needed is that the present epoch is none too particular but that the Magellanic Clouds have suffered repeated disruption due to tides. Magellanic Streams form whenever the spin orientations at closest approach are most favourable to disruption.

(The Figure is adapted from Ref. 5, with permission).

NGC 3992 - A GALAXY WITHOUT A MASSIVE HALO

S. T. Gottesman, J. H. Hunter, Jr., and J. R. Ball
Department of Astronomy, University of Florida, Gainesville,
Florida, U. S. A.

As a continuation of our earlier work (Gottesman and Hunter, 1982), we have reobserved the HI emission from the galaxy NGC 3992. We have combined all the data and produced new maps, at a significantly improved signal-to-noise ratio, of the gas density and velocity distribution with resolutions of ~ 22" and 25 km s^{-1}. The resultant, angle averaged, HI rotational velocity is shown in Figure 1 for the symmetric and nearly circular flows for r ≤ 3.35' from the center (r ≤ 14.0 kpc, assuming de Vaucouleurs' (1979) distance of 14.2 mpc for NGC 3992). Shown, also, in Figure 1 is a fit to the observations provided by a Toomre disk of index n = o. No attempt was made to fit the observational data within 1', in view of the low signal to noise.

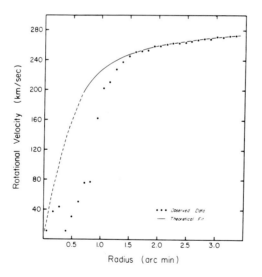

Figure 1. Angle Averaged Rotation Curve for the HI in NGC 3992

The surface density and gravitational potential of the n = o disk are obtained by <u>integrating</u> the density and potential for the n = 1 disk

E. Athanassoula (ed.), Internal Kinematics and Dynamics of Galaxies, 93–94.

(Toomre, 1963), Viz. $\sigma_0(a, r) = -\int \sigma_1(a, r)ada = [C_1^2/(2\pi G)](a^2 + r^2)^{-\frac{1}{2}}$, and $\phi_0(a, r) = -\int \phi_1(a, r)ada = -C_1^2 \ln[a + (a^2 + r^2)^{\frac{1}{2}}]$. The rotational velocity as a function of r and disk mass interior to r are given by $v(r) = C_1 [1 + a^2/r^2 + (a/r) (1 + a^2/r^2)^{\frac{1}{2}}]^{-\frac{1}{2}}$ and $M(r) = (C_1^2 r/G) [(1 + a^2/r^2)^{\frac{1}{2}} - a/r]$ respectively. With our fit, $C_1 = 293$ kms^{-1}, $a = 1.76$ Kpc, and $M(14.0$ Kpc$) = 3.0 \times 10^{11}$ M$_\odot$. (The mass has been augmented by 20% to allow for finite disk thickness.) By way of comparison our published Brandt model, which was based on half the present data, resulted in disk masses interior to r = 19.7 Kpc in the range $(1.6 - 4.3) \times 10^{11}$ M$_\odot$, with the uncertainty reflecting the allowable range in the Brandt function parameters. Serendiptiously, we observed HI emission from three nearby dwarf systems, UGC 6923, 6940, and 6969, which appear to be Magellanic cloud-like satellites of NGC 3992. The global line profiles of the objects are well resolved, and in Table 1 we list the velocity differences δV_i and projected separations S_i between the satellites and NGC 3992. Listed, also, are indicative masses for the primary galaxy, $q_i = 2.325 \times 10^5 (\delta V_i)^2 S_i$ M$_\odot$.

Table 1

Object	δV_i (km/sec)	S_i (Kpc)	$q_i (10^{10}$ M$_\odot$)
UGC 6969	64	44.2	4.2
UGC 6940	50	34.7	2.7
UGC 6923	16	59.5	0.55

Using the mass indicator of Bahcall and Tremaine (1981), $M = (24/\pi)$ $1/3 \sum_{i=1}^{3} q_i = 1.9 \times 10^{11}$ M$_\odot \pm 1.5 \times 10^{11}$ M$_\odot$. After allowing for observational errors, particularly uncertainties in the δV_i values, we conclude that the mass interior to 60 Kpc from the center of NGC 3992 cannot exceed about 4×10^{11} M$_\odot$.

Therefore, we conclude that NGC 3992 is unlikely to possess a massive halo. This confirms our earlier analysis, and is based on twice as much data and an alternative model for the mass distribution in NGC 3992. Indeed, if we invert the argument and suppose that the galaxy does possess a massive isothermal halo, then, at large radii, $V(r) \rightarrow C_1/\sqrt{2} = 175$ km s^{-1}. A χ^2 test for three degrees of freedom yields only a 3% probability that satellites moving in such a halo would exhibit velocity differences as small as those observed.

REFERENCES

Bahcall, J. N. and Tremaine, S. 1981, Astrophys. J., 244, 805.
de Vaucouleurs, G. 1979, Astrophys. J., 227, 729.
Gottesman, S. T. and Hunter, J. H. 1982, Astrophys. J., 260, 65.
Toomre, A. 1963, Astrophys. J., 138, 385.

GROUPS OF GALAXIES AND THE MISSING MASS PROBLEM

P. N. Appleton* and R. D. Davies

Nuffield Radio Astronomy Laboratories,
Jodrell Bank, England.

The existence of a massive dark component to the matter distribution of galaxies (the 'missing mass') is inferred from the now overwhelming evidence for flat rotation curves in galaxies. However observational data on the linear extent of such a dark component and its total mass contribution is usually restricted by the limited radial distance to which rotation curves of individual galaxies can be measured (typically < 100 kpc). The magnitude of the mass contained within a larger radius around a galaxy can in principal be inferred by studying the kinematics of small groups of galaxies and making assumptions about their dynamical stability (see Faber and Gallagher, 1979, for review). However, one of the major difficulties in such studies is the question of group membership. The inclusion of disrelated foreground or background galaxies into a dynamical calculation of mass obtained for example via the Virial Theorem, can lead to spurious results. The effects of varying membership criteria on the dynamical properties of groups is well illustrated by the work of Huchra and Geller (1982).

In order to investigate further the problem of the dynamics of groups, Appleton and Davies (1982) carried out a deep HI survey, in Ursa Major. The data has provided detailed information about the distribution of galaxies in a region $\sim$ 100 sq. degrees in area and covering a velocity range of 0 – 3000 km s^{-1}. The Virial Theorem can be used to investigate the dynamics of groups contained within such a survey. For a system in dynamical equilibrium $|2T/\Omega| = 1$, where T is the kinetic energy of the system and Ω is the gravitational potential energy. Here $T = 3/2 \, \Sigma m_i \, \Delta V_i^2$ and $\Omega = 2/\pi \, \Sigma m_i m_j / r_{ij}$ (Limber and Mathews, 1960), m_i is the mass of the ith galaxy, ΔV_i is velocity of the ith galaxy relative to the group barycentric velocity and r_{ij} is the apparent (linear) separation between the ith and j^{th} galaxies. The qualities m_i and ΔV_i can be determined from the HI profiles of galaxies detected in the survey.

In order to avoid making subjective decisions about group membership we have developed a new method of analysing the data obtained from galaxy

---

* Now at Department of Astronomy, University of Manchester, Manchester.

E. Athanassoula (ed.), Internal Kinematics and Dynamics of Galaxies, 95–96.
Copyright © 1983 by the IAU.

surveys (Appleton and Davies, in preparation). Many different combinations of galaxies are taken together to form putative groups and the ratio $|2T/\Omega|$ is calculated for each group. A frequency distribution function of values of $|2T/\Omega|$ is then constructed (called the virial discrepancy spectrum). Such a spectrum would be strongly peaked around $|2T/\Omega| = 1$, for many bound low mass groups in the sample but would peak at values > 1 for groups containing missing mass (assumping group stability). The algorithm has been extensively tested using N-body simulations to investigate the effects of contamination by foreground and background groups. Preliminary results of the application of the method to the Ursa Major data show two main peaks in the virial discrepancy spectrum. The first peak occurs at values of $|2T/\Omega| = 1$ to 2 and may correspond to bound group with no missing mass. The second significant peak is at $|2T/\Omega| = 15$ to 16 and is the result of the combination of the former groups into a larger single group. The physical reality of such a large system is uncertain. It could be interpreted as a single group containing large quantities of missing mass but is more likely the result of erroniously including a number of separate (bound) groups into a larger unphysical group. If the former view was accepted, it is difficult to interpret the existence of bound low mass substructure within a larger dynamical system. The results therefore suggest that massive dark haloes probably do not extend much beyond the HI dimensions of galaxies.

REFERENCES

Appleton, P. N. and Davies, R. D., 1982. Mon. Not. R. astr. Soc. (in press).

Faber, S. M. and Gallagher, J. S., 1979. Ann. Rev. Astron. Astrophys., 17, 135.

Huchra, J. P. and Geller, M. J., 1982. Astrophys. J., 257, 423.

Limber, D. N. and Mathews, W. G., 1960. Astrophys. J., 132, 286.

HI OBSERVATIONS OF THE INTERACTING GALAXIES - VV 371 AND VV 329

Edwin J. Grayzeck,
University of Nevada Las Vegas

1. INTRODUCTION

Recent evidence for the tidal interaction among neighboring galaxies in the form of HI bridges and streamers has been accumulated by Haynes(1981). The successful application of computer simulations to model these encounters has largely been restricted to the nearby systems such as the Milky Way. In this paper, I will report on preliminary observations of the HI distributions associated with two interacting systems - VV 371 and VV 329.

2. OBSERVATIONS

Initially, those galaxies classed by Vorontsov-Velyaminov(1977) as "Dwarf Satellites on a Stem" were selected for observations at the Arecibo Observatory using its 21 cm, dual circular feed, to search for HI emission. From the sample of 20 objects, 6 were selected for subsequent observations employing the flat feed which has a HPBW of 3.9; for detailed characteristics of this receiver system see Hewitt, et al. (1982). For the two galaxies in this report, the 1008 channel autocorrelator was used to provide a velocity resolution of 8 km/s (VV 371) or 4 km/s (VV 329). Grid positions were chosen around these two galaxies to sample the environs of both the parent and nearby companion objects. A typical observation consisted of a 5 minute total power integration both on the galaxy and at a position to the east providing the same zenith angle coverage; the resulting RMS is 4.5 MJy. From this collection of profiles, a partially sampled map has been accumulated and preliminary contour corresponding to column densities exceeding 10^{19} cm^{-2} have been constructed for both galaxies as shown in Figure 1a and 1b.

3. DISCUSSION

VV 371

The contour map in Figure 1a indicates that the HI peak associated with this galaxy shows a distortion to the east, with a secondary rise approximately 12' from the main galaxy. A sampling of positions near

97

E. Athanassoula (ed.), Internal Kinematics and Dynamics of Galaxies, 97–98.
Copyright © 1983 by the IAU.

this enhancement shows that the peak is centered on a faint galaxy(VV371c).

VV 329A(NGC7679) and VV 329B(NGC 7682)
        The HI emission in Figure 1b is strongest towards VV 329B, but extends to the west to include both objects in a common envelope. The former object, which is also a radio source (Mirabel 1982), shows a distorted two lobed profile that exhibits only a single feature at VV 329A.

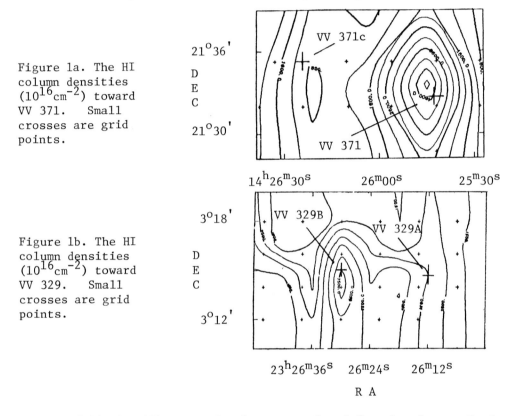

Figure 1a. The HI column densities $(10^{16}cm^{-2})$ toward VV 371. Small crosses are grid points.

Figure 1b. The HI column densities $(10^{16}cm^{-2})$ toward VV 329. Small crosses are grid points.

The following HI properties have been found for the above galaxies:

Table I

Galaxy	Type	V_H(21cm)	ΔV(0.2)	$\int$ S dV	Dist	Mass(HI)
VV371	SBb	1146 km/s	130 km/s	4.96 Jy-km/s	21 Mpc	0.50×10^9 M_0
VV371c		1130	80	2.19		0.21
VV329A	SB0	5080	150	2.76	92	5.5
VV329B	SBa	5120	240	4.22		8.4

4. REFERENCES

Haynes, M. 1981, Astron. J., 86, 1126.
Hewitt, J., Haynes, M., and Giovanelli, R. 1982, preprint.
Mirabel, F. 1982, preprint.
Vorontsov-Velyaminov, B.A. 1977, Astron. Astrophys. Suppl., 28, 1.

DISTRIBUTION AND MOTIONS OF ATOMIC HYDROGEN IN LENTICULAR GALAXIES

Hugo van Woerden, Wim van Driel and Ulrich J. Schwarz
Kapteyn Astronomical Institute
Groningen University
Groningen, the Netherlands

ABSTRACT

We report the results of HI observations of eleven gas-rich S0/S0a galaxies with the Westerbork Synthesis Telescope. The majority of these galaxies have most of their hydrogen outside the optical body, in annular configurations with diameters ~2 times the optical. These outer gas rings are often clumped, incomplete, and in approximately circular motion. They may represent the remnants of primordial, often warped, gas disks; or they may have formed from gas (or dwarf galaxies) accreted recently. Optical spectra could discriminate between these possibilities.

A few objects have filled gas disks, and several have inner HI rings with radii ~0.4 × optical. Two objects have peculiar distributions suggesting, respectively, tidal effects and stripping. However, stripping by intergalactic gas appears not to be a major current process.

1. INTRODUCTION

The gas content of lenticular (or, equivalently, S0) galaxies plays a key role in the problem of their origin and evolution.

Faber and Gallagher (1976) estimate that a galaxy of blue luminosity L_B should, through mass loss from evolving stars, in a Hubble time produce an amount of interstellar gas, M_g, given by $M_g/M_\odot \sim 0.1\ L_B/L_\odot$. Such amounts should be readily detectable, but are rarely observed in S0 galaxies. Faber and Gallagher consider several ways to remove the gas. Star formation would probably make the S0's bluer than they are observed to be. Stripping by collisions with other galaxies or with intergalactic gas might work in clusters, but not for field S0's. Faber and Gallagher conclude that hot, supernova-driven galactic winds are probably responsible for the lack of gas in S0's.

Larson, Tinsley and Caldwell (1980) calculate that star formation will exhaust the disk gas in most spiral galaxies in a fraction of the Hubble time. They suggest that spirals are replenished from gas-rich envelopes consisting of tidal debris, dwarf companions and/or remnants of primordial gas; and that S0's were formed from disk galaxies which early lost their envelopes and then consumed their gas by star formation. Larson et al. consider that gas produced later by evolving stars may be removed by intergalactic gas, by galactic winds, or by short bursts of star formation.

99

E. Athanassoula (ed.), Internal Kinematics and Dynamics of Galaxies, 99–104.

Thus, the absence of gas traditionally assumed in SO galaxies may be theoretically understood. But what is the observational situation?

Since 1975, major surveys of the HI content of SO and SO/a galaxies have been carried out at Arecibo, Green Bank and Parkes. Some 200 objects have been observed, and about 40-50 of these detected. The M_{HI}/L_B values measured range between <0.001 and 0.6. Clearly, the gas content of SO and SO/a galaxies varies over a wide range. This fact suggests that the HI component in these galaxies may be a transient phenomenon, and that we see them in different stages of evolution.

In an attempt to understand the origin and evolution of the gas in lenticular galaxies, we have measured the HI distribution and velocity field in a large number of such galaxies with the Westerbork Synthesis Radio Telescope. The present paper summarizes the current results of this program, and draws some tentative conclusions.

2. WESTERBORK OBSERVATIONS

For our Westerbork program, we selected galaxies of morphological type SO or SO/a (as listed in the Second Reference Catalogue, RC2), having a good HI detection in the recent surveys, positive declination, and sufficient angular size. Our sample spans considerable ranges in M_{HI}/L_B and in luminosity. Table 1 lists 11 objects for which maps are now available; it includes a few objects observed by colleagues at Groningen, who kindly made their results available for the present discussion. Several more objects have been observed, and the program continues. The Revised Shapley Ames (RSA) types listed show that most of our objects are genuine SO's or SO/a's.

The observations have ~40 km/s velocity resolution; the angular resolution is 0!4 in α and 0!4/sin δ in δ for most objects.

3. THE HYDROGEN DISTRIBUTIONS AND MOTIONS

In this section, we first describe the results obtained in two typical cases: NGC 4203 and 4262. Next we summarize and discuss the gas distributions and motions found in our sample of 11 galaxies.

3.1. NGC 4203 and 4262

NGC 4203 is a pure SO galaxy of optical diameter D_{25} = 3!6 and inclination i ~ 30°. The hydrogen in this galaxy has been mapped at Arecibo with 3' resolution by Burstein and Krumm (1981). We find (Fig. 1) most of the gas to lie outside the optical body, in an annular configuration of 6!5 diameter, i.e. almost twice the optical. This outer HI ring is incomplete and strongly clumped. If its true shape is roughly circular, it has an inclination of ~60° to the plane of the sky, and of 30° or 90° to the stellar disk. We also observe an inner ring, of dia-

Figure 1. Hydrogen distributions (left) and velocity
fields (right) for two galaxies. Column density contour
values are 1.0, 2.0, 3.0 ×
10^{20} atoms/cm^2 for NGC 4203;
and 1.4, 2.8, .. 7.0 × 10^{20}
atoms/cm^2 for NGC 4262. Velocities (in km/s) are heliocentric. The dashed ellipses are
25 mag/arcsec^2 isophotes from
RC2; the shaded ellipses show
the angular resolution of the
HI observations. Linear scales
are for distances based on
H_0 = 100 km s^{-1} Mpc^{-1}.

The HI distributions are
dominated by incomplete,
clumpy, tilted outer-ring
configurations. The velocity
fields indicate that these
rings are in approximately
circular rotation.

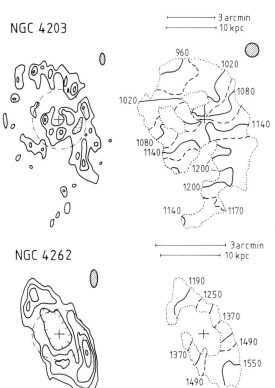

NGC 4203

NGC 4262

TABLE 1
Structural and dynamical properties of HI in lenticular galaxies

(1) Name NGC	(2) Morphol. type (RC2) (RSA)		(3) Features in HI distribution c r , D_r', χ (d) R, D_R' , χ T	(4) Notes	(5) hR_R kpc	(6) V_{rot} km/s	(7) hP_R 10^9yr	(8) hM_R $10^9M_\odot$	(9) M_R/hL_B $M_\odot/L_\odot$
1023	S B 0-	SB0 1	R,2.5 , < T	2 vel. systems	25	150	1.1	110	8
2273	SABrs0:		d R,0.8 , <		8	210	0.2	80	7:
2787	S Br 0+	SB0a	R,2.5:,60°						
3900	SA r 0+	S a(r)	r_0,0.4,< d	*	11	220	0.3	120	14
3941	S Bs 0+	SB0a1-2	r ,0.4,< R,1.6 , <		8	160	0.3	50	9
3998	SA r 0?	S 0 1	r? R?0.7:,70°	**	4:	280	0.1	70:	12:
4203	SAB 0-	S 0 2	r ,0.4, < (d) R,1.8 ,30°		11	150	0.4	60	10
4262	S Bs 0-?	SB0 2-3	R,1.7 ,40°	Virgo Cluster	7	240	0.2	90	27
4694	S B 0p	Amorph	c T	Virgo Cluster					
5084	S 0/	S 0 1	R,2.5 , <		27	310	0.5	600	60
7013	SA r 0a		r_0,0.4,25°: d R,0.85, <	***	6	170	0.2	40	7

Key to Table 1
(2)   Morphological types, from Second Reference Catalogue (RC2) and Revised Shapley Ames Catalog (RSA)
(3)   Features in HI distribution: c = concentration at centre; r = inner ring; r_0 = r coinciding with
       optical inner ring; d,(d) = filled, resp. partly filled gas disk; R = outer ring; T = tail. D_r',
       D_R' give ring diameters relative to D_{25} (from RC2); χ = inclination of ring (assumed circular)
       relative to optical disk; < means χ is small.
(4) *   NGC 3900: Data in columns (5)-(9) given for edge of HI disk.
    ** NGC 3998: Diameter and category of HI ring uncertain.
    *** NGC 7013: Inner ring tilted or motions non-circular.
(5)-(7) Linear radius R_R, rotation speeed V_{rot}, and period P_R of outer HI ring.
(8),(9) Total mass M_R within outer ring (assuming spherical distribution), and ratio to blue luminosity L_B.

meter 0.4 D_{25} and negligible tilt relative to the disk. A smoothed map
shows weak disk emission between both rings. The velocity field is somewhat irregular, but circular motions in a warped disk with a flat rotation curve give a reasonable fit.

NGC 4262 is an almost face-on (i < 30°), barred S0 galaxy of 2!2
diameter, 3° NW of the centre of the Virgo Cluster. Its hydrogen (Fig.
1) lies in an outer ring of 3!7 diameter (i.e. 1.7 × optical) and 60°
inclination (if circular), hence tilted >30° relative to the disk. The
ring is fairly complete, but quite uneven in brightness. The velocity
field bears the signature of solid-body rotation, indicating a circular
ring in circular motion.

3.2. A summary for 11 galaxies

Table 1 summarizes our results for 11 galaxies. Among these, 8 have
most of their hydrogen in outer-ring configurations (R), with diameters
between 0.8 and 2.5 times the optical (D_{25}). In general, these rings are
roughly concentric with the optical disk; however, if circular in shape,
several are strongly inclined to the disk. The density distributions in
these rings are quite uneven; some are strongly clumped, and often parts
of the ring are not detected. The velocities measured in the rings
closely follow the pattern of circular rotation; in NGC 1023 two velocity systems are observed. In section 4 we discuss the rotation speeds
and periods, and the galaxy masses derived.

Four galaxies have inner HI rings (r) of diameters about 0.4 times
the optical disk. Except in NGC 3900, these inner rings combine with
outer HI rings. In NGC 3900 and 7013, they approximately coincide with
optical (probably stellar) inner-ring features. In NGC 7013 the velocity
field indicates that the inner HI ring is tilted relative to the disk,
or that it possesses non-circular motions.

HI distribution and motions in NGC 3998 indicate a ring observed
edge-on and highly inclined to the optical body (cf. Knapp, this symposium). The ring diameter is uncertain because of the edge-on situation.

In four galaxies we observe a disk of hydrogen, in addition to an
inner and/or outer ring. In NGC 4203 we find a faint, extended disk. The
HI disks in NGC 2273, 3900 and 7013 are about as large as the optical
disks; these 3 galaxies, however, may be early-type spirals.

Two galaxies in Table 1 have exceptional HI distributions. NGC 1023
(Sancisi et al., in preparation) shows an incomplete outer ring in
circular motion, an anomalous velocity system, and a bright HI tail. The
tail and the anomalous-velocity system may be due to tidal action of NGC
1023 on a nearby, smaller late-type galaxy. NGC 4694 has a narrow,
clumpy HI tail and an unresolved central condensation of gas. This
unique distribution and the galaxy's location in the Virgo Cluster (5°
SE of the centre) suggest stripping by intergalactic ram pressure as a
possible cause.

4. ROTATION SPEEDS AND PERIODS; MASSES AND M/L RATIOS

The observed outer gas rings allow us to derive some dynamical
properties of gas-rich lenticular galaxies. The results (Table 1) are
given for distances based on a Hubble constant H_o = 100 h km s^{-1} Mpc^{-1}.

The outer HI rings have radii R_R ranging from 6 to 27 h^{-1} kpc.
Assuming circular ring shapes to derive inclinations, we find rotation
velocities V_{rot} between ~150 and ~300 km/s. The present sample shows no
Tully-Fisher relation between rotation velocity and luminosity, a
finding similar to that of Dressler and Sandage (1982, and this symposium) from optical rotation curves for another sample of SO's. The rotation periods of the outer HI rings range between 2 and 10 h^{-1} 10^8 years.

Assuming spherical mass distributions, we derive total masses
inside the outer HI rings, M_R = $V_{rot}^2 R_R/G$, ranging from 0.4 to 6 h^{-1} 10^{11}
solar masses. The total-mass-to-blue-light ratios cluster around 10h;
two values are much higher. We find no correlation between M_R/L_B and
relative ring diameter, D_R/D_{25}.

5. ORIGIN AND EVOLUTION OF THE GASEOUS COMPONENT

In this section we draw some tentative conclusions from the observational results summarized in section 3 and Table 1.

1) Our major finding is that in many gas-rich lenticulars most of
the hydrogen is located in the outer parts, or even outside, of the
luminous body. This strongly suggests that stripping by intergalactic
ram pressure is not a major current process in SO galaxies. For such
stripping should primarily remove the gas in the outer, rather than the
inner, parts. The only likely stripper in our sample is NGC 4694, a
galaxy in the Virgo Cluster. However, the other Virgo member, NGC 4262,
has all of its gas in a well-formed outer ring structure!

2) The frequent outer rings might well be interpreted as a stage in
the process of gas removal by galactic winds. However, it is unclear
that such winds would produce the observed uneven distributions in the
outer HI rings. Also, the circular motions observed may be inconsistent
with angular momentum conservation in an outward gas flow (although this
argument may be invalid in non-axisymmetric cases).

3) Another possibility to be considered is that of accretion of
intergalactic gas clouds, tidal debris or dwarf companions (cf. Silk and
Norman, 1979) by a formerly gas-poor lenticular galaxy. The uneven
structure of the outer rings may suggest dwarf capture. The close-to-
circular motions suggest that capture occurred a few periods, that is:
about 10^9 years, ago. The inner HI rings observed in a few cases might
be due to earlier accretion, cf. the Spindle galaxy, NGC 2685 (Shane,
1980).

4) Finally the outer rings may represent the remnants of primordial, often warped gas disks, of which the dense inner parts have been used up in star formation, while the tenuous outer parts never gave birth to stars. (For removal of the gas produced later in the inner parts by stellar evolution, a galactic wind may then still be required.) This scenario is partly similar to that of Larson, Tinsley and Caldwell (1980) summarized in section 1. However, most of our gas-rich SO's apparently do have gas-rich envelopes, which somehow may have failed to replenish the inner disks.

These are tentative suggestions, which will have to be analyzed in detail before they can lead to firm conclusions. However, an important observational test appears evident. Comparison of the sense of rotation in the outer gas rings with that in the inner, stellar disks should decide for or against scenario 3). Opposite senses of rotation would require accretion; if the senses always agree, recent accretion of dwarfs or intergalactic gas can probably be ruled out.

Another item for further study is what the gas distributions tell us about the place of SO galaxies in the morphological sequences. For this purpose, we have undertaken a companion program on Sa galaxies, as part of a comparison between lenticulars and spirals. Also, the relationship of the outer HI rings to the phenomenon of optical outer rings requires investigation.

ACKNOWLEDGEMENTS

First, we thank Jill Knapp for her vital contributions to this program. She has freely provided us with unpublished detections, and shared in many of our analyses and discussions. We further thank Renzo Sancisi and Seth Shostak for making their results on NGC 1023 and 2787 available, and for much help and discussion. Jay Gallagher and Nathan Krumm collaborated on individual galaxies. Beatrice Tinsley, Bob Sanders, James Binney and Scott Tremaine contributed through discussions.

Wim van Driel is supported by ZWO through the ASTRON foundation. Jill Knapp's visits to Groningen were financed by a ZWO fellowship. Our work on gas in early-type galaxies is further supported by NATO Scientific Affairs Division through grant RG 098.82.

The Westerbork Radio Observatory is operated by the Netherlands Foundation for Radio Astronomy with financial support from ZWO. The observations were analyzed using GIPSY at the Computing Centre of Groningen University. We thank the Foundation staff and our local colleagues for their share in this work.

REFERENCES

Burstein, D. and Krumm, N.A. 1981, Astrophys. J. 250, 517
Dressler, A. and Sandage, A.R. 1982, Astrophys. J., in press
Faber, S.M. and Gallagher, J.S. 1976, Astrophys. J. 204, 365
Larson, R.B., Tinsley, B.M., Caldwell, C.N. 1980, Astrophys. J. 237, 692
Sancisi, R., van Woerden, H., Davies, R.D., Hart, L. 1982, in preparation
Shane, W.W. 1980, Astron. Astrophys. 82, 314
Silk, J. and Norman, C.A. 1979, Astrophys. J. 234, 86

NEUTRAL HYDROGEN MAPPING OF THREE S0 GALAXIES

W.W. Shane
Astronomical Institute, Catholic University of Nijmegen

N. Krumm
Physics Department, University of Cincinnati

Synthesis maps of HI in three S0 galaxies are discussed. In one case several companions are observed. An evolutionary sequence based on accretion is suggested.

Neutral hydrogen in three S0 galaxies has been mapped using the Westerbork Synthesis Radio Telescope. Each galaxy was observed for one 12-hour period during 1978-79 using the digital back-end and with a system temperature of 90 K. The bandwidth after Hanning smoothing was 33 km/s, and the maximum baseline was 1500 m. Before presentation, the maps were convolved to a larger beam. The galaxies are listed in table 1. A Hubble constant of 100 km/s/Mpc has been assumed.

NGC 2655: The HI extends well beyond the prominent optical halo and shows a chaotic ring-like structure with many fragments and little correlation with the optical picture. The velocity field indicates rotation (east side approaching) but is also very irregular (figure 1 left). The global line profile is quite asymmetric.

NGC 4138: Again the HI extends well beyond the optical image. It is distributed roughly in two concentric rings, both apparently seen more face-on than the optical galaxy. The velocity field is quite regular; the rotational velocity decreases from the inner to the outer ring (figure 1 right). The lowest velocities were not observed.

Table 1. List of observed galaxies

Name	Type (from 2RCBG)	B_T^0	V(hel) (km/s)	M_H ($10^8 M_\odot$)	M_H/L_B ($M_\odot/L_\odot$)
NGC 2655	SAB(s)0/a	10.49	1374	8.5	0.035
NGC 2859	(R)SB(r)0+	11.36	1670	>1.2	>0.01
NGC 4138	SA(r)0+	11.97	875	>2	>0.09

E. Athanassoula (ed.), Internal Kinematics and Dynamics of Galaxies, 105–106.

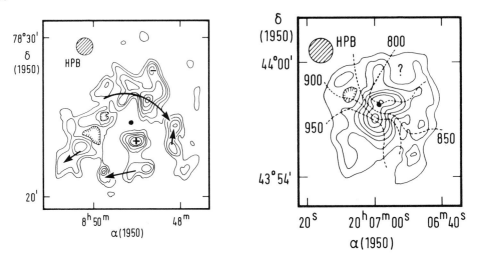

Figure 1. (left) Column density map of NGC 2655 with contour levels
1.8, 2.7, 3.7, ... x 10^{20} cm^{-2}. The arrows indicate positive velocity
gradients, the + and - regions of exceptional velocity and the dot the
optical center. (right) The same for NGC 4138 with contour levels
0.4, 0.9, 1.3, ... x 10^{20} cm^{-2} and with isovelocity contours added. The
question mark indicates the incompletely observed region.

     NGC 2859: The two HI peaks (n and f) lie oñ the optical ring but
may be part of a more diffuse cloud. Many known companions and two HI
clouds are detected (table 2). Some require confirmation (:).

     Attributing the HI in these galaxies to accretion events, the
following evolutionary sequence is suggested: A dwarf galaxy or HI cloud
is captured and fragmented (NGC 2655); the gas settles into a disk or
rings (NGC 4138); the denser regions are depleted by star formation and
produce an outer optical ring (NGC 2859).

Table 2. Objects in the field of NGC 2859

Name	HI Mass (10^6 M$_\odot$)	Velocity (km/s) hel	Velocity width	Name	M$_H$	V	ΔV
:Cloud 1	160	1754	50	:UGC 5011	>150:	>1848	>50:
:UGC 4988	70	1535:	50:	:Cloud 2	180	1725	100:
NGC 2859 n	40	1668	40	:UGC 5014	>400:	>1820	>120
NGC 2859 f	80	1573	65	UGC 5015	380	1644	120
UGC 5004	>90	>1836	>80	UGC 5020	520	1635	200
Positions (1950): Cloud 1 = 9 19.8, +34 46; Cloud 2 = 9 22.3, +34 33.							

II

SPIRAL STRUCTURE

THEORY OF SPIRAL STRUCTURE

Agris J. Kalnajs
Mount Stromlo and Siding Spring Observatories, Research School
of Physical Sciences, Australian National University

ABSTRACT

Spiral structure is associated with a slow redistribution of gas,
which may already be quite significant over time scales short compared
to the age of the galaxy. One has to worry about replacing the gas in
order to keep the structure alive.

1. INTRODUCTION

One of the more interesting discoveries of the past decade has been
the liveliness of self-gravitating stellar disks. Much of the zip can
be directly attributed to differential rotation, a feature which was
once thought to be inimical to any coherent structure. Disks show a
definite preference for bisymmetrical deviations from axial symmetry,
which also happens to be one of the more striking features of large
scale spiral structure. Thus there is little doubt in my mind that the
understanding of spiral structure is to be found in the study of the
dynamics of stellar disks, and what their oscillations can do to the
gaseous component.

An alternative approach to spiral structure is to view it as the
result of stochastic self-propagating star formation (Mueller and Arnett
1976; Gerola and Seiden 1978) - a galactic version of the computer game
Life, played in a shearing world. While the physical mechanisms invoked
by this theory are all plausible, it lacks one important ingredient: the
large scale organisation and symmetry. I like to illustrate the latter
with the help of a photographic trick. Figure 1 shows a well-known
galaxy on the left, togather with a symmmetrised version. The latter
was obtained from the same negative, except that after the first half of
the exposure the paper was rotated 180 degrees around the center and
then exposed again. The main thing to note is that the exposures were
so chosen as to cancel any feature which did not have a counterpart on
the other side of the center, e.g. the companion and the field stars
are gone, but the spiral structure remains almost in tact, even the
bifurcations in the arms in the 10 and 16 o'clock quadrants. It is the

E. Athanassoula (ed.), Internal Kinematics and Dynamics of Galaxies, 109–116.
Copyright © 1983 by the IAU.

explanation of this large scale symmetry and its permanence which should
be the primary task of a spiral theory, the irregularities I would
consider as secondary.

Figure 1.   M-51 (left) and its symmetric version (right).

The task of any reviewer of this topic has been lightened
considerably by the existence of a very comprehensive review (Toomre
1977). All I can hope to do is to add a few footnotes, pointing out new
results and changes in emphasis.

2.   STELLAR MODES, GAS, AND TIDES

The theory of spiral structure was born when people began to think
of it as a density wave phenomenon. The theory should tell us why
spirals

      (a) are long lived,
      (b) have a large scale, and
      (c) require both gas and stars.

A combination which satisfies the above criteria is a stable
stellar disk which can support a large scale mode, to which one adds a
bit of gas.

In order to be long lived, the mode has to be practically stable,
and therefore it can not be noticeably spiral. The spirality arises
from the inability of the gas to settle down in the periodic orbits of
the non-axisymmetric field, because they often intersect in space. As a
result the gas must bump into itself, and this gives rise to a
quasi-steady flow with shocks. The bumping is the feature which makes
the gas essential and distinguishes it from the stars.

For the above scenario to work there are many parameters such as mode shapes, amplitudes, pattern speeds, gas distribution, etc. which have to be determined, as well as observational constraints to be satisfied (Kormendy and Norman 1979).

The logical place to start is by choosing a disk and calculating the modes. The question is: which disk? All the seemingly reasonable disks turned out to be unstable, and stable ones unreasonable, usually because they were too hot. One way to make cooler disks is to suppose that they are only partly self-gravitating, by imbedding them in a halo. This I find a mixed blessing.

Many studies have by-passed this initial difficulty by simply assuming a mode or a bar (= finite amplitude mode), and proceeded to the second stage: to see what will happen to the gas. The results have been very encouraging, which should provide added incentive to tidy up the disk stability problem.

If we do not insist that spirals be long-lived, we may assume that both stars and gas have a spiral form, as in the case of tidally forced structure. While suggestions and hints that a tidal interaction could generate a fine spiral have been around for at least a decade, the convincing demonstration of how and why it works was provided only recently (Toomre 1981). The key ingredient turns out to be an effective swing-amplifier.

3. INSTABILITIES OF STELLAR DISKS

The stock of global modes of a variety of disks has shown a healthy growth since 1977, most of them demonstrating the well-known fact that disks are unstable, and that haloes can help to make them less so. More important, there has also been considerable growth in understanding of the causes of the fiercest instabilities (Toomre 1981). There appear to be two distinct non-axisymmetric amplifying mechanisms, giving rise to swing-amplified and edge modes. The former depends on the ability of shear to amplify leading outgoing waves while swinging them around and sending them inward as trailing ones. The cycle is closed, when the trailing wave passes through the center, and returns as a leading one. It has been possible to demonstrate the demise of this type of instability by simply placing an absorbing plug in the center of the disk. The same effect could be achieved by redistributing the mass so as to give rise to an inner Lindblad resonance, but the demonstration is not as convincing, since tampering with the mass distribution causes other parameters to change. (This is one of the frustrations inherent in trying to isolate and localise features of a global mode.) Of course we may also turn down the gain of the amplifier by decreasing shear, or by placing some of the mass in a halo. While stability can be achieved by turning down the gain or breaking the cycle, a possible distinction arises when the stable disk experiences a tidal field. Because by breaking the cycle we preserve the lively nature of the disk, it will still respond enthusiastically to any tidal forcing, although in a transient manner.

The edge modes depend on the gradients in angular momentum distribution for their growth. They do not care about central absorbing plugs. Here the obvious cure is to minimise the gradients.

The unstable modes seldom fall neatly into one or the other category, usually both mechanisms operate. However by judicious fiddling one can select almost pure examples to illustrate the two instabilities.

4. EQUILIBRIUM MODELS

The study of spiral structure would be greatly simplified if we knew the equilibrium structure of a disk galaxy. Under the most favourable conditions we may hope to discover the mass distribution from a combination of rotational data and photometry. That still leaves us with the problem of determining the velocity distribution. It is conceivable that some information about the latter could be obtained by requiring the absence of instabilities. The introduction of dark haloes to stabilise disks nullifies that, and certainly the rotation will no longer tell us much about the disk mass. Thus life would be simpler without haloes.

One fairly obvious way of generating stable disks is to let the instabilities run their course. Hohl (1971) showed that it is possible to get rid of the resulting stable bar by simply reshuffling the stars in azimuth. The symmetrised disk remained stable, but was rather hot in the sense that the axisymmetric stability parameter Q was around 4. By a more careful selection of the initial configuration, the final Q could be reduced to around 2. Again it is conceivable that the high Q's might have been due to the violence of the instability. To check this, Hohl tried cooling the disk and in so doing managed to reduce Q to the range 2 to 3 before a slowly growing bar started to heat it up again. It now transpires that the reported radial velocity dispersion was an azimuthal average and hence included all systematic motions. Insofar that the random motions are due to the decay of the systematic flow, the latter could dominate the azimuthal average. Already a reduction of Q by, say 1.5, would make it quite compatible with the (only known) value near the sun. Did we give up too soon?

5. EVOLVING GAS FLOWS

There are two somewhat distinct approaches to generating spiral structure in gas. One is by brute force: you assume a spiral forcing field and get a spiral response (Visser 1978; Roberts et al 1979). The WKBJ theory necessarily falls in this category. The other relies on finesse: you start with a non-spiral field such as a bar or oval distortion, and let the inhomogeneities of the disk - the resonances - give the gas a spiral form. The bar forcing has received most attention simply because the bar phenomenon is so widespread both in nature and theory.

There are several distinct regimes of spiral making. The first is the initial or turn-on transient. Since all calculations seem to start from a circular flow, there is a transition period of the order of a bar revolution. During this phase both stars and gas start out in a similar fashion, and continue until pressure forces part them. The gas remains spiral, while the stars begin to phase mix and finally adopt a distribution which reflects the symmetry of the force field.

Spirals can be generated by an exponentially growing field. As long as the orbital excursions remain small, both stars and gas will show spirality, although the shapes may differ.

Spirality can also be sustained in a steady field by dissipation, due to viscosity or shocks.

The common factor in the above three cases is evolution. Another way of saying this is to note that whenever the gas response to a bar is spiral, the torques exerted on different annuli will not vanish, implying that the angular momentum distribution has to evolve. The torques can be computed from the observed or computed mass distribution, and do not depend on how that distribution was produced.

Clearly the initial transient stage is not the answer. However practical considerations such as numerical stability and accuracy, discourage long integrations, and the slower evolution of the "quasi-steady" state has been largely ignored, or wrongly attributed to numerical effects.

The age of a typical "quasi-steady" gas spiral is about one to two bar revolutions, although in the most recent work (Schempp 1982) stops after half a revolution. The evolution during this shock dissipation phase can be quite rapid in regions close to the bar. Huntley (1980) provides sufficient data to estimate the rate at which the gas in the straight off-set shocks is losing angular momentum. If you assume that the angular momentum at his radius 10L is that corresponding to circular motion in the axisymmetric field, and divide it by the annular torque, the decay time turns out to be about half a rotation period or only 1/6 of a bar period! This is an alarmingly short time, and it may be that the "quasi-steady" pattern survives only because the gas which falls into the center is recycled. Huntley calls the infall "numerical leakage".

A similar rate of infall is evident from tracing the streamlines of the standard model of Sanders and Tubbs (1980). It can be seen that a streamline which leaves the shock at radius a, meets the next shock at half this value.

The above calculations use the beam scheme. The inward spiraling of the streamlines in the vicinity of the bar is also quite evident in the calculations using other hydrodynamical schemes (van Albada and Roberts 1981). The time scale for moving from one end of the bar (6.9 kpc) to the center (0.47 kpc) in their model is about one gigayear, or two bar revolutions. A comparison with the beam scheme shows that both behave the same way.

Because of the numerical viscosity inherent in the above schemes, it was not at all clear how much of the evolution should be attributed to computational artifacts, and how much is real. That shocks imply some evolution was pointed out by Pikel'ner (1970). The first clear demonstration that it was significant over a lifetime of a galaxy came from the work of Schwarz (1979; 1981). In contrast to most studies, he put his effort in trying to understand the flows around the outer Lindblad resonance. (One of the reasons for this choice is the fact that the model parameters become fewer: all bar fields and rotation curves begin to look the same at large distances.) Schwarz modeled the gas with inelastic clouds, and used a Lagrangian scheme to follow their evolution. Once you accept the cloud model, there is little room for numerical misgivings.

The initial evolution is quite similar to that of the continuum schemes. After two bar revolutions a quasi-steady state forms, with a nice spiral collision front. In a typical case, it then slowly evolves into a ring (actually a periodic eccentric orbit) around the outer resonance, over a time scale of ten bar revolutions, or 2.5 gigayears. Once the ring is formed, there are no more collisions, and all evolution stops. One can understand the formation of the shock in terms of intersecting periodic orbits (Kalnajs 1972), and its demise as the result of their drift outwards to the resonance. In order to keep the spiral structure alive, one has to repopulate the depleted orbits, for example by mass loss from stars or infall. (The incipient ring has a distinct shape which is quite common among galaxies, and this identification argues that the pattern speed should be high enough to give rise to an outer resonance.)

The evolution towards a periodic orbit around an eccentricity resonance appears to be quite general, and not peculiar to the outer resonance. There is a hint of this phenomenon at the end of the hydrodynamical calculations of Sorensen and Matsuda (1982).

Since it is the torques that appear to be responsible for the redistribution of the gas, it is natural to ask whether a tightly wrapped spiral field where the ratio of tangential to radial force is small, could slow the drift. Provided it was was tight enough, the answer must be yes. But in the case of M-81, which is very average in terms of the tightness of its arms, Visser's spiral lasts only about four bar revolutions before the redistribution destroys it (Schwarz 1979).

REFERENCES

Gerola, H., and Seiden, P.E. : 1978, Astrophys. J. 223,129.
Hohl, F. : 1971, Astrophys. J. 168, 343.
Huntley, J.M. : 1980, Astrophys. J. 238, 524.
Kalnajs, A.J. : 1973, Proc. Astron. Soc. Austr. 2, 174.
Kormendy, J., and Norman, C.A. : 1979, Astrophys. J. 233, 539.
Mueller, M.W., and Arnett, W.D. : 1976, Astrophys. J. 210, 670.
Pikel'ner, S.B. : 1970, Astron. Zh. 47, 752.
Roberts, W.W., Huntley, J.M., and van Albada, G.D. : 1979, Astrophys. J.
      233, 67.
Sanders, R.H., and Tubbs, A.D. : 1980, Astrophys. J. 235, 803.
Schempp, W.V. : 1982, Astrophys. J. 258, 96.
Schwarz, M.P. : 1979, Ph. D. Thesis, Australian National University.
Schwarz, M.P. : 1981, Astrophys. J. 247, 77.
Sorensen, S-A., and Matsuda, T. : 1982, Monthly Notices Roy. Astron.
      Soc. 198, 865.
Toomre, A. : 1977, Ann. Rev. Astron. Astrophys. 15, 437.
Toomre, A. : 1981, In The Structure and Evolution of Normal Galaxies, ed.
      S.M. Fall and D. Lynden-Bell, p. 111, (Cambridge : Camb. U. Press)
Van Albada, G.D., and Roberts, W.W. : 1981, Astrophys. J. 246, 740.
Visser, H.C.D. : 1978, Doctoral Thesis, University of Groningen.

DISCUSSION

VAN ALBADA : Using the FS2 hydrodynamic code described elsewhere in this
volume, a very long computational run was made of a model of M81. The
description of the potential was taken from M.P. Schwarz's thesis (1979) ;
a velocity of sound of 15 km/s was chosen. The model was computed on a
80 by 160 zone half grid with 200 pc cells ; it was continued to a model
time of 4.3 Gyr, or approximately 13 pattern revolutions. As is the case
in Schwarz's cloud-particle computations (which were run for approximately
4 pattern revolutions) most of the gas has been removed from the spiral
arm region by the dissipation in the shocks. However, the similarity ends
here. Instead of narrow rings of matter, the following structures can be
identified :
1) An annulus around corotation, attributable to the fact that the gas
streaming velocities relative to the pattern are too low to cause shock
dissipation in this region. The relatively high gas density in this annulus is due to the large extent of the assumed initial density distribution.
2) A very wide inner annulus having an enhanced gas density caused by the
gas transported from the depleted regions somewhat further out in the
plane. The width of this annulus is probably real as its appearance remained unaltered in a continuation of the computation on a very high resolution grid. It appears to be caused by the gradual disappearance of shocks
near the inner Lindblad resonance.
3) Arms clearly delineated by shocks, having peak densities comparable to
those in the corotation annulus. The occurence of shocks in this region,
whatever the local gas density, is a direct consequence of modelling the
ISM as a perfect gas.

At first sight we might state that Schwarz's and my computations represent
two extremes of the possible behaviour of the gas in M81. The ISM is evidently not as viscous and inelastic as modelled by Schwarz, nor is it as
free of viscosity as it is assumed to be in my computation. The observed
gas distribution, however, does not resemble a cross between Schwarz's and
my results. The HI is predominantly concentrated in the region of spiral
arms and not in some kind of inner ring, as predicted by both computations. The result presented here is more realistic in that the arms are
delineated by shocks. Star formation processes are presumably important
in shaping the observed distribution of the gas in M81, preventing the
buildup of an inner gas annulus. The observed gas distribution may also
be influenced by the acquisition of new gas, as implied by Tinsley and
Larson's models of spiral galaxies. Alternatively, we may not exclude the
possibility that the observed strong and regular spiral structure in M81
is wholly temporary and caused by the recent interaction with M82 and
NGC 3077 evident from the extended HI distribution.

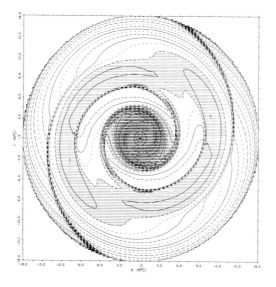

W.W. ROBERTS : The shortness of the time scale suggested by the N-body
gas/particle calculations, quoted from Schwarz, for the destruction of
gaseous spiral structure and the pile up of gas into a ring, may in part
be attributed to the high dissipation adopted for the cloud-cloud collisions in his code. If the dissipational processes in real galaxies are
not so severe, then the spiral structure may be expected to persist over
longer time scales, and considerably more so if dissipational processes
are partially offset by sources of energy input to the cloud system, such
as supernovae and stellar winds.

KALNAJS : I agree that decreasing the dissipation in the collision process for particle calculations, or making the gaseous shocks milder by
increasing the sound speed, slows the secular radial migration of the
gas. But at the same time the spiral arms become broader and less distinct

QUASI-STATIONARY SPIRAL STRUCTURE IN GALAXIES

C.C. Lin
Massachusetts Institute of Technology
Cambridge, Massachusetts 02139 USA

The hypothesis of *quasi-stationary spiral structure* in galaxies was explicitly formulated in the early 1960's in papers of Bertil Lindblad and of Lin and Shu. It asserts that the grand design observed in spiral galaxies may be described by the superposition (and interaction) of a *small* number of spiral *modes*. (See Lin and Bertin, 1981 for a fairly extensive review of the theory.) We wish to re-affirm the correctness of this hypothesis in the present contribution. Early numerical experiments by P.O. Lindblad and by Miller, Prendergast and Quirk demonstrated that spiral structures occur naturally in certain models of stellar systems, although it was difficult to control the morphological types of galaxies simulated. We are now able to simulate galaxies of various morphological types in a controllable manner. Numerical fluid-dynamical codes developed by Pannatoni (1979) and improved by Haass (1982) have been used to calculate normal modes of various spiral types (Haass, Bertin, and Lin 1982) in the morphological classification of Hubble, Sandage, and de Vaucoulers. Furthermore, the processes that govern the maintenance and the excitation of these modes simulating both normal spirals and barred spirals, can be understood by using analytical theories which are closely related to the local dispersion relationship, as Bertin will describe in his paper at this conference. Understanding these mechanisms enables us to choose the parameters and the distribution functions in our models more properly in order to exhibit the desired characteristics in the computed modes. Such an approach also has important implications on observational studies. Much of the previous work on comparison between theory and observations in normal spirals used only the short trailing waves. A mode must consist of at least two waves propagating in opposite directions. It has been found that, at least in normal spiral modes, the long wave branch provides essentially only a *modulation* of the amplitude along the short wave branch, which accurately describes the phase. Previous calculations are thereby justified. [These points were not adequately covered in the previous paper reviewing theory of spiral modes.]

I wish now to turn to the discussion of another point of great

E. Athanassoula (ed.), Internal Kinematics and Dynamics of Galaxies, 117–118.
Copyright © 1983 by the IAU.

significance to observers. The powerful "swing" amplification process
suggests that the spiral structure might be rapidly varying, on the
time scale of an epicyclic period and *not* quasi-stationary. Actually,
this perception of swing amplification only holds for wave *packets*.
In the context of steady wave *trains* or *modes*, it can be shown that an
exactly equivalent amplification mechanism can be described in terms
of a process of wave amplification via stimulation of emission of
radiation (WASER, see figure). The same analytical and numerical
calculations are involved; only the physical interpretation of the
symbols are different. Indeed, the amplification factor at the corotation circle of a mode has been systematically calculated and published
by Drury (1980) for two standard rotation curves. We have now further
clarified the comparison of the two approaches by deriving a second
order differential equation for the Fourier transform of the perturbation of the gravitational potential, with a time-*independent* wave
number ξ as the independent variable. The parameter ξ then takes the
place of the time-*dependent* wave number in the swing formalism.
Otherwise, the equations to be solved are identical in form to Eq.
(35a) [homogeneous part] and Eq. (35b) in the 1978 paper by Goldreich
and Tremaine and to Eqs. (18) and (20) in the 1981 paper of Toomre.
The numerical calculations required are also the same. The WASER is,
however, more *appropriate* for the description of modes or steady wave
trains. Indeed, Goldreich and Tremaine (1978) already demonstrated the
equivalence of the swing process with the WASER processes in the
special case studied earlier by Mark, who incorporated the amplification mechanism into the asymptotic theory of normal modes (WASER I).
Now the equivalence is demonstrated for all spiral modes. Incorporation of the WASER process into the theory of bar modes (WASER II) will
now be explained by Bertin.

Figure 1. *WASERS OF TYPES I AND II:* As the incident wave is refracted,
there is also stimulated an emission of radiation. This strengthens
the returned wave, thereby providing amplification in the feedback cycle.

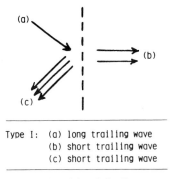

Type I: (a) long trailing wave
         (b) short trailing wave
         (c) short trailing wave

Type II: (a) (short) leading wave
         (b) (short) trailing wave
         (c) (short) trailing wave

REFERENCES

Drury, L.O'C.: (1980), M.N. 193, 337–343.
Goldreich, P. and Tremaine, S.: (1978),
    Ap. J. 222, 850–858.
Haass, J., Bertin, G. and Lin, C.C.: (1982),
    Proc. Nat. Acad. Sci. USA 79, 3908–3912.
Lin, C.C. and Bertin, G.: (1981), in *Plasma
    Astrophysics*, eds. Guyenne, T.D. and Levy,
    G. (European Space Agency, Noordwijk,
    Netherlands), SP-161, 191–205.
Pannatoni, R.F.: (1979), Ph.D. Thesis MIT,
    Cambridge, Massachusetts, USA.
Toomre, A.: (1981), in *The Structure and
    Evolution of Normal Galaxies*, eds. Fall,
    S.M. and Lynden-Bell, D. (Cambridge
    University Press, London), 111–136.

DYNAMICAL MECHANISMS FOR DISCRETE UNSTABLE SPIRAL MODES IN GALAXIES

Giuseppe Bertin
Scuola Normale Superiore,
I - 56100 Pisa, Italy

Progress in understanding physical mechanisms for the excitation and maintenance of spiral structure has considerably benefited from investigations of tightly wound spiral density waves (e.g., see Bertin 1980). These studies have identified the existence of four basic kinds of density waves (trailing and leading waves, and in each case short and long waves) with different propagation properties. In addition, they have led to the conclusion that some realistic galaxy models can support self-excited global normal spiral modes. These owe their maintenance to the presence of trailing waves with opposite propagation properties and are excited mostly as a result of a WASER (superreflection) mechanism at corotation. In discussing the dynamics of spiral structure and in comparing theory with observations a number of important issues should be kept in mind (Lin and Bertin 1981). Here we just recall that the calculation of spiral modes is being pursued by many researchers, using different methods. In general the structure and the growth rates of the dominant modes are determined by the radial distributions of the active disk density, the differential rotation, and the dispersion speed through the dimensionless functions ε_o, j, and Q (Haass, Bertin and Lin 1982).

Analytical methods of investigation, based on certain asymptotic studies of properties of steady wave trains, prove to be very useful tools in understanding the physics of spiral modes. In the Figure we show propagation diagrams and density contours for two different regimes, one typical of normal spiral modes and the other characteristic of certain barred spiral modes. The difference in the regimes of the basic parameters is clearly reflected in the change of topology of the propagation diagrams. This in turn results in the different structure of the first two-armed mode. The relevant channels for wave reflection, wave refraction, and WASER are well illustrated in the diagrams (upper frames) which "diagnose" the numerically computed modes (lower frames). The main new result presented here is contained in the right frames, where we have tested an analytical model for barred modes. In the regime of higher Q and j considered, a new asymptotic analysis reveals the possibility of a WASER of type II (to be distinguished from

119

E. Athanassoula (ed.), Internal Kinematics and Dynamics of Galaxies, 119–120.

the WASER of type I, found by Mark, which operates in the regime
illustrated in the left frames). This mechanism is characterized by a
leading wave impinging on corotation and producing an amplified trailing
wave traveling back to the galactic center and a transmitted trailing
wave which is eventually absorbed in the outer regions. The same
asymptotic analysis reveals the possibility of a "reverse swing" close
to the galactic center which completes the loop necessary for the
maintenance of the barred spiral mode. For the case shown, there is
quantitative agreement between the structure and the growth rate of the
mode computed numerically and the predictions of our simplified analytical model based on a local dispersion relation. The equilibria investigated and the modes shown are part of an extensive exploration made by
Haass (1982).

Clear results for extreme regimes, such as those presented above,
encourage us to extend and apply concepts borrowed from asymptotic analysis to interpret more complicated transition cases. In this regard
power spectra (Haass, Bertin and Lin 1982) and propagation diagrams are
very useful and indicate that all different kinds of waves and loops
participate in the structure of transition modes. Studies of this kind
can form the basis for a dynamical approach to the classification of
spiral galaxies.

Bertin, G.: 1980, Phys. Reports, 61, pp.1-69
Haass, J.: 1982, Ph. D. Thesis, MIT, USA
Haass, J., Bertin, G., and Lin, C.C.: 1982, Proc. Natl. Acad. Sci. USA,
    79, pp.3908-3912
Lin, C.C., and Bertin, G.: 1981, in "Plasma Astrophysics", T.D. Guyenne
    and M. Levy, eds ESA SP-161, pp.191-205.

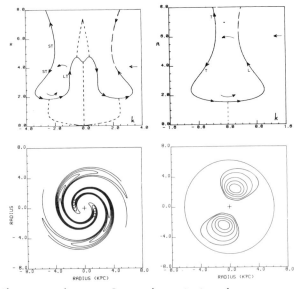

Propagation diagrams (upper frames) and density contours (lower frames)
for a normal spiral mode (left) and a barred spiral mode (right).

GLOBAL INSTABILITIES OF A GALAXY MODEL

Jon Haass
Department of Mathematics
Massachusetts Institute of Technology

A simple two-component fluid model of a galaxy is analyzed numerically.
For this equilibrium configuration a large number of unstable spiral
modes is found. It is of particular interest that some of these modes
are well described by the asymptotic theory developed for tightly wound
trailing spirals, while others are best understood in terms of the
swing formalism which includes both leading and trailing waves.

The model consists of a Toomre disk of order 5 and length scale
12, plus a Plummer sphere with length scale 2 containing half as much
mass. Only the disk is dynamically active. The sphere, regarded as
frozen, affects only the total gravitational potential. This combination produces the rotation curve shown in Figure 1. To complete the
description I specify the stability function as

$$Q(r) = 1 + \exp\left(\frac{-r^2}{2}\right) . \tag{1}$$

In other words, I assume that the innermost portion of the disk (residing more or less within the sphere) is quite "hot", whereas its exterior
is just warm enough to avoid Jeans instability. These two conditions
were adopted purposely to favor - and thereby test - the refraction and
amplification of the short and long trailing waves involved in the
asymptotic theory of Lau, Lin and Mark (1976).

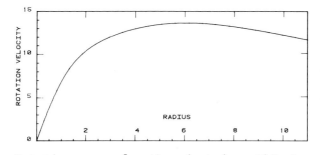

Figure 1. Rotation curve for the adopted equilibrium model.

E. Athanassoula (ed.), Internal Kinematics and Dynamics of Galaxies, 121–124.
Copyright © 1983 by the IAU.

The dynamics are assumed to be governed by the linearized equations
of motion for the system. I concentrate on the unstable spiral modes
using methods similar to those developed by Pannatoni (1979), and find
that there are plenty of them. Figure 2 reports the modal "spectrum"
for the basic model in a form suggested by Toomre. It obviously offers
a lot of information. For example, the third two-armed mode (2C) has a
growth rate 0.512 and corotation radius 5.91 corresponding (with the
help of Fig. 1) to a pattern speed 2.305. That this basic equilibrium
is quite unstable makes it all the better to illustrate two distinct
sources of instability.

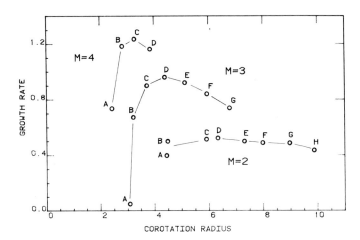

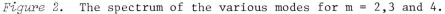

Figure 2.   The spectrum of the various modes for m = 2,3 and 4.

One feature seen in the spectrum is the nearly identical pattern
speed and growth rate for the first two m=2 modes. This "degeneracy"
is now known to be one of the signs that the model (for these modes) is
approaching the asymptotic regime where the instability can be traced
to a short-long trailing wave feedback cycle first discussed by Lau,
Lin and Mark (1976). As evidence for the correctness of this description, I have computed the mode based on the asymptotic second order
equation and found agreement in pattern speed and growth rate to
within 5% and 15% respectively. Further, the eigenfunctions produced
by the different methods are nearly indistinguishable. (See Fig. 3.)

The spectrum in Figure 2 also cautions, however, that these
all-trailing modes are not the whole story even in this favorable
setting. Notice that several of the m=3 modes grow about twice as
rapidly as the m=2 modes we have just been discussing. And although
the first two m=3 modes are close in pattern speed, their structures
and growth rates are not. These modes owe their instability to a feedback loop quite different from the aforementioned cycle. This possibility was first recognized by Bardeen (1976) in his own gas-disk calculations. He realized that the regularly spaced interference patterns
typical of these modes (see especially mode 3D in Fig. 3) signify a
superposition of trailing and leading waves of similar length and

amplitude. This intuition is now supported by a solid theory. Given
the pattern speed and model, Toomre's (1981) method for calculating the
growth rate from group transport and swing amplification yields agreement with the "exact" values to within 10% for modes 3D, 3E, etc.
Again the essence of the modes seems to have been grasped.

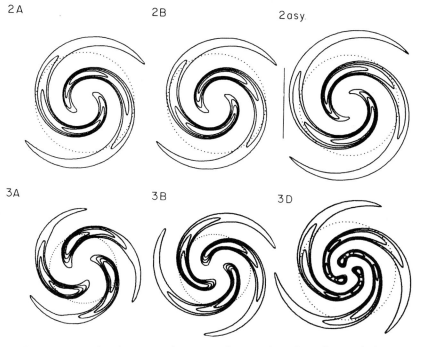

Figure 3. The perturbation density for modes 2A, 2B and 2-asymptotic
is shown in the top row. Modes 3A, 3B and 3D are along the bottom.

     Alas, most galaxies or their modes are not so simple as either of
these two pure cases. But surely an understanding of these building
blocks is a prerequisite for the analysis of the muddier situations
where both cycles may be operable. Parameter variations, other modes,
fluid effects and resonances have not been mentioned. (See Haass, 1982.)
A detailed description of the numerical method and some variations of
the model is in preparation.

REFERENCES

Bardeen, J.M.: 1976, unpublished modal calculations.
Haass, J.: 1982, Ph.D. Dissertation, MIT.
Lau, Y.Y., Lin, C.C. and Mark, J.W.-K.: 1976, Proc. Natl. Acad. Sci.
   USA Vol. 73, pp. 1379-1381.
Pannatoni, R.F.: 1979, Ph.D. Dissertation, MIT.
Toomre, A.: 1981, The Structure and Evolution of Normal Galaxies, S.M.
   Fall and D. Lynden-Bell eds., Cambridge University Press, pp. 111-136.

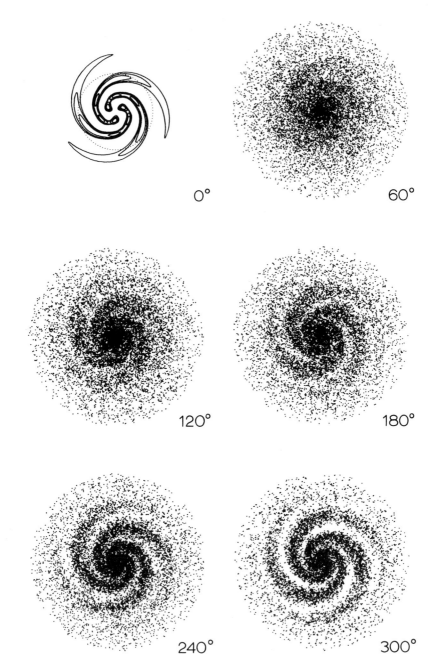

PS from Haass: 10,000-dot visualization of the rapidly-growing mode 3D
from Figure 3, shown here at 60° intervals of pattern rotation.

GLOBAL STABILITY ANALYSIS OF AN S0 GALAXY NGC 3115

Masanori Iye[1], Tatsuo Ueda[2], Masafumi Noguchi[2] and Shinko Aoki[1]
[1]Tokyo Astronomical Observatory, Mitaka 181 JAPAN
[2]Department of Astronomy, University of Tokyo, Tokyo 113 JAPAN

The global modal analysis is applied to a family of models which are consistent with recent dynamical data of NGC 3115. A halo component at least as massive as the disk component is required to explain the apparent lack of spiral structures in NGC 3115.

NGC 3115 is an isolated edge-on galaxy of type S0 (Figure 1). The luminosity profile along the major axis is highly symmetric and smooth. Figure 2 shows the rotation velocity profile $V(r)$ and the velocity dispersion profile $\sigma(r)$ along the major axis of NGC 3115.

We construct a family of gas dynamic models which are reproducing the observed $V(r)$ and $\sigma(r)$. Each model consists of a disk component and a halo component and is characterized by a halo-mass fraction s and by a dimensionless halo radius r_h.

We suppose that the apparent lack of prominent spiral structures in NGC 3115 means that this galaxy is dynamically stable and has no growing spiral modes of oscillation. The global modal analysis (Aoki, Noguchi, and Iye 1979) is applied to these models of NGC 3115 to check this hypothesis. The use of gas dynamic models is justifiable for studying global modes of long waves. Figure 3 shows the growth rate of the most unstable mode for a series of models with $r_h = 1$.

It is concluded that a halo component at least as massive as the disk component ($s > 0.5$) is required to interpret the apparent lack of unstable modes in NGC 3115 in terms of the modal analysis. This lower limit of the halo mass is obtained based not only on the equilibrium condition but also on the stability analysis of the observed galaxy.

REFERENCES
Aoki, S., Noguchi, M., and Iye, M.: 1979, *Publ. Astron. Soc. Japan* 31, pp.737-774.
Illingworth, G. and Schechter, P. L.: 1982, *Astrophys. J.* 256, pp.481-496.
Rubin, V. C., Peterson, C. J., and Ford, W. K., Jr.: 1980, *Astrophys. J.* 239, pp.50-53.

E. Athanassoula (ed.), Internal Kinematics and Dynamics of Galaxies, 125–126.
Copyright © 1983 by the IAU.

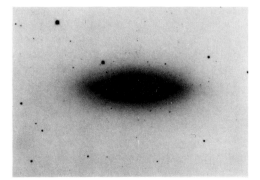

Figure 1.   NGC 3115 taken at
the Las Campanas 254-cm telescope (by courtesy of Dr. K.
Wakamatsu).

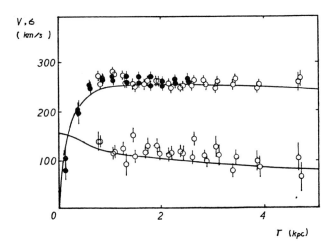

Figure 2.   Rotation velocity profile $V(r)$ and the velocity dispersion
profile $\sigma(r)$ of NGC 3115.   Open dots (Illingworth and Schechter 1982)
and filled dots (Rubin *et al*. 1980) are observed values.   Solid lines
are model profiles corrected for the line-of-sight effect and are
common to every model analyzed in the present paper.

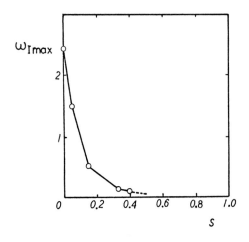

Figure 3.   The growth rate
ω_i of the most unstable growing
mode for a series of models of
NGC 3115 with various halo-mass
fractions s.   The halo radius
of this series of models is
fixed ($r_h = 1$) and it is equal
to the disk radius.

SPIRAL INSTABILITIES AND DISC HEATING

R.G. Carlberg and J.A. Sellwood
Institute of Astronomy, Madingley Road, Cambridge, CB3 OHA.

Heating-Cooling Equilibrium of a Disc

Many simulations of disc galaxies exhibit a succession of transient spiral patterns when sufficient halo is included to suppress the global bar instability (Hohl 1970, Hockney and Brownrigg 1974, Sellwood and James 1979). This behaviour causes the velocity dispersion to rise secularly which in turn reduces the intensity of the patterns. Spiral activity usually ceases altogether as Q tends asymptotically to a value somewhat greater than 2.

Such simulations however, take no account of star formation which continually injects low velocity dispersion stars into the disc. In our current simulations, we mimic this process by adding particles at a constant rate. The new particles are chosen randomly from the same density distribution as the original disc and their initial velocities are purely tangential at the local circular speed. The mass in the stabilising halo, initially 70% of the total, is steadily reduced to conserve mass. The halo density profile was chosen in such a way that the rotation curve is independent of the disc to halo ratio in order that the mass transfer does not affect the centrifugal balance of the disc.

The Q of the disc rises from its initial value of 1, quickly reaching a steady equilibrium value after 2 or 3 rotation periods. Spiral structure is continuously present for the duration of the experiments, 35 rotation periods. However, as in previous work, any one pattern lasts typically for less than a rotation period. The patterns are large scale and have a fair degree of symmetry, often two armed but frequently more. A logarithmic spiral Fourier analysis of the particle distribution shows that leading features are continually being "swing amplified" (Toomre 1981) into strong trailing waves which propagate for a short time before fading.

The equilibrium Q of the disc is dependent on the rate of addition of stars. From three different experiments we find Q = 1.9, 1.7 and 1.7 for f = .006, .015 and .03 respectively, where f is the fraction of the

E. Athanassoula (ed.), Internal Kinematics and Dynamics of Galaxies, 127–128.

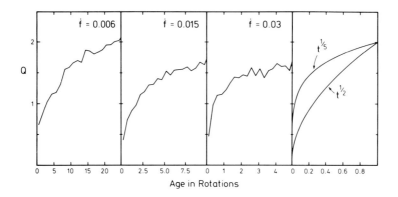

Figure 1.   The age-velocity dispersion relation for stars in three models
having different star formation rates.   The fourth plot gives two
fiducial curves indicating the range of power laws spanned by the models.

total mass added to the disc per rotation period, measured at the half
mass radius.   The mass of the disc reaches twice its initial value after
50 rotation periods when $\dot{f}$ = .006, which represents the best scaling to
a roughly constant star formation rate in the Galaxy.

Thus the steady addition of cool stars balances the heating caused
by the transient spirals at an equilibrium Q low enough for spiral
activity to be maintained practically indefinitely.

The Age-Velocity Dispersion Relation

Particles are divided into groups by their age, that is the time
since they were inserted into the simulation.   The local velocity
dispersion of each group is computed and expressed as a Q, i.e. a
fraction of the local $\sigma_{u,min}$.   The measured Q is found to be almost
independent of radius, so we average the whole disc together and obtain
the relations shown in figure 1.

The experimental results substantially agree, both in form and
magnitude, with that observed in the solar neighbourhood (Wielen 1974).

References

Hockney, R.W. & Brownrigg, D.R.K. (1974) M.N.R.A.S. 167, 351.
Hohl, F. (1970) NASA TR R-343.
Sellwood, J.A. & James, R.A. (1979) M.N.R.A.S. 187, 485.
Toomre, A. (1981) "Normal Galaxies" Eds. Fall, S.M. & Lynden-Bell, D.
                        Cambridge Univ. Press.
Wielen, R. (1974), Highlights of Astronomy 3, 395.

ON THE EVOLUTION OF PERTURBED GAS DISKS

Søren-Aksel Sørensen,
Department of Computer Science,
University College London, U.K.

Numerical experiments have shown that axisymmetric shearing gas disks will develop a spiral pattern when a non axi-symmetric contribution is introduced. This mechanism initially seems to be a prime candidate as the source of galactic spiral arms. The experiments have, however, failed to explain the formation of more than the most open systems and the smooth transition in the morphological sequence does not indicate any change in the physical mechanism between Sa and Sc types.

In order to study the evolution and persistence of these spiral arms, experiments have been performed on a disk of gas rotating in equilibrium in a Toomre potential.

$$(1) \qquad \phi = c^2/a \ (a^2+r^2)^{-3/4}$$

This disk is perturbed by a point mass passing along a parabolic trajectory.

$$(2) \qquad r = p/(1+\cos f)$$

where p is the semilatus rectum and f the true anomaly. If p is assumed to be independent of the mass of the perturber, the system can be determined by five independent parameters: The Toomre constants (a,c), the isothermal sound speed in the gas and two parameters for the perturber (p,m) where m is the mass ration perturber/disk. In the extremes this model has well known solutions. For small values of p the perturbation is reduced to a short tidal jolt and the spiral system is formed when differential rotation. winds up material arms. For large p and m values the system will resemble the well studied case where the potential has a bar superimposed [Sanders 1977; Matsuda and Isaka 1980; Sørensen and Matsuda 1982], and a quasi-stationary open pattern, correlated with the major resonances will result. Neither of these types of arms are entirely satisfactory as candidates for the observed spirals, being either too short lived or too open to fit the observations.

Here we concentrate on a third type which previously has been discussed only briefly [Sørensen 1979]. In the perturbation experiments tightly wound arms forms in the density distribution. Although the

E. Athanassoula (ed.), Internal Kinematics and Dynamics of Galaxies, 129–130.

amplitude of the pattern depends on the strength of the perturbation, the rest of its attributes including the degree of winding is independent of both p and m. The principal wavefronts have been traced in both the density distribution and the velocity field. In the velocity field the basic mode, simple epicyclic waves, can easily be identified. These wave fronts have no simple geometrical form for this type of potential and their pitch angle varies with radius as well as time [Nelson, 1976]. For illustration purposes, however, a mean pitch angle was defined by fitting the best logaritmic spiral to each arm, and the measured variation in pitch angle with time is shown in fig 1 (solid) with the evolution of a retrograde epicyclic wave (dotted). A similar procedure was followed for the density. Due to non-linear effects the density response is quite different from that of the velocity field and the arms fit a logaritmic spiral to a very high degree as seen in fig 2. The density pattern seems considerably tighter wound and less prone to further winding than the velocity mode as can be seen from fig 1 (dashed). Further experiments indicate that the the two modes follow the same evolutionary track although their position on this track is displaced. The evolutionary track is closely correlated with the radial variation in the potential, represented here by the Toomre a-constant, being progressively steeper with increasing central concentration. The pattern persists until the winding of the velocity response matches that of the density response.

References
Matsuda, T. and Isaka, H.,1980. Prog. Theor. Phys., 64, 1265.
Nelson, A.H.,1976. Mon. Not R. astr. Soc., 177, 265.
Sanders, R.H.,1977. Astrophys. J., 217, 916.
Sørensen, S.-A.,1979. In Photometry, Kinematics and Dynamics of
        Galaxies, ed. Evans, D.S.
Sørensen, S.-A. and Matsuda, T.,1982. Mon. Not. R. astr. Soc., 198, 865.

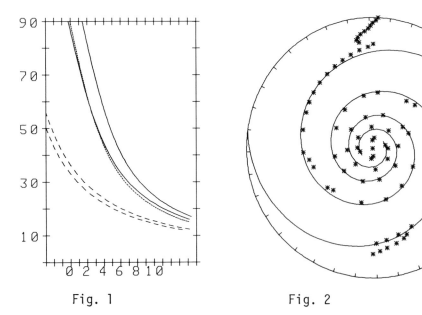

Fig. 1   Fig. 2

CLOUD–PARTICLE DYNAMICS AND STAR FORMATION IN SPIRAL GALAXIES

W. W. Roberts, Jr., M. A. Hausman, and F. H. Levinson[1]
University of Virginia, Charlottesville, Virginia

We study the gas in a spiral galaxy with a cloud–dominated "stellar association-perturbed" interstellar medium from the standpoint of a cloud-particle model. Through N-body computational simulations, we follow the time evolution of the system of gas clouds and the corresponding system of young stellar associations forming from the clouds. Basic physical processes are modeled in a three-step cyclic procedure: (1) dynamical propagation of the clouds and young stellar associations, (2) simulation of cloud-cloud collisions, and (3) formation of new associations of protostars that are triggered by the local mechanisms of cloud-cloud collisions and cloud interactions with existing young stellar associations.

From its initially-uniform distribution, the cloud system evolves in time, driven by the prescribed spiral gravitational field of the model galaxy. Figure 1 shows a photographic color-intensity display of the distribution of 20,000 clouds in one representative disk of 25 kpc diameter at a representative time of 250 Myr after approximate "steady state" has been reached. The light blue regions have the highest concentrations of clouds. The gaseous spiral structure on the global scale exhibits characteristics of a global density wave, galactic shock wave manifestation. The local, small-scale disorder is a consequence of the turbulent nature of both the collisional and supernova processes. We are continuing to investigate to what extent the local, turbulent state might be associated with the corresponding raggedness and degree of disorder on the small scale that are often observed as characteristics permeating the global spiral structures of many galaxies.

Figure 2 illustrates the nature of the computed distribution of gas clouds and flow characteristics. Plotted versus phase of the spiral field are the components of flow velocity, the normalized number density distribution, and the velocity dispersion for clouds in a representative annulus at a radius of 10 kpc. Each plus (+) represents about 10 clouds on the average. The galactic shock, characterized by the rapid decline in the normal component of flow velocity (bottom panel), occurs over a width of the order of several mean free paths. The density enhancement measured from maximum to mean is of the order of 3:1.

131

E. Athanassoula (ed.), Internal Kinematics and Dynamics of Galaxies, 131–132.
Copyright © 1983 by the IAU.

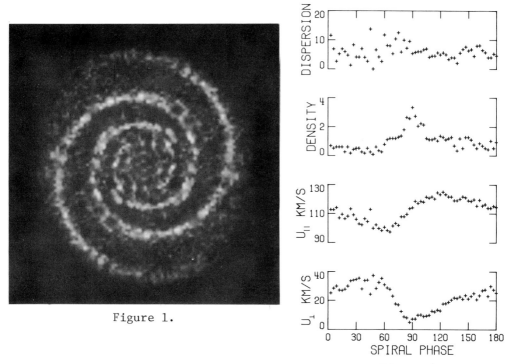

Figure 1.

Figure 2.

In order to determine characteristics with respect to the z dimension, we are studying special three-dimensional flows within the context of this cloud-particle picture. Figure 3 provides a photographic intensity display of the computed distribution of clouds (projected in the x-z plane) for a case of three-dimensional flow (in x, from left to right) in the presence of a sinusoidal gravitational field. The regions near zero z height contain the highest concentrations of clouds. A shock occurs about midway in x along the flow; there, the compression to higher density cloud conglomerations takes place, as those clouds entering with super- sonic velocities (from the left) are decelerated and exit subsonically.

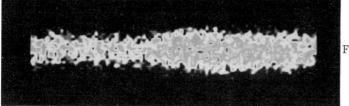

Figure 3.

It is possible to complement these N-body numerical computations with quasi-analytical theory. We are currently in the process of doing so.

This work has been supported in part by the National Science Foundation under grant AST 7909935.

[1]Now at Bell Laboratories, Allentown, Pennsylvania.

THE BALLISTIC PARTICLE MODEL

Frank N. Bash
University of Texas at Austin

THE MODEL

Bash and Peters (1976) suggested that giant molecular clouds (GMC's) can be viewed as ballistic particles launched from the two-armed spiral-shock (TASS) wave with orbits influenced only by the overall galactic gravitational potential perturbed by the spiral gravitational potential in the arms. For GMC's in the Milky Way, the model predicts that the radial velocity observed from the Sun increases with age (time since launch). We showed that the terminal velocity of CO observed from $\ell \simeq 30°$ to $\ell \simeq 60°$ can be understood if all GMC's are born in the spiral pattern given by Yuan (1969) and live 30×10^6 yrs. Older GMC's were predicted to have radial velocities which exceed observed terminal velocities.

In Bash, Green and Peters (1977) we assume that stars form in the GMC's, after some delay time, and that the stars continue in ballistic orbits which can be integrated over times less than a relaxation time. We made observations of CO in molecular clouds connected to young star clusters and used the cluster's main-sequence turn-off to determine its age. We found that very young clusters (earliest star is an O-star) still are inside the molecular cloud from which they were born while older clusters (earliest star is a B-star) show no evidence of their molecular cloud. This confirmed the suggestion of our earlier paper that GMC's disappear abruptly at a certain age. The lifetime determined above and the main-sequence lifetime of a B0 star allows the determination of the delay between GMC formation and star formation in the GMC.

Bash (1979) examined the optical surface brightness shapes of spiral arms predicted to result from stars born in our ballistic GMC's. The predicted surface brightness shapes agree well with observed ones and the agreement allows the initial-mass-function of spiral arm stars to be determined. Bash (1979) also determined, from observations of the Milky Way and M81, that the best agreement with the model is achieved when the molecular clouds are launched at the post-shock velocity and live for 40×10^6 yrs.

E. Athanassoula (ed.), Internal Kinematics and Dynamics of Galaxies, 133–134.
Copyright © 1983 by the IAU.

FURTHER TESTS

Bash, Hausman and Papaloizon (1981) suggest that ballistic molecular
clouds will hit smaller interstellar clouds at relatively high
velocities ($\sim$ 15 km s^{-1}) and that those collisions are capable of
continuing to stir the GMC's internal turbulence.

Bash and Visser (1981) combined Visser's model for the HI gas in M81
with the ballistic particle model for GMC's and spiral arm stars. The
resulting model predicts an optical surface brightness which resembles
photographs of M81 and predicts HII region velocities which can be
checked by observation.

Bash (1981) showed that observed Milky Way giant HII region radial
velocities and galactic longitudes agree with predictions of the
ballistic particle model which assumes that the Galaxy has two spiral
arms and that the HII regions do not move on circular orbits.

Finally, Hilton and Bash (1982) showed that the observed vertex
deviation of the velocity ellipsoid for B0 and B1 stars near the Sun
is predicted by the ballistic particle model. The cause of the amount
and the sense of the deviation is the initial post-shock velocity of
the GMC's from which the stars were born and the fact that the stars
are young enough to still "remember" their initial velocities.

Present work on testing the ballistic particle model focuses on
comparing a set of detailed VLA and optical measurements of M81 to the
predictions of our model and on a theoretical examination of how our
model predicts the properties of the Hubble Types of spiral galaxies.

REFERENCES

Bash, F. N. 1981, Astrophys. J., 250, 551.
Bash, F. N., and Visser, H. 1981, Astrophys. J., 247, 488.
Bash, F. N., Hausman, M., and Papaloizou, J. 1981, Astrophys. J., 245, 92
Bash, F. N. 1979, Astrophys. J., 233, 524.
Bash, F. N., Green, E., and Peters, W. L. 1977, Astrophys. J., 217, 464.
Bash, F. N., and Peters, W. L. 1976, Astrophys. J., 205, 786.
Hilton, J. L., and Bash, F. N. 1982, Astrophys. J., 255, 217.
Yuan, C. 1969, Astrophys. J., 158, 871.

RELATION BETWEEN STAR FORMATION AND ANGULAR MOMENTUM IN SPIRAL GALAXIES.

L. Carrasco and A. Serrano
Instituto de Astronomía
Universidad Nacional Autónoma de México

Abstract. We derive the radial distribution of the specific angular momentum j=J/M, for the gas in M31, M51 and the galaxy, objects for which well observed unsmoothed rotation curves are available in the literature. We find the specific angular momentum to be anti-correlated with the present stellar formation rate, i.e. minima of spin angular momentum correspond to the loci of spiral arms. We find that the stellar formation rate is an inverse function of j. We derive new values of Oort's A constant for the arm and interarm regions in the solar neighborhood.

High quality Hα or 21-cm rotation curves of galaxies give rise to the angular velocity, Ω, as a function of radius. This unsmoothed rotation curve refers to a particular direction to the galaxy. Through numerical differentiation one obtains the curve of Oort's A constant $A(r) = -r/2 \; [d\Omega(r)/dr]$. Since A is the vorticity, the specific angular momentum j of a cloud, due to differential rotation, is given by

$$j = A \, R^2$$

where R is the radius of the cloud. In this way, A is proportional to j, for "standard" clouds of a given R. A comparison of A(r) and the stellar formation rates (the latter determined by the emission measure of radio recombination lines) is shown in Figure 1 for the Galaxy. One can easily notice that minima of j correspond to an efficient star forma-

TABLE 1

Object	k (adopted)	s (derived)	correlation coefficient
The Galaxy	1	1.21±0.3	0.67
	2	1.26±0.3	0.66
M31	1	0.7 ±0.4	0.57
	2	0.12±0.3	0.74
M51	1	0.66±0.3	0.77
	2	0.51±0.3	0.68

$$<s> \; = \; 0.91 \pm 0.3$$

135

E. Athanassoula (ed.), Internal Kinematics and Dynamics of Galaxies, 135–136.

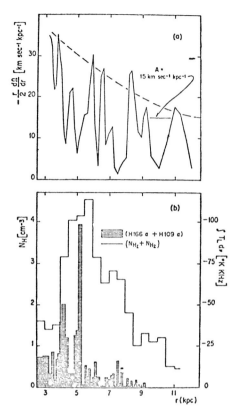

tion. This indicates that the gas located in spiral arms is rotating -to a first approximation- as a rigid body. Equality of radial an tangential velocity dispersions for OB stars is consistent with a value A=0. The rotational data used in this paper is that of Burton and Gordon (1978) and Jackson et.al. (1979). While the recombination line data are those by Lockman (1976) and Hart and Pedlar (1976). From Figure 1 one can derive new local values of Oort's A constant for the arm and interarm regions

$$A(\text{interarm}) = 18 \ km \ s^{-1} Kpc^{-1}$$
$$A(\text{arm}) = 3.5 \ km \ s^{-1} Kpc^{-1}$$

We propose a stellar formation value of the form

$$dN_* / dt \propto \rho^k \ j^{-s}$$

A least squares analysis of the Galaxy M31 and M51 is summarized in Table 1. From Figure 1 it is clear that gas density alone cannot describe in detail the radial trend of the star formation rate. The results presented in Table 1, for the exponent s, indicate that the stellar formation rates in spiral arns indeed vary as the inverse of the specific angular momentum j of the clouds.

Figure 1. The derived values of Oort's A constant are plotted as a function of galactocentric distance r for the Milky Way. The dashed curve represents the value of A associated with a differentially rotating disk with σ=8 km s⁻¹. IAU's value of A for the solar vicinity is plotted as reference. 1b.- Histograms of the radial distribution of gas (solid line, Gordon and Burton 1976), and of the radial distribution of H II regions (hatched area) for the Galaxy. The latter being proportional to the stellar formation rate.

REFERENCES

Burton, W.B., Gordon, M.A: 1978 Astron. Astrophys. 63, 7.
Gordon, M.A., Burton, W.B. 1976, Astrophys. J. 208, 346.
Hart, L., Pedlar, A. 1976. Monthly Notices Roy. Astron. Soc. 176, 547
Jackson, P.D., Fitzgerald, M.P., Moffat, A.F.J., 1979, IAU Symp. No. 84, ed. W.B. Burton (Dordrecht: D. Reidel)
Lockman, F. 1976, Astrophys.J. 209, 429.

MODEL CALCULATIONS ON THE LARGE SCALE DISTRIBUTION OF BUBBLES IN GALAXIES

J. V. Feitzinger[x] and P. E. Seiden[xx]
[x] Astronomical Institut, Ruhr-University, Bochum, FRG
[xx] IBM Watson Research Laboratory, Yorktown Heights, NY 1o598, USA

Introduction

One main characteristic of the ionized and neutral component of the interstellar medium in galaxies is the appearance of shells and bubbles in all sizes up to 1 kpc diameter (Hodge,1974; Heiles, 1979). Between different galaxy types differences in size and radial distribution of these bubbles are observed as a function of galaxy type. The stochastic self-propagating star formation model is able to simulate such distributions on a global galactic scale.

The Model

The model includes the distribution of the stars as well as a two component gas (Feitzinger, Glassgold, Gerola, Seiden, 1981; Seiden, Schulman, Feitzinger, 1982). We track of both active and inactive gas for each cell in the model. Active means that the gas is ready for star formation (i.e. cool). Inactive gas is not ready for star formation because it has been heated or thrown out of the plane of the disk by previous star formation events. Creation of a star cluster (association) modifies a cell by converting the active into inactive gas by action of <u>ionization fronts,</u> <u>stellar winds</u> and <u>supernova explosions</u>. The inactive gas is allowed to return to the active state with a given time constant. The bubbles are created and maintained by this star-forming process.

In Fig. 1 the star and gas distributions are shown for a typical time step. The cold gas disk (black) shows many various sized bubbles (white). For a whole sequence of models we have determined(maximum rotation velocity is taken as a crude measure of the galaxy type) the sizes and radial distributions of bubbles. The decreasing shear (lower differential rotation) and the decrease of the gas density as function of the galaxy radius causes an increase of the diameters of the bubbles as function of the distances from the center of the galaxies. This depends on the bubble evolution time scale and the shear time scale.

Comparison with Observations

In Fig. 2 we compare the results of our model simulations with the bubble radius versus distance distribution for the Milky Way. The general trend

E. Athanassoula (ed.), Internal Kinematics and Dynamics of Galaxies, 137–138.
Copyright © 1983 by the IAU.

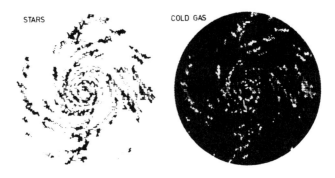

Fig. 1 The star and gas disk at a typical time step (Milky Way simulation)

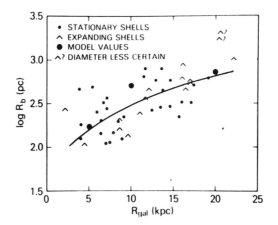

Fig. 2 Bubble radius versus distance for the Milky Way; data after Heiles
       (1979), adopted from Bruhweiler et al. (1980). The line represents
       our model simulation.

of increasing radius of the bubbles is very well reproduced by our models. The same accordance between observations and model simulations is
obtained for the size distribution of the bubbles. The results of Bruhweiler et al. also agree well with our simulations.

References
Bruhweiler, F., Gull, T.R., Kafatos, M., Sofia, S., 1980, Ap. J. 238, L27
Feitzinger, J.V., Glassgold, A.E., Gerola, H., Seiden, P.E., 1981, Astro-
    nomy Astrophys. 98, 371
Heiles, C., 1979, Ap. J. 229, 533
Hodge, P.W., 1974, PASP 86, 845
Seiden, P.E., Schulman, L.S., Feitzinger, J.V., 1982, Ap. J. 253, 91

HI-SHELLS IN M31

E. Brinks and E. Bajaja
Leiden Observatory, Leiden, The Netherlands

1. INTRODUCTION

Close inspection of the channel maps which make up the high resolution M31 HI-data set (Brinks, this volume) resulted in finding numerous
shell structures and bubble-like regions. To avoid any bias in the interpretation we shall call holes, in the course of this contribution, all
those structures where we find a deficiency of neutral gas. Some of the
holes can be identified directly in a total surface density map of M31,
but due to blending of the various velocity components, e.g. the warp,
holes tend to loose contrast. Therefore the best way to search for the
holes is by inspection of the full resolution channel maps ($\Delta\alpha \times \Delta\delta \times \Delta V =$
$24'' \times 36'' \times 8.2$ km s^{-1}).

2. RESULTS

At present some 300 holes are listed, of which some fraction might
be spurious, i.e. not due to structure in the HI surface density but
caused by kinematical effects. The sizes of the holes range from about
1 kpc to less than the beam size, i.e. smaller than 80 pc. The definition
of a hole is subjective and strongly dependent on the contrast in the
channel maps. A prerequiste for a hole to be listed is that it is present
in a range of channel maps and that it is stationary.

Figure 1(a) shows the channel map centered at -370.8 km s^{-1}, as a
typical example, in which several holes can be seen. Figure 1(b) is a
position-velocity diagram along a line parallel to the minor axis of M31
centered on the hole indicated by an arrow in figure 1(a), which is located at 15.4 minutes of arc to the south of the center along the major
axis and 12.3 to the east along the minor axis. This particular hole
shows an expanding (or contracting) shell-like structure. The spectrum
at its center (figure 1(c)) indicates that there is no HI at the velocity
of the channel map shown.

The expansion velocity of about 14 km s^{-1} and the amount of HI in
the shell derived from this velocity profile would indicate, assuming

139

E. Athanassoula (ed.), Internal Kinematics and Dynamics of Galaxies, 139–140.
Copyright © 1983 by the IAU.

spherical symmetry, a kinetic energy of about 6×10^{50} erg.

One hole has been studied in detail (Brinks, 1981). It coincides with NGC 206, a bright OB-association. This hole is about 400 x 800 pc in size and about 2.10^6 $M_{\odot}$ of HI is deficient, assuming a uniform background HI density. The stars of the association are a strong source of UV-emission and they are embedded in a low surface brightness HII-region. The energy required to produce this hole is of the order 10^{52} ergs. The NGC 206 hole is quite similar to the (super)shells as discussed by Heiles (1979) in our own galaxy and to the superbubbles in the Large Magellanic Cloud discussed by Meaburn (1980). Although much work still has to be done, the holes seem to be related to regions with young stars and recent star formation.

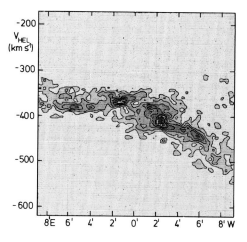

Figure 1a: Channel map at -370.8 km s^{-1}. The cross indicates the center of M31. East is at left; North at top.

Figure 1b: Position-velocity diagram along a line parallel to the minor axis of M31 and centered on the hole indicated in (a) by an arrow. The first contour levels are 2.5, 10, 20, ... K.

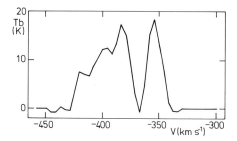

Figure 1c: Velocity-profile at the center of the hole indicated in (a).

REFERENCES

Brinks, E.: 1981, *Astron. Astrophys.* *95*, L1
Heiles, C.: 1979, *Astrophys. J.* *229*, 533
Meaburn, J.: 1980, *Monthly Notices Roy. Astron. Soc.* *192*, 365

SELFPROPAGATING STOCHASTIC STAR FORMATION AND SPIRAL STRUCTURE

J. V. Feitzinger[x] and P.E. Seiden[xx]
[x] Astronomical Institut, Ruhr-University, Bochum, FRG
[xx]IBM Watson Research Laboratory, Yorktown Heights, NY 1o598,
    USA

Spiral structure in galaxies can arise from both dynamic and non dynamic phenomena: spiral density waves and stochastic selfpropagating star formation. The relative importance of these effects is still not known. Deficiences of the original selfpropagating star formation model (where only stars are taken into account) are overcome by explicitly considering the stars embedded in and interacting with a two-component gas (Seiden and Gerola, 1979; Seiden, Schulman and Feitzinger, 1982; Seiden and Gerola, 1982). The two-component gas is essential because it is the means by which we get feedback in the interaction between stars and gas. The coupling between stars and gas regulates and stabilizes star formation in a galaxy. Under proper conditions this model can give good grand design spirals (Fig. 1).

The inclusion of a realistic radial gas profile eleminates the hard edges of the model galaxies and a smooth fading out of the spiral structure can be observed as in real galaxies (Fig. 2). This means that by a quite natural method problems of the density wave theory,where the spiral structure ends and the difficulties at the resonances, are eliminated. Since the driving mechanism for the stochastic spiral structure is differential rotation trailing spirals are always generated and no mode problems arise. Galactic dynamic is taken into account implicitly in the models via the rotation curve (i.e. the mass distribution).

Propagating star formation can generate the observed galactic forms and realistic color and luminosity distributions. It can generate the SAa - Sm Hubble sequence, two-armed symmetric global structures, flocculent many-armed structures as well as the patchy structure of irregular galaxies like the Large Magellanic Cloud (Feitzinger et al. 1981). Also arm spurs, the irregular interspiral arm structures, can be interpreted as material features built up by stochastic star formation processes (Feitzinger and Schwerdtfeger, 1982).

E. Athanassoula (ed.), Internal Kinematics and Dynamics of Galaxies, 141–142.
Copyright © 1983 by the IAU.

References:
Feitzinger, J.V., Glassgold, A.E., Gerola, H., Seiden, P.E., 1981,
    Astron. Astrophys. 98, 371
Feitzinger, J.V., Schwerdtfeger, H., 1982, Astron. Astrophys. in press
Seiden, P.E., Gerola, H., 1979, Ap. J. 233, 56
Seiden, P.E., Schulman, L.S., Feitzinger, J.V., 1982, Ap. J. 253, 91
Seiden, P.E., Gerola, H., 1982, Fund. Cosmic Physics 7, 241

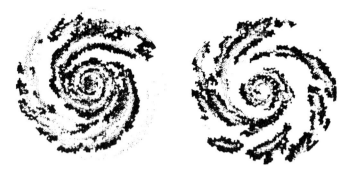

Fig. 1 Examples of two-armed symmetric spiral patterns

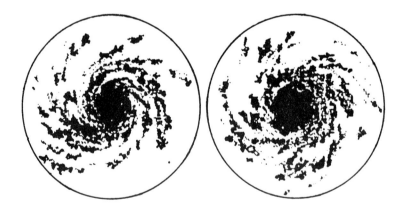

Fig. 2 Examples of galaxies whose gas density decreases exponentially
    with radius. The circle indicates the size of the simulation
    array.

VELOCITY STRUCTURES IN THE VERTICAL EXTENSIONS OF SPIRAL ARMS

J. V. Feitzinger, J. Spicker
Astronomical Institut, Ruhr-University, Bochum, FRG

Yuan and Wallace (1973) explained the "Rolling motion" in galactic spiral arms by geometrical means as apparent, but not actual motions.
Strauss and Poeppel (1976), however, could demonstrate, that these geometrical effects are not sufficiently large to produce the rolling motions. The aim of this investigation is to explain the remaining part of
the rolling motion effect with the galactic fountain model (Bregman 1980).

Data
We use two 21-cm line surveys in the form of $(v,b)_1$ contour maps of
brightness temperature; the Maryland Green Bank survey and the Berkeley
survey. A two armed spiral (Simonson 1976) was assumed and the three-
component galactic mass model of Rohlfs and Kreitschmann (1981) adopted
for the calculation of the radial velocity field. We investigated the
Perseus arm and measured the slope dv/db on the contour maps as well as
the location of (v,b_0) of the spiral arm center and the extension Δb of
the arm.

Discussion
After correcting for the geometrical effects (displacement, warp), we
find dv/db $\neq$ 0 for $70° < 1 < 160°$ (dv/dz > 20 km/sec/kpc); there are also
large regions with dv/db $\approx$ 0 (Fig. 1). From these residuals a net velocity $v_z = (dv/dz)$ dz (velocity difference between the velocity at the
height z and z = 0) was calculated as a measure of the rising and falling
motions. The net velocities (Fig. 2) could be compared with the velocity
of a mass falling free from some height z to the galactic plane (Feitzinger, Kreitschmann 1982). They group reasonably well around the free
fall velocities for different z heights. The distribution of star-forming sites reavels concentrations of young objects in regions with great
v_z velocities, so that a model with rising and falling gas (Bregman 1980)
seems to be adequate. The remaining part of the apparent rolling motion
can be attributed to such phenomena.

E. Athanassoula (ed.), Internal Kinematics and Dynamics of Galaxies, 143–144.

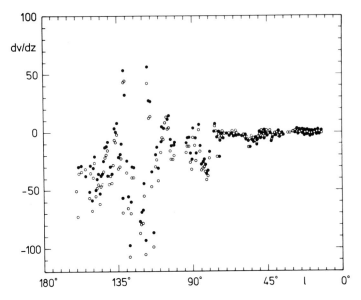

Fig. 1 Observed velocity gradients dv/dz in the Perseus arm with (dots)
and without (circles) geometrical corrections.

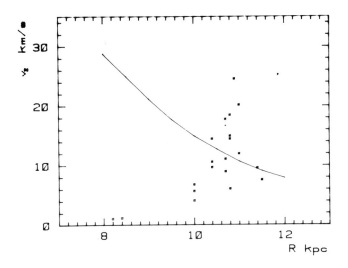

Fig. 2 Velocity distribution v_z(R) for z = o.3 kpc; the solid line is
the free fall velocity

References:
Bregman, J.E., 198o, ApJ 236, 577
Feitzinger, J.V., Kreitschmann, J., 1982, Astron. Astrophys. 111, 255
Rohlfs, K., Kreitschmann, J., 1981, Ap. Sp. Sci. 79, 289
Simonson, S.C., 1976, Astron. Astrophys. 46, 261
Strauss, F.M., Poeppel, W., 1976, ApJ 2o4, 94
Yuan, C., Wallace, L., 1973, ApJ 185, 453

VERTICAL STRUCTURE OF SPIRAL SHOCKS IN A CORRUGATED GALACTIC DISC

A.H.Nelson[1], T.Matsuda[2], T.Johns[1].
1) Dept. of Applied Maths.   2) Dept. of Aeronautical
        & Astronomy,   Engineering,
   University College,   Kyoto University,
   Cardiff, U.K.   Kyoto, JAPAN.

Numerical calculations of spiral shocks in the gas discs of galaxies (1,2,3) usually assume that the disc is flat, i.e. the gas motion is purely horizontal. However there is abundant evidence that the discs of galaxies are warped and corrugated (4,5,6) and it is therefore of interest to consider the effect of the consequent vertical motion on the structure of spiral shocks. If one uses the tightly wound spiral approximation to calculate the gas flow in a vertical cut around a circular orbit (i.e the θ-z plane, see Nelson & Matsuda (7) for details), then for a gas disc with Gaussian density profile in the z-direction and initially zero vertical velocity a doubly periodic spiral potential modulation produces the steady shock structure shown in Fig. 1. The shock structure is independent of z, and only a very small vertical motion appears with anti-symmetry about the mid-plane.

Now if instead of an initially flat gas disc we impose an initial perturbation of density and velocity derived from linear corrugation wave theory (8), and assume that the corrugations are spirals congruent to the potential spirals, then the response of the gas to the same spiral potential as for Fig. 1 is shown in Fig. 2. Note that there is now no steady state since the corrugation will in general have a different phase velocity from the stellar density wave producing the potential, hence the corrugation moves in the rest frame of the spiral potential. The main feature produced by the corrugation is a splitting of the main shock peak into two peaks as in Fig. 2, a phenomenon which appears and disappears periodically. The period between succesive splittings in this case is 10^8 years, and the duration is 4×10^7 years. The reason for the splitting is that the corrugation carries its own doubly periodic density modulation (8), and when the corrugation and spiral shock are out of phase the total density modulation becomes quadruply periodic around a full galactic orbit.

The main conclusion to be drawn from this calculation is that, while the overall doubly periodic density modulation due to a two-armed stellar spiral (or bar) persists, the detailed structure of the density is significantly modified by a corrugation of the gas disc. This is true even if the corrugation amplitude is relatively small (maximum gas ridge z deviation ≈ 30 pc, and z velocity ≈ 6 km/sec in Fig. 2), and hence the existence of higher harmonics in spiral galaxies (9), leading to such features as

E. Athanassoula (ed.), Internal Kinematics and Dynamics of Galaxies, 145–146.

secondary filaments and gaps and asymmetries in arms, might easily be
explained by corrugations.

References

1) Shu, F.H., Milione, V., & Roberts, W.W.: 1973, Ap. J. 183, p. 819.
2) Nelson, A.H. & Matsuda, T.: 1977, M.N.R.A.S. 179, p. 663.
3) van Albada, G.D. & Roberts, W.W.: 1981, Ap. J. 246, p. 740.
4) Sancisi, R.:in "The Structure and Evolution of Normal Galaxies", eds.
   Fall & Lynden-Bell, pub. Cambridge University Press, 1981, p. 149.
5) Bosma, A.: Ph.D. Thesis 1978, Groningen University.
6) Quiroga, R.J.: 1974, Astrophys. & Space Sci. 27, p. 323.
7) Nelson, A.H. & Matsuda, T.: 1980, M.N.R.A.S. 191, p. 221.
8) Nelson, A.H.: 1976, M.N.R.A.S. 177, p. 265.
9) Iye, M., Hamabe, M. & Watanabe, M.: in "Photometry, Kinematics and
   Dynamics of Galaxies", ed. Evans, pub. University of Texas Press, 1979.

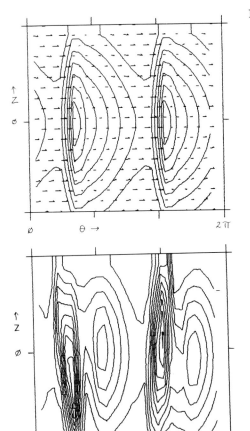

Fig. 1 Contour plot of gas density in
the θ-z plane in response to a
spiral potential alone. The
contours start at a density
(relative to the unperturbed
ridge density) of 0.4 and the
step is 0.2. The arrows show
the velocity component perpendicular to the spirals. The z
thickness of the box is 200 pc,
and the θ length is 31 kpc.

Fig. 2 Contour plot of gas density in
the θ-z plane in response to a
spiral potential and an initia
corrugation of the gas layer.
Contours are as for Fig. 1.

RECENT TAURUS RESULTS ON Hα VELOCITIES IN M83

R.J. Allen[1], P.D. Atherton[1], T.A. Oosterloo[1], and K. Taylor[2]
[1]Kapteyn Astronomical Institute,
University of Groningen, The Netherlands
[2]Royal Greenwich Observatory,
Herstmonceux Castle, Hailsham, U.K.
[2]Anglo-Australian Observatory
Epping, N.S.W., Australia

SUMMARY:

Preliminary Hα observations with the TAURUS imaging spectrometer confirm a pattern of systematic radial motions in a section of spiral arm in M83. The velocity gradients are not consistent with those predicted for the neutral gas.
Non-circular motions have also been discovered in the central regions of the galaxy.

1. INTRODUCTION:

For some time now the existence of large-scale patterns of non-circular motions in the interstellar gas near spiral arms has been accepted as evidence for the presence of underlying density waves in the mass distribution of the disk. The well-known example of the distribution and motions of HI in M81 as observed by Rots (1975) and analysed by Visser (1980a, b) has remained unique in its clarity for many years.

The kinematics of the ionized gas may be different from that of the neutral gas (see for example Bash and Visser, 1981), and these differences may be of great interest for the study of the mechanisms and time scales for the formation of massive stars out of the interstellar HI in spiral arms. With this general motivation we have examined a section of spiral arm near the minor axis of M83 using an Hα data cube which was available more by accident than by design. We have also studied the velocity field in the central regions of the galaxy with the hope of finding kinematic evidence for a bar of about 1 arc-minute size which has been reported as being present there (Comte, 1981).

E. Athanassoula (ed.), Internal Kinematics and Dynamics of Galaxies, 147–150.

2. OBSERVATIONS AND DATA REDUCTION:

The observations were made with the TAURUS imaging spectrometer
(Taylor and Atherton, 1980; Atherton et al., 1982) during one of the
first test runs at the Anglo-Australian Observatory in the spring of
1980. M83 had been selected as an interesting candidate for a number of
reasons, and an approximately one hour exposure on the 3.9-meter telescope produced a data cube of 180 × 180 pixels at 91 spacings of the
Fabry-Pérot interferometer. Unfortunately, the pre-filter defining the
overall passband of the system was not optimally centered on the
systemic velocity of M83, and the Hα emission from the southwestern side
of the galaxy was very weak. Partially for this reason the data have not
been photometrically calibrated. The transformation from Fabry-Pérot
spacing to velocity was carried out using the STARLINK computer at the
Royal Greenwich Observatory; subsequent spatial smoothing and velocity
profile analysis of selected regions of the galaxy have been done with
the Groningen Image Processing System (GIPSY). The final spatial
resolution was 3 arcseconds for the spiral arm region, and 6 arcseconds
(Gaussian FWHM) for the central areas of the galaxy. The velocity
resolution is 35 km s^{-1} (Lorentzian FWHM).

3. RESULTS:

In the accompanying figure we show a contour map of the Hα emission
in the northeastern side of M83; as remarked earlier, the southwestern
side is virtually absent and the map is not photometrically calibrated.
Isovelocity loci are drawn over the contours in several places. The
broken curve indicates the position of a prominent dust lane.

a) The Eastern Spiral Arm:

A region of this spiral arm near the minor axis has been chosen
for study, since any radial streaming velocities should be greatest
there. If we assume that the spiral arms are trailing, then the SE-side
in the figure is the far side of M83. Since the dust lane is on the
inside of the arm, we expect corotation to be located at some larger
radial distance. The models of neutral gas flow (e.g. Visser, 1980b)
under these conditions predict that just outside the shock (dust lane),
the gas radial velocity $V_r < 0$ indicating infall, and furthermore
$\delta V_r / \delta r > 0$ as we encounter streamlines which are further downstream
from the shock.

If we accept the minor axis position angle ϕ of 135° (indicated by
the straight line segments in the figure) and a systemic velocity V_s
of 505 km s^{-1} (e.g. Comte, 1981) we encounter the first contradiction,
viz. the radial velocity of the ionized gas is greater than the
systemic velocity in the region where the minor axis crosses this arm.
This would indicate outflow, were it not for the uncertainties in the
two parameters ϕ and V_s. However, quite independent of the exact
choice of these parameters, we see from the figure that $\delta V_r / \delta r < 0$
since a velocity gradient of approximately -10 km s^{-1} in a radial

increment of +20 arcseconds is clearly present across the spiral arm.
The isovelocity contours have a slope opposite to that predicted by the
neutral gas flow models.

The observational results in the SE spiral arms (and elsewhere in
M83) are consistent with the earlier photographic Fabry-Pérot study
carried out by Comte (1981). Similar photographic observations have been
recently made by De Vaucouleurs, Pence and Davoust (1982); their results
do not appear to show the flow velocities described above, although
their sensitivity and spatial averaging methods make a direct comparison
difficult.

Because the photographic techniques result in a spatially
undersampled velocity field it could be that peculiar velocities of the
HII regions compared to the surrounding diffuse emission may account for
some of this discrepancy, depending upon the regions which were sampled.
The possibility that these peculiar velocities exist was first pointed
out in a comparison of the HI and HII velocities in M101 by Viallefond,

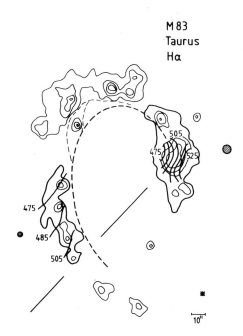

M83
Taurus
Hα

Distribution and radial motion of the Hα emission in the eastern
arm of M83. The contours are Hα intensity in linear equal increments
after subtraction of the continuum emission, but not photometrically
calibrated. The angular resolution is shown by the two hatched circles,
one to the right of the nucleus (giving the resolution of 6" for the
central regions), and the other to the left of the eastern spiral arm
(3"). Isovelocity loci in these two regions are also shown, as is the
dust lane (broken line). The symbol directly above the 10" horizontal
bar is a star.

Allen, and Goss (1982, see for example NGC 5447). Since TAURUS can obtain reliable Hα velocities for both the HII regions and the diffuse gas, we can now make a comparison for the ionized gas alone. We find that there is indeed considerable small-scale structure in the observed velocities, or order 10 km s^{-1}. However, these velocities appear insufficient to explain the discrepancy between the map of De Vaucouleurs, Pence and Davoust on the one hand, and that of Comte (1981) and the present results.

b) The Central Regions:

The velocity contours in the central areas of the galaxy show a strong gradient and a general S-shaped morphology which is reminiscent of the kinematic effects associated with the presence of a bar in other galaxies, for instance NGC 5383 (Peterson et al., 1978; Sancisi et al., 1979; Sanders and Tubbs, 1980). This result was not available in previous Fabry-Pérot work on M83 owing to saturation effects on the photographic plates.

ACKNOWLEDGEMENTS

This work has been partially supported by a NATO research grant no. 052.81. We acknowledge with thanks the assistance of R.N. Hook and C.D. Pike of the RGO in the data reduction.

REFERENCES

Atherton, P.D., Taylor, K., Pike, C.D., Harmer, C.F.W., Parker, N.M.,
    and Hook, R.N.: 1982, MNRAS, in press
Bash, F.N. and Visser, H.C.D.: 1981, Ap. J. 247, 488
Comte, G.: 1981, Astron. Astrophys. Suppl. Ser. 44, 488
De Vaucouleurs, G., Pence, W.D. and Davoust, E.: 1982, preprint
Peterson, C.J., Rubin, V.C., Ford, W.K., Jr., Thonnard, N.: 1978, Ap. J.
    219, 31
Rots, A.H.: 1975, Astron. Astrophys. 45, 43
Sancisi, R., Allen, R.J. and Sullivan, W.J., III: 1979, Astron.
    Astrophys. 78, 217
Sanders, R.H. and Tubbs, A.D.: 1980, Ap. J. 235, 803
Taylor, K. and Atherton, P.D.: 1980, MNRAS 191, 675
Viallefond, F., Allen, R.J. and Goss, W.M.: 1981, Astron. Astrophys.
    104, 127
Visser, H.C.D.: 1980a, Astron. Astrophys. 88, 149
Visser, H.C.D.: 1980b, Astron. Astrophys. 88, 159

POSSIBLE EVIDENCE FOR LIN'S THREE-WAVE INTERACTION MECHANISM AT COROTATION IN THE GALAXY NGC 1566

G. Comte
Observatoire de Marseille
2 Place Le Verrier
13248 Marseille Cedex 4   (France)

In a previous paper, (Comte and Duquennoy, 1982), we have reported observations of the distribution and kinematics of the ionized hydrogen in the large southern spiral NGC 1566 (SAB(r?s)bc I). We found evidence for a severe warping of the galactic disk, and discussed the spiral distribution of 477 detected HII regions. The *observed* spiral structure, deduced from the positions of the regions deprojected in a $(\ln r, \theta)$ diagram, shows a four-armed design (Fig. 1):

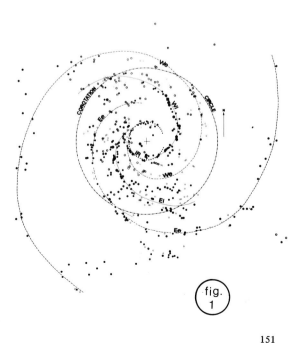

fig. 1

- 2 "principal, inner" arms (WI and EI on Fig. 1), containing very bright HII regions, with pitch angle 28°, finishing abruptly near $r \simeq 130''$.

- 2 "outer" arms, fainter in Hα, of pitch angle 15° till $r = 130''$ then opening slightly to 20°, winding away around the galaxy into a "pseudo-outer-ring". (WE and EE on Fig. 1).

- An uncomplete inner ring located at $r \simeq 35''$.

From a preliminary velocity field obtained by means of Hα Fabry-Perot interferometry, a rotation curve was drawn and fitted by a simple two-populations (bulge + disk) mass model, based on de Vaucouleurs' (1973) surface photometry and the method described by Monnet and Simien (1977). Possible non-circular velocities exist near the inner ring zone.

E. Athanassoula (ed.), Internal Kinematics and Dynamics of Galaxies, 151–152.

The model rotation curve was used to derive the angular velocity Ω, the epicyclic frequency κ and the quantities $\Omega \pm \kappa/2$ (Fig. 2).

The possible existence of spiral waves of different pitch angle in a same pattern (i.e. rotating with a unique velocity Ωp), was suggested first by Lin (1969) as a natural replenishment mechanism of the spiral through 3-wave interaction at corotation: a *long* (*open* spiral) wave interacts with two *short* (tight spiral) waves; detailed calculations were made by Mark (1976). In NGC 1566, we may locate the corotation zone at r = 130" where the main inner open spiral vanishes. This gives $17.5 < \Omega p < 19$ km s^{-1} kpc^{-1} for Δ = 15.3 Mpc; the inner Lindblad resonance (ILR) is then just inside the inner ring and the outer resonance at 300" < r < 350". (The outermost HII region is observed at r = 325"). Using the Lin-Shu-Toomre dispersion equation for density waves, we compute two wavenumbers which lead one to the tight spiral, the other to the open spiral (dotted and dashed lines on Fig. 3). The agreement with the observed pitch angles is good for 4.5 < r < 7.5 kpc.

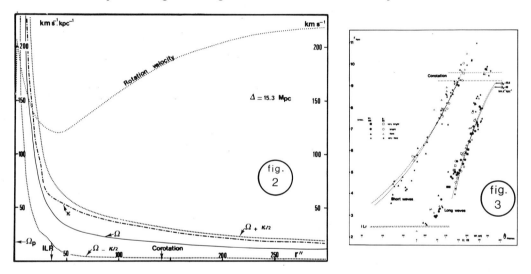

From Mark's calculations, the short wave should be amplified with respect to the long one, in terms of carried angular momentum. We observe a high-contrast long wave, and a low-contrast short wave in terms of stellar formation. However, computation of $w_{lo} = (\Omega - \Omega p)r \sin i$ for both waves shows (Roberts, 1969) that a shock may be generated at the front of the long wave, with subsequent triggering of stellar formation (supersonic flow of gas), but is likely to be unimportant at the front of the short wave.

REFERENCES
Comte, G., Duquennoy, A.: 1982, Astron. Astrophys. (in press).
Lin, C.C.: 1969, I.A.U. Symposium n° 38.
Mark, J.W-K: 1976, Ap. J. 203, 81 and 205, 363.
Monnet, G., Simien, F.: 1977, Astron. Astrophys. 56, 173.
Roberts, W.W.: 1969, Ap. J. 158, 123.
Vaucouleurs, G. de: 1973, Ap. J. 181, 31.

SPECTRAL ANALYSIS OF THE LUMINOSITY DISTRIBUTION OF SEVEN GALAXIES IN THE VIRGO CLUSTER

Masanori Iye
Tokyo Astronomical Observatory, Mitaka 181   JAPAN

An objective method involving a Fourier analysis is applied to seven galaxies for measuring the strength of spiral and bar components.

A spectral analysis of the spiral patterns of galaxies is proposed by Iye et $al.$(1979).   In this scheme, we define a Fourier transform $A(\alpha,m)$ of the intensity distribution $I(u=ln$ r, $\theta)$ of a galaxy by

$$A(\alpha,m) = \frac{1}{2\pi} \int_0^{2\pi}\int_{-\infty}^{\infty} I(u,\theta) \exp\{-i(\alpha u + m\theta)\} \, du d\theta \qquad , \qquad (1)$$

and the normalized power spectrum $P(\alpha,m)$ by

$$P(\alpha,m) = |A(\alpha,m)|^2 / \sum_{m=1}^{\infty} \int_{-\infty}^{\infty} |A(\alpha,m)|^2 d\alpha \qquad . \qquad (2)$$

The application of this method to an asymmetric spiral galaxy NGC 4254 revealed that the spiral pattern of NGC 4254 is a superposition of five-armed and three-armed spiral components on the main one-armed spiral (Iye et $al.$ 1982).

In the present paper, the same method is applied to seven galaxies NGC 4303 (ScI), NGC 4321 (ScI), NGC 4340 (SBa), NGC 4472 (E4), NGC 4501 (Sb$^+$I), NGC 4535 (S(B)cI:), and NGC 4579 (Sbn).   Figures 1, 2, and 3 show the V-band image, the power spectrum $P(\alpha,m)$, and the pattern of the primary spectral component, respectively, of two spiral galaxies NGC 4303 and NGC 4535.   The primary spiral component of NGC 4535 is very strong and this singly reproduces the overall spiral feature of NGC 4535 (Figure 3b).   On the other hand, the spiral feature of NGC 4303 is composed of a few spiral components of which the strongest is shown in Figure 3a.

REFERENCES

Iye, M., Hamabe,. M., Watanabe, M., and Okamura, S.: 1979, in $Photometry,$ $Kinematics$ and $Dynamics$ of $Galaxies$, ed. D. S. Evans, University of Texas, pp.407-410.
Iye, M., Okamura, S., Hamabe, M., and Watanabe, M.: 1982, $Astrophys.$ $J.$ 256, pp.103-111.

E. Athanassoula (ed.), Internal Kinematics and Dynamics of Galaxies, 153–154.

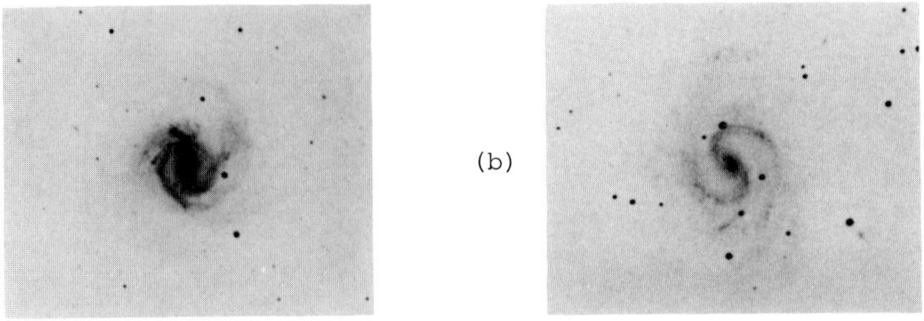

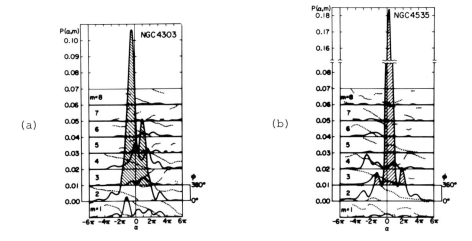

Figure 1.   V-band images of (a) NGC 4303 and (b) NGC 4535 taken using the
105-cm Schmidt telescope at the Kiso Observatory.   North is at the top.

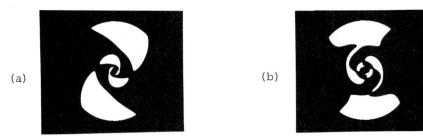

Figure 2.   Normalized power spectra $P(\alpha,m)$ of (a) NGC 4303 and (b) NGC
4535.   Spectra for $m=1,2,..,$ and 8 are shown by successively shifting
0.01 in the ordinate.   Strong components with $P(\alpha,m)\geq0.01$ are hatched.
Dots are the phase angles ϕ of the corresponding $A(\alpha,m)$.   The primary
component of NGC 4303 is peaked at $P(-0.45\pi, 2)$ and has a FWHM of 1.1π.
The primary component of NGC 4535 is peaked at $P(0.25\pi, 2)$ and has a
FWHM of 0.9π.

Figure 3.   Patterns of the primary spectral components of (a) NGC 4303
and (b) NGC 4535.

FOURIER ANALYSIS OF SPIRAL TRACERS. A PROGRESS REPORT.

S. Considère and E. Athanassoula
Observatoire de Besançon

Fourier analysis of the observed distribution of spiral tracers (e.g. HII regions) can yield a wealth of information about the spiral structure of a given galaxy. A description of the method has been given by Kalnajs (1975) and, in brief, involves decomposing the observed distribution into components with given angular periodicity m. Component m = 0 corresponds to the axisymmetric part, m = 1 to either a one-armed spiral or an asymmetry, m = 2 to a spiral or a bar, etc. Each component is then analysed into a superposition of logarithmic spirals. This does not imply that the observed spirals are assumed to be logarithmic, but rather that logarithmic spirals are convenient building blocks.

We have set out to apply this method to a sufficiently large number of galaxies to permit a statistical treatment. The database for this type of study has been successfully accumulated by various astronomers over many years. There are approximately 50 galaxies with sufficiently complete catalogues of HII regions to permit a Fourier analysis (see Hodge 1982 for a compilation of the existing literature). To this already long list should be added the available information on other spiral tracers. For instance Iye (this volume) has applied it to V-band plates.

There are many interesting applications of this study. Norman (this volume) has noted its relevance for theoretical models. Correlations of various properties of the spiral arms like angular periodicity, pitch angle or maximum amplitude with Hubble type and luminosity class could help towards a more quantitative classification of galaxies. Correlations with observational parameters that influence the dynamics of the galaxy (e.g. the form of the rotation curve) could provide constraints for theoretical studies. The influence of observational constraints (Kormendy and Norman 1979) and of the environment (Elmegreen and Elmegreen 1982) can be made more quantitative. Finally it would be interesting to compare the spiral structures of various tracers in the same galaxy.

We have made only a very small first step in this long-term project. We have tested our method on four galaxies (Considère and Athanassoula 1982). The first two, M51 and M33, were chosen to test the method.

155

E. Athanassoula (ed.), Internal Kinematics and Dynamics of Galaxies, 155–156.

Indeed in M51 we found the well-known two-armed spiral in addition to
the observed N-S asymmetry. The inner 15' of M33 revealed a clear m = 2
component of which the southern arm shows both better fit and better
continuity than the northern one. Thus these results gave us confidence
in the method but, at the same time, brought forth the problem of
choosing the orientation parameters (see Considère and Athanassoula 1982
and Bosma, this volume for a discussion). However, there could be no
particular merit to the method if it only found already obvious results.
Thus the two other galaxies chosen, NGC 2997 and M31, were more controversial. In NGC 2997 we found both an m = 2 component and a, somewhat
weaker, m = 3. Note that m = 3 is a favoured periodicity when the percentage of halo mass is large (Toomre 1981). For M31 we too found the
one-armed leading spiral first proposed by Kalnajs (1975). In addition
we delineated a faint two-armed leading pattern for 45' < r < 60' and
a faint two-armed trailing one for r > 65'. In all four galaxies we
found that few components can very well represent the observed distribution. The sum of these components defines an area of some 15-35 % of
the galaxy surface (within a radius defined by the outermost region)
which contains 75-90 % of the observed HII regions.

The analysis of several more galaxies is underway.

REFERENCES

Considère, S. and Athanassoula, E. : 1982, Astron.'Astrophys. 111, 28.
Elmegreen, D. and Elmegreen, B. : 1982, Monthly Notices Roy Astron.
      Soc., in press.
Hodge, P.W. : 1982, Astron. J. 87, 1341.
Kalnajs, A.J. : 1975, in "La dynamique des galaxies spirales", ed.
      L. Weliachew, p. 103.
Kormendy, J. and Norman, C. : 1979, Astrophys. J. 233, 539.
Toomre, A. : 1981, in "Structure and Evolution of Normal Galaxies", ed.
      S.M. Fall and D. Lynden-Bell, Cambridge, p. 111.

LARGE SCALE MAGNETOHYDRODYNAMICAL CONSIDERATIONS IN SPIRAL GALAXIES

E.Battaner, M.L.Sánchez-Saavedra

Instituto de Astrofísica de Andalucía.CSIC.Apdo.2144.Granada
and Dpto.Física Fundamental.Universidad de Granada.(Spain)

ABSTRACT.–A magnetohydrodynamical result is deduced, which could contri
bute to our understanding of spiral and ring structures in galaxies.The
usual expressions for the continuity, momentum and induction equations
are adopted for the gas of a galaxy, and the following simplifying hypo
tesis are made :   a) Steady state conditions, b) Axisymmetry, c) A velo
city field given by (π=0, θ=θ(r), Z=0) for the interstellar gas (where
π,θ and Z are the radial, azimuthal and vertical to the galactic plane
components and r is the distance from the galactic center).Then, the di
rection of magnetic field must be azimuthal and the plasma distribution
is compatible with ring structures.

The aim of this work is to determine the distributions of magnetic field
and gas in a steady state galaxy, in which the gas corotates with the
stellar field. The equations of continuity, momentum and induction must
be integrated. The form adopted for these equations are standard in ga
lactic gas dynamics. The above mentioned hypothesis are made and cilin
drical coordinates are used. Then the following results are obtained, af
ter a lengthy but easy derivation

$$B_r = 0 \tag{1}$$

$$B_z = 0 \tag{2}$$

$$\frac{\partial p}{\partial z} + \rho \frac{\partial F}{\partial z} + \frac{\partial}{\partial z}\left(\frac{B_\phi^2}{8\pi}\right) = 0 \tag{3}$$

$$\frac{\partial p}{\partial r} + \frac{\partial}{\partial r}\left(\frac{B_\phi^2}{8\pi}\right) + \frac{2}{r}\left(\frac{B_\phi^2}{8\pi}\right) = 0 \tag{4}$$

where F is the gravitational potential, and p the gas pressure due main
ly to clouds turbulence. Some of the equations give trivial results, and
are not included. Equation (2) has been obtained from $\nabla.\vec{B}$=0, and it has
been assumed that intergalactic magnetic fields are negligeable. The fo

E. Athanassoula (ed.), Internal Kinematics and Dynamics of Galaxies, 157–158.

llowing conclusions arise directly :

1) Equations (1) and (2) indicate that only the azimuthal component B_ϕ of $\vec{B}$ can be different from zero. Therefore the magnetic field lines in a steady state galaxy should be circles around the rotation axis. This is a severe restriction to the possible directions of $\vec{B}$. Measurements of $\vec{B}$ in the Galaxy, seem to agree with this restriction.

2) When B_ϕ is removed in (3), the equivalent width obtained agrees reasonably with the observations. The magnetic force is probably not so important in deciding the vertical distribution of gas.

3) Equation (4) cannot be solved to obtain both p and B_ϕ. Near the center $B_\phi^2/r \gg dB_\phi^2/dr$ and p should decrease approximately exponentially. In outer regions $B_\phi^2/r \ll dB_\phi^2/dr$ and the sum of magnetic pressure and gas pressure should remain approximately constant. A ring structure is compatible with equation (4). Rings should be favoured to form in the outer parts of the galaxy and the magnetic field should have lower values within the rings.

Active galaxies are clearly not in a steady state, and the above conclusions should be related to normal galaxies, mainly ellipticals and spirals, which are in conditions probably not very far from steady. Some facts may indicate the identification of galaxies having an azimuthal magnetic field ; second, ellipticals have less gas content and magnetization should be less important. Speculations might also be made if current ejection theories could explain the slight deformation of ring patterns into spirals in expanding galaxies slightly separated from the steady state.

More details of the calculations can be found in Battaner and Sanchez-Saavedra (1982).

REFERENCE

Battaner, E. and Sánchez-Saavedra, M.L.: 1982, Astron. and Space Sci.
    86, 55.

MAGNETIC FIELDS AND SPIRAL STRUCTURE

R. Beck
Max-Planck-Institut für Radioastronomie, Bonn, FRG

Interstellar magnetic fields are known to be a constraint for star formation, but their influence on the formation of spiral structures and the evolution of galaxies is generally neglected. Structure, strength and degree of uniformity of interstellar magnetic fields can be determined by measuring the linearly polarised radio continuum emission at several frequencies (e.g. Beck, 1982). Results for 7 galaxies observed until now with the Effelsberg and Westerbork radio telescopes are given in the table. The Milky Way is also included for comparison.

Galaxy	Type	λ	Distance	Linear resolution element	$<p_n>$	$<B_u/B_r>$	$<B_t>$	M_B	Observers
		(cm)	(Mpc)	(kpc²)	(%)		(µG)		
Milky Way	Sbc II	21.2–73.5	—	—	—	∿0.5	∿4	−20.1	Spoelstra & Brouw (in prep.)
NGC 224 (M31)	Sb I-II	11.1	0.7	0.9 x 4.2	20 ± 3	0.9 ± 0.1	4 ± 1	−21.61	Beck et al. (1980)
NGC 253	Sc(p)	2.8	2.5	1.0 x 4.9	∿10	∿0.4	∿10	−20.72	Klein et al. (in prep.)
NGC 598 (M33)	Scd II-III	11.1	0.7	0.9 x 1.6	13 ± 4	0.4 ± 0.1	3 ± 1	−19.07	Beck, Berkhuijsen, Wielebinski (unpubl.)
NGC 3031 (M81)	Sab I-II	6.3	3.2	2.3 x 4.5	17 ± 3	0.5 ± 0.1	8 ± 3	−20.75	Beck, Klein (in prep.)
NGC 5194 (M51)	Sbc I	6.0	9	3.2 x 3.6	∿20	∿0.6	∿10	−21.60	Segalovitz et al. (1976)
NGC 6946	Scd I	2.8	7	3.1 x 3.6	10 ± 3	0.3 ± 0.1	12 ± 4	−20.30	Klein et al. (1982)
IC 342	Scd I-II	6.3	4.5	3.2 x 3.5	20 ± 4	0.5 ± 0.1	7 ± 2	−21.4	Gräve, Beck (in prep.)

Galaxy types and absolute magnitudes are mostly taken from the Second Reference and Shapley-Ames catalogues. The linear resolution at the distance of each galaxy is given along the major and minor axis. The thermal contribution to the total flux density was estimated with help of the spectra (Klein and Emerson, 1981) and optical data if available (Beck and Gräve, 1982; Klein et al., 1982). The mean ratio of the uniform to random field strengths $<B_u/B_r>$ follows from the mean degree of polarisation of the total nonthermal flux density $<p_n>$ using the formulae given by Segalovitz et al. (1976) and Beck (1982), with the inclination and the nonthermal spectral index as input parameters. $<B_u/B_r>$ refers to the linear resolution element. The mean strength of the total field $<B_t>$ was computed from the total nonthermal flux density assuming equipartition between cosmic ray and magnetic field energy densities. As the total flux density varies with almost the fourth power of the field strength, $<B_t>$ is believed to be accurate to ∿ 30% despite the large uncertainties of the input parameters.

A dependence of some property of the magnetic field on galaxy type would be of great importance for the theories of field origin. The limited sample of the present observations does not allow statistically founded conclusions, but two tendencies appear in the data:

159

E. Athanassoula (ed.), Internal Kinematics and Dynamics of Galaxies, 159–160.

1. The strength of the total magnetic field $<B_t>$ seems to vary with
    luminosity class: Galaxies of class I contain a field of $\sim$ 10 μG
    strength, galaxies of class I-II only $\sim$ 6 μG. M33 representing
    class II-III has the weakest field.
2. For linear resolution elements of the same size, the mean ratio of
    the uniform to random field strengths $<B_u/B_r>^*$ (a measure of the
    degree of uniformity of the magnetic field) increases with increas-
    ing luminosity. $<B_u/B_r>^*$ is smallest in M33 and highest in M51. On
    a scale of 3×3 kpc^2, the ratio between the strengths of the uniform
    and random fields $<B_u/B_r>$ is typically 0.5, but can reach values
    around 2 locally.

The distribution of the polarisation vectors in galaxies, corrected for
Faraday rotation, reveals the direction of the magnetic field lines.
The present data allow the following conclusions:

3. Magnetic field lines generally follow the spiral arms.
4. Large-scale magnetic field lines seem to be closed within the plane
    of the galaxy. No field reversals are observed.

Closed magnetic field lines suggest the action of a dynamo mechanism.
Maximum deviations from circular field lines are ±10° in IC 342 and ±20°
in M31 and NGC 6946. A coupling of the field lines to the streaming
lines of the gas in a density wave potential is indicated. Shock fronts
are able to align the field (Beck, 1982); strong shocks are expected in
galaxies of high luminosity. In density wave theory, magnetic fields
may also play an active rôle because they could control the growth rate
of the waves.

Magnetic fields may also directly influence the star formation
rate: Galaxies with bright, massive spiral arms contain the strongest
fields. This can be understood in the framework of the stochastic theory of star formation (SSPSF; see Seiden and Gerola, 1982) where field
lines give a preferential direction to the propagation of star formation. Hence it becomes inevitable to consider magnetic fields for theories of the spiral structure and evolution of galaxies.

LITERATURE

Beck, R.: 1982, Astron. Astrophys. 106, pp. 121–132
Beck, R., Berkhuijsen, E.M., Wielebinski, R.: 1980, Nature 283,
    pp. 272–275
Beck, R., Gräve, R.: 1982, Astron. Astrophys. 105, pp. 192–199
Klein, U., Emerson, D.T.: 1981, Astron. Astrophys. 94, pp. 29–44
Klein, U., Beck, R., Buczilowski, U.R., Wielebinski, R.: 1982, Astron.
    Astrophys. 108, pp. 176–187
Segalovitz, A., Shane, W.W., Bruyn, A.G. de: 1976, Nature 264,
    pp. 222–226
Seiden, P.E., Gerola, H.: 1982, Fund. Cosm. Phys. 7, pp. 241–311

SPIRAL STRUCTURE: DENSITY WAVES OR MATERIAL ARMS?

S.V.M. Clube
Royal Observatory, Edinburgh

Recent studies (Frenk and White 1980, 1982) of the (x, y, z) and (Π, θ) distributions of the Galactic globular cluster population have given R_o = 6.8±0.8 kpc and Π = 51±26 kms^{-1} for the inner halo. The observations leading to these solutions are well illustrated by the ρ vs. x plot in figure 1 (Clube and Watson (CW), 1979) for the globulars with the best determined data within $|1|$, $|b|$ < 20°. The independently determined R_o value clearly divides the distribution in such a way that most of the objects on the nearside of the G.C. approach the Sun while those on the farside recede, the probability that this arrangement arises by chance from a "stationary" distribution being $\sim$ 0.002.

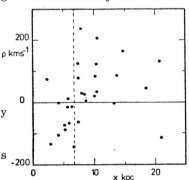

The average value of Π for the inner halo (R < 4 kpc) based on figure 1 is 76±15 kms^{-1} in broad agreement with Frenk and White though the formal significance may be underestimated because of an inappropriate choice of rest frame. Since the globular population is equally represented on either side of the G.C., the best current choice may be considered the average of the ρ's within R < 4 kpc, and we thus obtain $\Pi \sim$ 76±14 kms^{-1} while $\Pi_\odot \sim$ 13±15 kms^{-1}. (The average of the extrema gives $\Pi_\odot \sim$ 42±30). It has previously been shown (Clube 1978) that the 3 kpc and +135 spiral arms of the HI disc, interpreted as an optically thick, kinematically bisymmetric system in the core of the Galaxy, give corresponding values of Π = 95(±1) and $\Pi_\odot$ = 40(±1) kms^{-1}. The evidence therefore, if correctly interpreted, seems to indicate both the inner halo and central disc expand together, presumably because of similar recent dynamical histories which may not in the circumstances be attributed to any kind of gas dynamical action. Indeed, should gravity be the most likely driving force, material spiral arms are probably implied and the observed phenomena presumably derive from a recent pre-existing steady state halo and centrally condensed disc through effects that are in some way equivalent to suddenly removing or introducing mass in the galactic centre (e.g. Clube 1980). The weight of evidence overall may not demand this kind of

E. Athanassoula (ed.), Internal Kinematics and Dynamics of Galaxies, 161–162.
Copyright © 1983 by the IAU.

explanation at present, but the possibility that spiral arms are recurrently produced, short-lived material bodies first moving outwards from the nuclear regions and then inwards under the influence of the Galactic potential field, should not be overlooked. It is therefore premature to interpret observed spiral arm streaming in the outer parts of external galaxies as exclusive evidence for density waves (e.g. Visser, 1980): it may equally be the residual motion of spiral arms injected into the outlying disc.

These conclusions are currently being tested in two ways. First by observing inner halo RR Lyrae radial velocities to check the globular cluster results. Spectra have so far been obtained (Rodgers 1977, CW 1980) for 31 RR Lyraes in the Plaut fields, and basing their distances on M_v = 1.0 corresponding to the above R_o value (Clube and Dawe 1980), it is already clear that negative values outnumber the positive values 4.2 to 1 on the nearside of the GC, giving much the same mean value and spanning the same range as the globular clusters. Although these results add weight to the globular cluster picture, RR Lyraes in the low obscuration windows in the G.C. will clearly provide the most direct determination of $\Pi_\odot$. Observations of $\sim$40 such stars have now been obtained and should soon throw further light on this question.

The second test is based on a new study of early type stars in the solar neighbourhood, r < 2 kpc (CW 1982). Eliminating Gould's Belt, it has been shown that the more concentrated spiral arm O, B stars represent a significantly different kinematic population from the remaining more widely dispersed disc O, B stars. The actual space motions of these two groups turn out to be very similar to those of Drifts I and II respectively occurring among later type stars within r < 200 pc. Drift I has an outward motion $\sim$30 kms^{-1} relative to the better mixed Drift II, and if the identities have been correctly established, it is impossible in the case of our Galaxy to attribute spiral arm streaming to density waves: indeed the picture of widely distributed material arms moving through an underlying disc substratum seems increasingly appropriate. If both these Drifts are represented by HI clouds, the disc overall at $R \sim R_o$ will appear to be expanding at some average velocity of the two Drifts in general accord with Kerr's expansion 7-10 kms^{-1}.

S.V.M. Clube, 1978, Vistas Astr. 22, 77.
S.V.M. Clube, 1980, Mon. Not. R. astr. Soc., 193, 385.
S.V.M. Clube and J.A. Dawe, 1980, Mon. Not. R. astr. Soc., 190, 591.
S.V.M. Clube and F.G. Watson, 1979, Mon. Not. R. astr. Soc., 187, 863.
S.V.M. Clube and F.G. Watson, 1980, Observatory, 98, 124.
S.V.M. Clube and F.G. Watson, 1982, in preparation.
C.S. Frenk and S.D.M. White, 1980, Mon. Not. R. astr. Soc., 193, 295.
C.S. Frenk and S.D.M. White, 1982, Mon. Not. R. astr. Soc., 198, 173.
A.W. Rodgers, 1977, Astrophys. J., 212, 117.
H.C.D. Visser, 1980, Astr. & Astr. 88, 149.

CONFLICTS AND DIRECTIONS IN SPIRAL STRUCTURE

C.A. Norman
Institute of Astronomy, Madingley Road, Cambridge.
Sterrewacht, Huygens Laboratorium, Wassenaarseweg 78, Leiden.

ABSTRACT

This is a critical and selective review of current theoretical and observational work related to spiral structure in disk galaxies. Productive future areas of research are suggested.

I. INTRODUCTION.

This area of astronomy contains some very interesting interrelated physical approaches that need to be discussed and evaluated in the context of the whole grand attempt to understand spiral structure. I have chosen to illustrate this schematically in Figure 1.

In the top right hand corner we have the physics of travelling wave amplifiers in inhomogeneous systems including the semiclassical WKBJ technique, the theory of instabilities in a shearing, rotating fluid and the phenomena of wave-particle resonant interactions. Moving clockwise we encounter the old percolation problem in a new guise: a shearing disk with the population effect being related to the probability that a burst of massive star formation will induce further massive star formation in neighbouring regions. The gas dynamics studies embody the usual one-fluid continuum approach with various techniques employed to treat the almost inevitable shocked response. Another currently fruitful mode of calculation of fluid response is to use a simulation of clouds or sticky particles that collide inelastically. Underpinning most of these gas dynamic calculations is some forcing potential such as a bar, a spiral or an oval. More sophisticated multiphase media are now under study. The global modes of stellar and gaseous disks present a classic problem in astrophysics that is currently severely ill-posed by our ignorance of that dynamically important component often referred to as dark stuff. This murky business I shall return to later. Global modes are now at least partially understood in terms of the local approximation used in the wave amplifier work but detailed simulations will assist greatly in clarifying this modal physics. The modes may often be a forced response

163

E. Athanassoula (ed.), Internal Kinematics and Dynamics of Galaxies, 163–174.

Figure 1

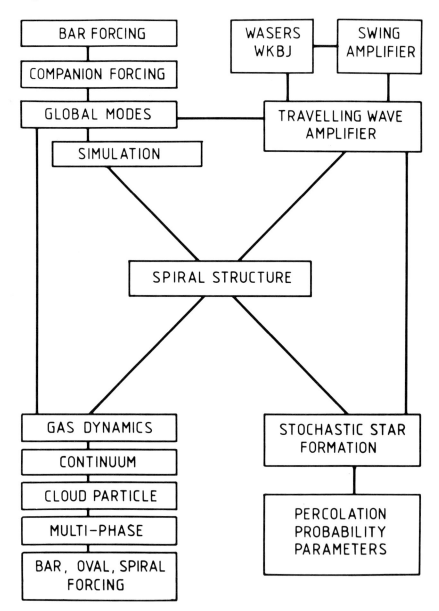

to an external driving agent, such as a bar or companion, that produces
a transient ringing of the global modes of the disk.

II. TRAVELLING WAVE AMPLIFIER

I cannot understand what all the fuss is about! The views of the two
principal protagonists are shown in Figure 2. In both cases the cause of
the wave generation is that galaxies wish to transfer their angular
momentum outwards. The WKBJ approach has been strongly advocated by CC
Lin and coworkers (Lin, Haass, Bertin the conference). In the vicinity
of corotation there is a three wave interaction between a long trailing
wave propagating outwards towards corotation, a short trailing wave
propagating inward away from corotation towards the outer Lindblad
resonance where it will be damped and consequently allowing the galaxy
to effect its radially outward flow of angular momentum. To close the
feedback cycle the WKBJ theory has a reflection point outside the inner
Lindblad resonance where *short* inwardly propagating trailing waves are
reflected as *long* wavelength trailing waves. This does depend rather
crucially on how steeply the velocity dispersion, as measured by the
Q factor (Toomre, 1977), rises in the inner region of a given galaxy.
Furthermore unless Q is of order unity near corotation a WKBJ barrier
has to be penetrated by tunelling processes. All the waves are trailing
waves in this picture. The original estimates of the growth factor in
amplitude per traversal of a cycle is of order $\sim \sqrt{2}$. A more detailed
analysis of shearing effects near corotation, incorporating the physics
of the Goldrich–Lynden–Bell (1965) instability, gives a growth factor
of ~ 4 per cycle.

The second detailed approach is due to Toomre and is significantly
behoven to the Goldreich–Lynden–Bell instability. A masterful presentation is given in Toomre (1981). The process feeds on shear and self-
gravity. In the region of corotation a short leading wave swings around
to become a short trailing wave propagating inwards and emits a shot
trailing wave propagating outwards that is damped at the outer Lindblad
resonance. Here the feedback cycle is closed in a rather different way.
There is assumed to be no inner Lindblad resonance so that the short
trailing wave reflects off the centre as a short leading wave propagating outwards. The growth factor can be very large $\sim 10^1$–10^2 per
transit.

In general, the two theories now before us have been brilliantly
developed over the last two decades. It all looks basically correct and
solved. Probably both processes occur, and swing amplication *could* be
incorporated as a type II waser, as Drury (1980) has indicated, or
wasering *could* be incorporated as a type II swing amplifier, particularly when the shearing Goldreich–Lynder–Bell response is included,
but I suggest that we stick to the nomenclature used in this paper to
avoid further subclassification into the mysteries of swinging or
percolating wasers!

III. GLOBAL MODES

Here I shall comment on a selection of recent global mode calculations where some attempts have been made to understand the physical growth mechanism in terms of the specific processes discussed in the previous section. The Gaussian disk (Erikson, 1974) has been analysed by Toomre and Kalnajs and the unstable modes have been interpreted as being due to swing amplifier effects. The growth rates are estimated by summing the travel time from the centre to the swing point, the swing time, and the reflection time, and also estimating the amplification factor per feedback cycle. There is agreement to within 10-20% between these estimates and the numerical work. The mode-shapes substantially agree. An important point to note is that if the central region is made into an absorber the whole instability goes away, which is quite consistent with the loss of the feedback cycle. The only other mode is a non-swinging edge mode generated on the sharp edge of the gaussian disk. A beautiful simple interpretation is given by Toomre (1982). The gaseous Kuzmin disk studied by the Iye (this conference) and Aoki et al. (1979) does not yet have a clear identification of the physical mechanism but Toomre (1981) has tentatively identified it with swinging.

Perhaps the most fruitful development in the near future will come from the N-body simulations such as the ones developed by Sellwood (this conference). These have good dynamic range, large N, and good diagnostics for the modes. The physics of travelling wave packets should be obvious if a simple initial disturbance can be propagated that travels, swings, reflects and refracts.

IV. STOCHASTIC STAR FORMATION.

As shown in Figure 3, we *start* with the idea of sequential star formation orginally proposed by Blaauw and summarised in his 1964 review; further developments were added by Elmegreen and Lada (1977). *Then*, add shear and choose a star-burst coherence length or equivalently a probability for continuing the star forming sequence at each point. This is a percolation problem. Simulate this on a large (IBM!) machine (see the review by Seiden and Gerola, 1982) and obtain, for a small coherence length, pictures that are quite reasonable for the large Magellannic clouds, star-bursts in dwarf galaxies such as IZW 18, IIZW 40, spurs in our own galaxy and filamentary structures in 2841-type galaxies.

Now, before considering the large coherence length studies, I should say that since the major proponents of this theory are not here at this meeting, I will not be so hard on the theory as I might otherwise have been! To continue, the large scale percolation phenomenon when applied to the grand design does *not* explain (i) smooth-armed SO's with no young stars (ii) the smooth neutral hydrogen and magnetic field compression correlated with dust lanes (iii) barred and oval galaxies (iv) the two-armed grand design (v) the arm symmetry even in the fine scale structure in M51 (Kalnajs, this conference). However, when there is an underlying

Figure 2

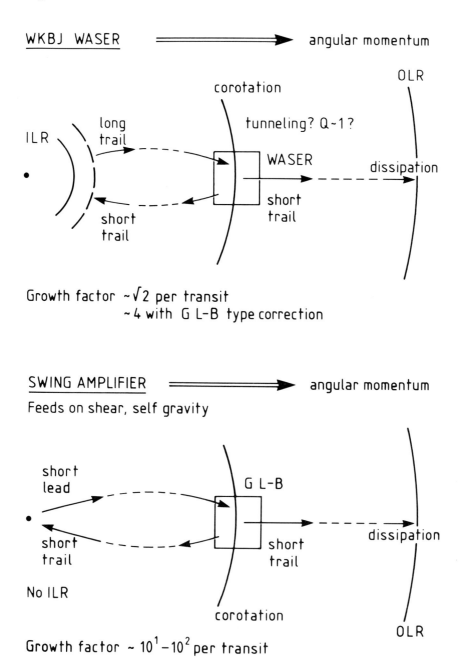

WKBJ WASER ⟹ angular momentum

OLR

corotation

tunneling? Q~1?

ILR

long trail

WASER

dissipation

short trail

short trail

Growth factor ~√2 per transit
~4 with G L–B type correction

SWING AMPLIFIER ⟹ angular momentum

Feeds on shear, self gravity

short lead

G L–B

short trail

short trail

dissipation

No ILR

corotation

OLR

Growth factor ~ 10^1–10^2 per transit

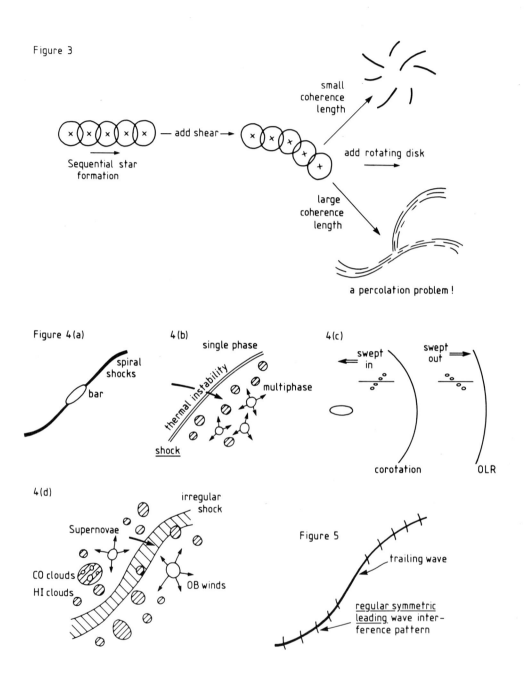

Figure 3

Sequential star formation

— add shear →

small coherence length

add rotating disk →

large coherence length

a percolation problem !

Figure 4(a)

spiral shocks

bar

4(b)

single phase

thermal instability

multiphase

shock

4(c)

swept in

swept out

corotation   OLR

4(d)

irregular shock

Supernovae

CO clouds

HI clouds

OB winds

Figure 5

trailing wave

regular symmetric leading wave inter-ference pattern

spiral wave such considerations of triggered sequential star formation are extremely germane, as they also are to filamantary 2841-type galaxies.

V. GAS DYNAMICS

This complex field will be partially reviewed by Prendergast (this conference) so I shall restrict myself to noting four aspects. Firstly it is now clear that the one-fluid models responding to an underlying forcing such as a bar, oval or spiral (Fig. 4a) describe very well the overall observed structure of the HI compression, the magnetic field compression and the dust-lane correlation. There is flow of gas into the central regions forced by the dissipation and the angular momentum out-flow carried by the spiral response. The efficiency of this process decreases as the effective internal sound speed of the gas is raised.

Marochnik et al. (1983) have sent me details of an extensive multicom-ponent fluid calculations incorporating thermal balance and selfgravity at the shock where they have isolated the condition for a phase transi-tion to occur at the shock from a one component to a two or multicompo-nent medium (Fig. 4b).

The calculational tool of simulating the gas response by using an N-body simulation with inelastic cloud collisions has been extensively explored by Schwarz (1979, 1981). As shown in Fig. 4c, the clouds can be regarded as test particles with a small but finite drag. Inside corotation, the particle response leads the bar forcing and the clouds lose angular momentum to the bar and move inwards. Outside corotation the clouds gain angular momentum from the bar and move outwards. The sweeping out time is short ~ 3-10 rotation periods.

More detailed simulations (Fig. 4d) involving a multiphase medium have been done by Roberts and collaborators (Roberts, this conference, Van Albada and Roberts, 1981). The most obvious effects are a longer mean free path with a corresponding larger shock width and weaker shocks with significant fluctuations in amplitude. Post shock heating occured because of the energy input from supernovae and stellar winds from early type stars. Such detailed models are most important for modelling the interstellar medium in M31; say, with its holes, bubbles, varied fine-scale kinematic structure, and also for modelling the detailed distri-bution and kinematic of the CO distibution, star forming regions and HI in an near spiral arms (Lieisawitz and Bash, 1982, Kennicutt and Hodge, 1982).

V. OBSERVATIONS.

The relation between morphological properties and dynamic properties has been extensively studied by Kormendy and Norman (1979) with more recent updates by Rubin et al (1980) and Elmegreen and Elmegreen (1982). The

basic point is that many grand design spirals have bars and companions
and these are quite plausibly associated with driven spiral structure.
However, apart from filamentary armed 2841-type spirals where there is
no clear forcing of the grand design, there is a class of apparently
isolated and non-barred galaxies which show grand design. There are
instances of galaxies that could be purely global growing modes (i.e.
not forced) due possibly to the absence of an inner Lindblad resonance
rendering their liable to pure swing amplifications. There may also be
very weak bars such as in NGC 1566 that are clear in the infrared
(Hackwell and Schweizer, 1982) but are difficult to detect optically
that can play the role of an almost invisible forcing agent.

The Fourier decomposition into logarithmic spirals is an useful tool for
analysis of the observed mode structure of a disk. It has serious
difficulties since the arms are not regular and large inhomogeneities
can generate anomalous structure. A start to such a program is represented in the work shown at this conference by Athanassoula and
Considère on NGC 2997 (trailing 3 arms) and M31 (2 trailing and 1
leading); Comte on NGC 1566 (three wave interaction at corotation); Iye
on a sample where he finds odd modes; and finally the eyeball decomposition by Haass of M100 (3 trailing arms). Interesting as such studies
are, they should be regarded as an initial stage in a long and sophisticated program. I wish to emphasize that swing gives us leading arms
although of relatively small amplitude and that regular symmetric
interference patterns should be found (Fig. 5). It is quite plausible
that such an interference pattern in the underlying potential field
could lead to regions of enhanced density in the gas resulting in the
periodic beads on a string of OB stars and also possibly regular
structure in the positions of giant molecular cloud complexes.

In a similar vein, the film shown of M31 by Brinks (this conference)
should inspire the type of detailed investigation I am advocating. The
kinematical and spatial information on HI, CO, holes, Z-structures and
stellar populations are all a very, rich field for study related to the
spiral structure problem. The Fabry Perot work from machines such as
Taurus can and will give important detailed constraints on the ionised
gas flows (Allen, this conference).

Disk structures as observed by van der Kruit and Searle (1982) with a
sharp cut-off at $4 \alpha^{-1}$ (where α^{-1} is exponential scale of the disk) over
a scale $<< \alpha^{-1}$ should, in fact, give an edge mode if sufficient self
gravity is present in the disks at such large radii. Observations of
velocity dispersions in disk that can be unambiguously related to Q
values such as in NGC 916 (Kormendy, this conference) will give radial
Q-profiles that will be useful discriminators in differentiating hard or
soft (i.e. leading or trailing) reflection in the central parts. The
final observation I advocate is to observe simulations for wasering,
swiging, edge moding and other!

VI. DIRECTIONS AND CONCLUSIONS.

From an immense array of possibilities for the future development of this field, I have selected the few areas that I consider will be most fruitful between now and the next IAU meeting.

Firstly, global mode calculations and simulations should be understood completely in terms of local effects and consequently more non-linear problems such as disk heating (Carlberg and Sellwood, this conference) and mode saturation can be pursued with confidence.

Secondly, at least some spirals may not have significant dark halos. The fits to the rotation curves shown by Kalnajs (this conference) with constant M/L ~ 2-6 over the whole disk out to the Holmberg radius do not require invisible dynamical dark stuff. The new understanding of disk stability (Toomre, 1982, Sellwood, this conference, Kalnajs, this conference) means that the remarkable Ostriker-Peebles criterion for disk stability may not be so relevant. Self-consistent disks *could* be constructed without dark material for at least some realistic galaxy models. There is no evidence for dark material in ellipticals so why should it not be absent in at least some spirals. The beautiful simulations of T.S. van Albada (this conference) fit the radial density dependence and the velocity dispersion profiles for ellipticals with no dark material. Certainly dark stuff exists, if one inspects the mass-radius relationship on scales of ~ kpc to ~ 10 Mpc (Rubin, this conference) but most of it could be *hot* dark stuff, that will not be associated with all individual galaxies, with Mpc scales intermediate between groups and clusters. There is little hard evidence for dark stuff either in groups with their long crossing times or binaries with their poor statistics. Defending such an hypothesis may be an example of theoretical overshoot but the evidence may be sifted more thoroughly as a direct consequence.

Evolutionary considerations are possibly *the* major new area with incorporation of effects such as large scale radial gas flows, infall and detailed star formation physics. Large scale secular stellar dynamical evolution should be analysed; bars may dissolve as they are torqued to become more and more slender up to the point of a dynamical catastrophe; secular evolution could redistribute the mass sufficiently to cut off the central reflection by repositioning or eliminating the inner Lindblad resonance; hot anisotropic lenses may result from dissolving bars; and a host of other related phenomena.

Finally, associated with the detailed observations of galaxies we have the Fourier decomposition technique; the search for leading arms or at least symmetric regular interference patterns along the arms and the detailed correlation of the HI/HII/CO and stellar population data with realistic multiphased gas dynamics models. There are all excellent bases for the next generation of detailed theories.

ACKNOWLEDGEMENT

It is a pleasure to acknowledge the significant input to this paper from my colleagues at both the Institute of Astronomy, Cambridge and the Sterrewacht, Leiden when we discussed many of these questions in depth. I thank the many scientists who sent me preprints on this subject. Especially helpful were E. Athanassoula, A. Bosma. L. Blitz, E. Brinks, B. Burton, R. Carlberg, D. Lynden-Bell, G. Efstathiou, J. Kormendy, H. Listzt, R. Kennicut, J. Sellwood, A. Toomre and R. Walterbos. I dedicate this paper to the memory of my brother Murray, who liked a good story and enjoyed living at the edge.

REFERENCES

van Albada, G.D. and Roberts, W.W., 1981, Ap. J. 246, 740.
Aoki, S., Noguchi, M. and Iye, M., 1979, Publ. Astr. Soc. Japan, 31, 737.
Blaauw, A., 1964, Ann. Rev. Astron. Astrophys. 2, 213.
Drury, L.O.C., 1980, MNRAS, 193, 337.
Elmegreen, B.G. and Lada, C.H., 1977, Ap. 214, 725.
Elmegreen, D. and Elmegreen, B., 1982, MNRAS, in press.
Erikson, S.A., 1974, Ph.D. Thesis, M.I.T.
Goldreich, P. and Lynden-Bell, D., 1965, MNRAS, 130, 125.
Kennicutt, R. and Hodge, P., 1982, Ap. J. 253, 101.
Kormendy, J. and Norman, C., 1979, Ap. J. 233, 539.
Lieisawitz, D. and Bash, F., 1982, Ap. J. 259, 133.
Marochnik, L.S., Berman, V.G., Mishuror, Y.N. and Suchkov, A.A., 1983, Astr. Space, Sci., 89.
Rubin, V.C., Ford, W.K., Jr. and Thonnard, N., 1980, Astrohys. J. 238, 471.
Schwarz, M.P., 1979, Ph.D. Thesis, Australian National University.
Schwarz, M.P., 1981, Astrophys., J. 247, 77.
Seiden, P.E. and Gerval, H., Fund. Cosmic Physics, 7, 241.
Toomre, A., 1977, Ann. Rev. Astr. Ap. 15, 437.
Toomre, A., 1981, in "Structure and Evlution of Galaxies", ed. S.M. Fall and D. Lynden-Bell, Cambridge, p. 111.
Toomre, A., 1982, in preparation.
Van der Kruit, P.C. and Searle, L., 1982, Astron. Astrophys. 110, 61.

GENERAL DISCUSSION ON SPIRAL STRUCTURE

BLITZ : One thing should be kept in mind when using the results of Blaauw and Elmegreen and Lada as justification for the stochastic star formation model. Both sets of authors have argued that star formation can infect neighboring regions on a scale of tens of parsecs. But for the stochastic star formation model to work, it is necessary for some mechanism to allow the infection to spread from one giant molecular cloud to another —a length scale near the sun of about 500 parsecs—

with a propagation time consistent with the theory. This process, which Seiden calls the "microphysics" of the theory, is hidden in the theoretical quantity P_{stim}. It will thus be necessary for proponents of the theory to show that there is a physical mechanism which can and will propagate star formation from one giant molecular cloud to another.

NORMAN : I agree, but Blaauw indicated that a coherence length of $\sim$ 500 kpc was not unreasonable from the observations of the solar neighbourhood.

IYE : Could you elaborate on your third conclusion or implication namely that large scale evolution may cut off the swing amplification ?

NORMAN : No swing amplification can occur if the mass distribution of the galaxy slowly changes so that the Inner Lindblad Resonance absorbs the inward travelling trailing wave before it can be reflected into an outward propagating leading wave. This effect can occur if, for example, the mass distribution becomes more centrally concentrated, possibly due to strong radial inflow or merging with dwarf companions over a Hubble time.

SIMKIN : Dr. Lin has noted that a Fourier analysis of the "swing amplified" mode shows that its Fourier coefficients can be treated in two different ways, one of which is time independent and the other time dependent. The differential equations (in both cases) are related to the potential distribution in the galaxy (i.e. the density distribution). Thus we should be able to predict what types of density distributions will yield growing modes by analysis of these equations. These predictions can then be compared with "reality" (galaxies). Has this been done ?

LIN : The investigation of the amplification mechanism near the corotation circle is a local discussion, whether it is approached as a swing process or as a WASER process. To fit this mechanism into the description of the global modes is indeed a primary concern of the theoreticians. This has been done, in varying detail, in a number of papers. Specifically, a discussion of the mode calculated for M81 may be found in the references cited at the end of my abstract (Lin and Bertin, 1981 ; Haass, 1982).

IYE : I have Fourier analysed the luminosity distribution of several galaxies including both early and late types and presented the Fourier spectra and the patterns of primary Fourier components of these galaxies in a poster paper in this symposium. We find that the coexistence of trailing and leading spiral components and the coexistence of one-armed, two-armed, and multi-armed spiral components takes place very often in actual galaxies.

RENZ (to LIN) : To decide whether a stationary spiral structure can really develop you have to consider global modes and not just the local mode at corotation. Furthermore, after at best 10^9 years, the growing global modes reach an amplitude where linear theory breaks down and

nonlinear effects have then to be taken into account. Without doing
nonlinear calculations you cannot pretend to have shown that a stationary
spiral structure is possible.

LIN :   There appear to be several misunderstandings on the part of the
questioner. Specifically, regarding the last statement, we only claim to
have demonstrated that quasi-stationary spiral structure is possible and
plausible through the calculation of unstable global spirals modes of
both the tightly wound type (associated with WASER of Type I) and of the
type involving leading waves (associated with WASER of Type II). These
have been discussed in our existing publications, in my paper just presented, and in the papers of Dr. Bertin and Dr. Haass presented at this
conference. We do not claim that the existence of stationary spiral
structure is mandatory. Indeed, it would be wrong to do so. (See Oort's
description of the issues on spiral structure in Interstellar Matter in
Galaxies, (Benjamin, 1962), p. 234). The occurence of vascillatory large
scale modes in an unstable basic state of flow has been repeatedly discussed in classical literature on turbulent flow, weather patterns, etc.
Our experience in these subjects strongly suggests that the same conclusions hold in the present case, even though the nonlinear theory has not
yet been developed.

     The questioner appears to have also misunderstood the purpose of
our local analysis. It is intended to clear up possible misunderstandings
of the role of "swing amplification". The dramatic transient amplification
of wave packets does not imply a corresponding strong amplification of
modes or stationary wave trains. The net growth rate is usually much more
moderate. Generally speaking, a mechanism of self-regulation of the magnitude of the dispersion of stellar velocities is needed to arrive at favorable conditions for the existence of quasi-stationary spiral structures.

RENZ (to BERTIN) :   To select a slowly growing mode as a dominant one
over a long time scale, as you argued, you need a mechanism which your
equations do not contain. Namely you must assume that fast growing modes
will disappear very quickly by enlarging the velocity dispersion. Can you
write down the basic equations from which you can obtain both the slowly
growing spiral modes and this necessary mechanism ?

BERTIN :   The equilibria that we are interested in are subject only to
mild instabilities. In choosing such equilibria we use the physical
argument that violently unstable disks should be discarded because they
would rapidly evolve so as to heal the strong instabilities. Linear
stability analysis generally indicates which parameters (such as Q)
should be affected in the saturation process, even though detailed equations for nonlinear evolution are not easy to write.

III

WARPS

THEORIES OF WARPS

Alar Toomre
Department of Mathematics
Massachusetts Institute of Technology

The large-scale warps of disk galaxies remain a challenge for theorists. Recent speculations about their origins or longevity have tended to focus on the possible benefits of halos. After some remarks about modes, this review will do likewise. It will conclude that any applause even there can as yet be only scattered rather than thunderous.

1. WHY NOT MODES?

Much of the sad state of this subject can be traced to the old rumor by Hunter and Toomre (1969, hereafter HT) which alleged that no isolated thin disk without an implausibly sharp outer edge admits discrete $m = 1$ modes of bending, let alone any that would permit it to remain warped for very long in a steadily precessing integral-sign shape. In retrospect, I have often wished this rumor were false. Life would indeed have been sweeter for warp theorists if HT had confirmed rather than contradicted Lynden-Bell's (1965) idea that the self-gravity of bent disks might enable them to avoid the differential precession to oblivion that was emphasized already by Kahn and Woltjer (1959). Faced with all the reports from radio astronomers since the mid-1970's that warped outer disks are remarkably common even in galaxies without close companions, we could now be smiling "we told you so" instead of scrambling around looking for excuses!

Could the mode idea have been right after all, and its dismissal too hasty? Anyone tempted to ponder this important question should remember that HT found plenty of discrete modes when their cold, self-gravitating disks were chosen sufficiently sharp-edged. What was it about the outer edges that made such a drastic difference?

To understand that sensitivity in modern terms, start with a ring of test particles in orbit near the periphery of a galaxy, each bobbing up and down with frequency $\kappa_z(r)$ about the plane $z = 0$ while circling with angular speed $\Omega(r)$. Much like Lindblad's kinematic waves in that same plane, any $\sin m\theta$ or $\cos m\theta$ vertical distortions of this ring will then advance in longitude θ either with the <u>fast</u> pattern speed

E. Athanassoula (ed.), Internal Kinematics and Dynamics of Galaxies, 177–186.

$$\Omega_F(r) = \Omega(r) + \kappa_z(r)/m \qquad (1)$$

or else with the <u>slow</u> pattern speed

$$\Omega_S(r) = \Omega(r) - \kappa_z(r)/m \quad . \qquad (2)$$

In the $m = 1$ context where the ring merely tilts as if rigid, the first type of motion is essentially an Eulerian nutation or wobble; as Lynden-Bell pointed out, it can occur even when a ring is flabby. The second kind of $m = 1$ motion will probably seem more familiar, especially when $\kappa_z > \Omega$, as must be true near any oblate object that provides a restoring torque back toward $z = 0$; it represents the slow <u>retrograde</u> drift or "regression" of the line of nodes of our tilted gyroscope.

To build a mode in a disk consisting of many rings, one requirement is obvious: Somehow or other, these rates of regression or nutation must be adjusted to exact constancy from radius to radius. In practice, this means using the gravity from neighboring rings to augment κ_z — since there can be little doubt that any local self-gravity tends to <u>increase</u> the vertical stiffness, and the more so the more corrugated we imagine the disk to be. The mathematics can get tedious when dealing with warps as gradual as the integral signs that are our prime concern, but as usual it simplifies greatly in the short-wave or WKBJ approximation. Then the self-attraction of a corrugated thin disk of projected density μ and radial wavenumber k can easily be shown to boost its effective κ_z^2 by the amount $2\pi G\mu|k|$, or to modify the precession rates to

$$\Omega_p(r;k) = \Omega(r) \pm \frac{1}{m} [\kappa_z^2(r) + 2\pi G\mu(r)|k|]^{1/2} \quad . \qquad (3)$$

As this formula indicates, the $m = 1$ rates of regression or nutation of outlying material in a thin disk can probably be increased enough by local forces to keep pace with the more rapidly precessing interior. However, we also see that such insistence may well penalize a disk with excessive corrugations in any broad outer zone of low density.

HT used essentially the above reasoning to conclude that one should expect distinct and well-behaved $m = 1$ modes of bending only in those unlikely disks whose reciprocal density $1/\mu$ remains integrable all the way to a postulated sharp outer edge. Much as I wish otherwise, this old argument still seems valid and it is buttressed, as we are about to see, by an even more blatant difficulty caused by the <u>group velocity</u>

$$c_g = m \, \partial\Omega_p/\partial k = \mathrm{sgn}(k) \cdot \pi G\mu(r) / (\Omega_p - \Omega) \quad . \qquad (4)$$

In plain English, equation (4) cautions that all local properties of a bending wave of <u>trailing</u> spiral planform ($k > 0$) tend to drift radially inward if that wave happens to be slow ($\Omega_p < \Omega$) and outward if it is fast ($\Omega_p > \Omega$), whereas for <u>leading</u> waves those senses of propagation are just the opposite. Now it may seem exotic to talk of spiral warps when mainly interested in stable modes which themselves must exhibit straight nodal lines and hence no spiral preference. But such talk is less farfetched than it sounds. Even decent bending modes — like various bar modes in the plane — can be regarded as superpositions of leading and trailing waves in equal measure. And granted that, it is only sensible to ask what processes can possibly replenish those constituent waves that drift radially despite our labors to fine-tune their pattern speeds.

The answer is easy enough for any proposed slowly-regressing $m = 1$ mode. Its inward-drifting trailing waves will in due course presumably transmit/reflect from the interior as outward-drifting leading waves, and thus they will indeed replenish briefly the leading waves that exist now. But the crisis will come as those new waves continue to travel outward with the speed c_g that decreases roughly like the density $\mu(r)$: Unless there is a sharp edge nearby to reflect them back as trailing waves to complete the cycle, we see that it can take literally forever for such wave information to drift out more and more sluggishly, never to return. A similar fate awaits a would-be fast mode — just swap the words leading and trailing — as illustrated vividly by the impulse response shown in Figure 8 of HT. In this mechanical sense, the outer parts of any gently tapered disk must therefore act very much like an absorbing beach to all $m = 1$ waves that might have sought to slosh up and down within it.

Modes aside, some beautiful examples of periodically _forced_ bending waves that lend themselves almost ideally to WKBJ reasoning were reported recently by Shu, Cuzzi and Lissauer (1982). The foremost of these is a slow $m = 4$ trailing spiral wave pattern that propagates inward from the radius where it continues to be excited thanks to a 5:3 resonance with a slightly inclined satellite. The name of that satellite is Mimas, and the locale is the A ring of Saturn.

2. FOUR RECENT SUGGESTIONS

In the absence of clearcut modes, at least four ideas have surfaced in print lately seeking to explain the observed warps of galaxies with the active help or passive acquiescence of halos. The "active" ideas have stressed two possible sources of fresh excitation, whereas the "passive" speculations have focused more on ways to reduce the bothersome rates of differential precession. Let me deal first with the activists.

2.1 Mathieu instability

Binney (1978, 1981) pointed out that a steadily revolving bar-like halo or a genuine central bar may spontaneously excite z-oscillations in a galactic disk near certain resonant radii, via the classic phenomenon known as Mathieu instability. What matters crucially here is that some significant portion of the vertical stiffness contributed by that bar or halo should appear to a given star or gas atom to vary periodically with time at almost exactly _twice_ the intrinsic frequency κ_z (ignoring the local self-gravity for the moment) with which that item and its neighbors would prefer to bob up and down. If this resonance condition can be met,

then it will be possible for each vibrating particle to acquire just such
a phase of oscillation that itᶜ vertical spring rate will be stronger than
average whenever it approaches the plane z = 0 , and weaker than average
when it recedes. Barring losses, the net result should be a steady gain
of amplitude — much as achieved by an energetic child on a swing.

The condition for Mathieu resonance at radius r follows from noting
that the bar-like potential needs to advance or recede relative to the
orbiting particle by exactly two cycles, or by one full turn, during one
intrinsic vertical period. Hence that relative speed must itself equal
$\pm \kappa_z$, or the imposed bar or halo must revolve in space simply with one
or the other of the elementary m = 1 speeds $\Omega \pm \kappa_z$ appropriate to
that radius. Assuming the resulting amplitudes to be independent of θ ,
it is easy to see that those motions again correspond to the aforesaid
fast nutation and slow regression, respectively.

Unfortunately, the addition of local self-gravity to this incomplete
story uncovers what may well be a fatal flaw. I am referring here <u>not</u>
to my own distaste for postulating any retrograde-moving triaxial halos,
nor to the lack of evidence as yet that the resulting wave propagation
(here presumably outward, for fast and slow waves alike) even from a fast
bar resonance can yield warps resembling those observed. My worry is
simply this: Unlike that production of fresh m = 4 bending waves by
Mimas — from scratch at a known rate — the Mathieu process manages only
to <u>amplify</u> by a finite and modest factor per revolution whatever m = 1
vibrations already exist near its resonance radius. And if the group
velocity always carries waves away and never returns them, there may soon
be precious little left to be amplified. Thus, I am afraid what seemed
"a major breakthrough" to Saar (1979) in the last IAU review of this
subject may yet fizzle away to nothing.

2.2 Flapping instability

Flapping instabilities as such have been known for a long time, but
until recently they were not thought relevant to the observed major warps
of galaxies. It was presumed that they arise only in circumstances that
are either too small-scale — like the two-stream or "hose" instability
recently revisited by Bertin and Casertano (1982), who insist that the
scales may be sizable after all — or else too preposterous, such as the
pair of superposed and <u>contra</u>-rotating Maclaurin disks which HT (p.752)
knew to be unstable for m ≥ 2 . Both of these related instabilities
seemed to demand a fairly stiff coupling between the oncoming streams,
to ensure that they deform vertically almost like one.

Bertin and Mark (1980) were the first to recognize that such lateral
stiffness is unnecessary, and that even a single thin sheet of matter that
slides rapidly through a hot background of stars, neutrinos, or whatever
may gravitationally amplify its corrugations — somewhat like a flag
flapping in the wind. Their original derivation of this important result
still frightens me, but by another route I concur that such amplification
is indeed possible, at least at moderate wavelengths. I also agree that

the local speeds of any growing corrugations must be intermediate between those of the disk and the spheroid/halo. And lastly I commend Bertin and Mark for realizing at once that their amplification — like the Mathieu kind just discussed — is convective, and thus requires wave feedback.

Alas, I doubt for two reasons that even these interesting disk-halo interactions have anything to so with the observed m = 1 warps:

1. It remains very much to be demonstrated that warps as gentle in the vital θ direction as our m = 1 tilted rings can amplify in this manner. One should not forget that even those furiously contra-rotating Maclaurin disks cited by HT showed no m = 1 instabilities.

2. As Bertin and Mark understood, their need for steady feedback postulates a discrete m = 1 mode that rotates forward more rapidly than the spheroid/halo but more slowly than the bulk of the disk. Earlier we saw why discrete ring-tilting modes remain so elusive, and also why any slowly advancing m = 1 mode would amaze gyroscopists. Evidently Bertin and Mark thought otherwise, judging from all their talk about corotation radii and a "type of bending mode different from that of HT". But their reasoning there rested on a blunder. In their dispersion relation (II-3), they simply forgot the frequency κ_z due to distant matter.

Ironically, HT themselves (p.762) reported that some m ≥ 2 modes of bending remain discrete even in disks with very blurry edges. Hunter (1969), who studied them extensively, noted that those modes generally do drift forward with modest speeds Ω_p . Hence I think they consisted essentially of slow multi-armed waves trapped between the center and the radius at which $\Omega_p = \Omega - \kappa_z/m$. Hunter and I have long wondered quietly whether such modes or their kin might be partly responsible for the minor ripples known to exist in our Galaxy interior to the Sun — and well they might be, if the Bertin-Mark process indeed manages to amplify them.

2.3 Nearly spherical potentials

Turning now to two proposals where the halo remains passive, I shall be briefer since they are more self-explanatory. For instance, on the premise that massive halos might be almost spherical, Tubbs and Sanders (1979) cautioned that one should not worry too much about differential precession. They stressed that the gaseous warps of galaxies like NGC 5907 (Sancisi 1976) or M83 (Rogstad, Lockhart and Wright 1974) occur far enough outside their optical disks that it remains at least arguable that such shapes could still be the (mangled?) relics of very ancient warps.

I think Tubbs and Sanders had a marginally valid point with those two galaxies and some others. To recapture the gist of their argument, remember first that those gas disks are huge: Out there near r = 40 kpc in NGC 5907, for instance, one revolution alone takes about 1.3×10^9 years. Even so, one probably needs to insist that any nodal regression rates be almost ten times slower; otherwise in 10^{10} years the rings at, say, r = 30, 35, 40 and 45 kpc would have drifted hopelessly out of

phase. Tubbs and Sanders noted correctly that the quadrupole potential
from the known stellar disks is unlikely to create much trouble at such
radii, and they also knew that any outlying dark matter needed to produce
the relatively flat rotation curves of those galaxies cannot remotely be
as flattened as the gas disks lest it cause too much regression. Assuming
for simplicity that the halo potential is constant on similar spheroids
of axis ratio c/a , it is easy to confirm that this tolerable flattening
is indeed rather small: In that special case, the vertical frequency κ_z
about the equatorial plane is everywhere just a multiple a/c of the
local angular speed Ω , and it follows from the regression rates Ω_s =
$\Omega - \kappa_z$ that the potentials had better be round to about one part in ten.
Halo densities, to be sure, can be somewhat more oblate, but even they
ought probably not exceed E2 shapes to leave some margin for disk self-
gravity. In short, uncomfortable maybe, but not impossible.

The situation remains grim, however, with the outer warp of our own
Galaxy — and by analogy probably also those of M31 and M33. Simply the
far-field formulas cited by Tubbs and Sanders imply that the regression
rates Ω_s in a pure exponential disk should amount to -0.147, -0.102 and
-0.074 times $\Omega(r)$ at radii equal to 4, 5 and 6 scale lengths. At
those rates and radii, I reckon that rings of test particles would regress
some 130°, 65° and 35° , respectively, every 10^9 years, assuming that
the orbital period equals 2.5×10^8 years at the distance of 3 scale
lengths. And these are but underestimates,
ignoring all extra nuisance from local gravity. Hence I still maintain
that the survival of our fairly clean and hardly spiral warp would become
remarkable in less than 10^9 years if we continue to pretend that our
Galaxy consists only of a disk, and after at most 3×10^9 years if we
adopt the most generous disk-halo composite. What a pity that the near-
perpendicular orientation and the large tip velocity of the Magellanic
Stream now exclude even the LMC from coming to our aid.

2.4 Increasingly oblate halos

Petrou (1980) added a nice logical twist to the reasoning by Tubbs
and Sanders. She urged that, rather than deplore any pronounced oblate-
ness of halos, we should actually welcome it — provided such flattening
increases outward rapidly enough to compensate for the decreasing $\Omega(r)$
by a steady increase of the relative stiffness κ_z/Ω about the imagined
common equatorial plane of the disk and the halo. Petrou reminded us all
that the goal is not to avoid the gyroscopic regression as such, but only
to make sure that its speed $\kappa_z - \Omega$ is almost identical at every radius.
She also illustrated with some examples that this goal can in principle
be met handily for several known huge warps including that of NGC 5907.

As implied above, I nonetheless doubt that even Petrou's idea can
ensure for 10^{10} years the survival of the relatively close-in HI warp
of our Galaxy against the torques from our minimal known stellar disk.
Even if it can, there remains one philosophical worry: Kinematic theories
concerned only with survival do not explain why warps should have begun
with their nodal lines so neatly aligned from radius to radius.

3.   STEADY FORCING BY A TILTED HALO?

To end on an upbeat note, let me conclude with a few tentative words about another idea involving halos that has long struck me (a) almost as outrageous astronomically, and yet (b) almost as delightful dynamically, as it would be to postulate that an invisible ring of massive satellites encircles a disk galaxy in a nearby and inclined orbit. The tidal force of that ring would certainly ensure that the central galaxy gets and stays warped about a single (though precessing) line of nodes. And even better, as HT (p.768) noted already, such <u>steady</u> forcing would yield much the same results regardless of whether the victim disk has a sharp or fuzzy edge.

Alternatively, as a classic example, imagine that our precessing Earth were surrounded by a disk of debris even more extensive than that around Saturn. Ever since Laplace it has been known that the equilibrium shape of such a disk would be dominated nearby by our equatorial bulge, and far away by the Sun and the Moon — with the warped outcome shown in Figure 1. Notice the shift in allegiance that would occur near $r_{crit} \simeq$ 7.7 Earth radii: It is so rapid because the ratio of the outer versus quadrupole torques varies here like r^5 for equal angles.

Along with Dekel and Shlosman (this volume), who had the same notion, I have been wondering lately whether something analogous might not still afflict disk galaxies that lacked the good manners to be born with spin vectors exactly parallel to the short axes of the oblate massive halos in which they may now be embedded. As said above, I used to regard such tilted geometries as unlikely indeed. However, the recent eloquence of several cosmologists to the effect that halos probably formed first, and galaxies slightly later by crashing gaseous infall, has persuaded me that it is at least conceivable that massive disks deep within halos started out misaligned from their hosts typically by some tens of degrees. And even without logic, desperate men try many things!

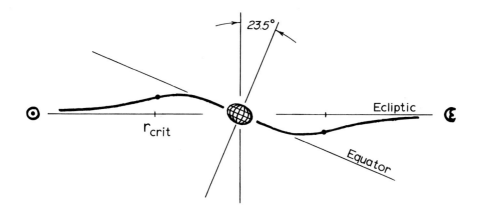

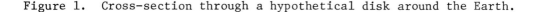

Figure 1.   Cross-section through a hypothetical disk around the Earth.

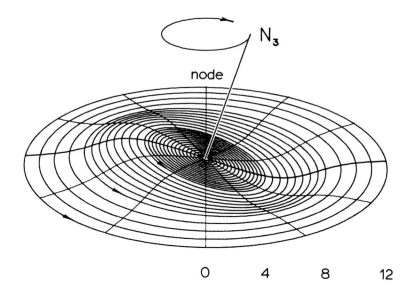

Figure 2. Perspective view of a tilted exponential disk regressing slowly
under the tidal influence of a distant massive ring imagined horizontal.
The disk slopes 1-in-3 at its center, and its first four scale lengths
are shaded darkest. The forcing was taken to consist only of the linear
part of the tidal field; its strength matches that of a ring of unit disk
mass and of a radius equal to 10 disk scale lengths.

One key question about this kind of proposal is how long the central
disk can thereafter remain tilted despite dynamical friction against the
doubtless flabby halo. Another serious concern is whether the tidal force
from a plausibly massive and flattened halo can compete <u>deeply</u> enough with
torques from the disk — i.e., whether the halo can impose a warp already
from about 4 scale lengths outward, as in this Galaxy.

In their own paper, Dekel and Shlosman offer some early encouragement
on the first topic, whereas for this review I concentrated on the second.
I took an exponential disk and divided it into some 50 equally-spaced
rings, ranging from quite massive ones in the interior to virtual test-
particle rings in the outer half. Next I calculated all the inter-ring
torques. Then I exposed this collection to a linear tidal force from
that imagined massive distant ring of $(R/10)^3$ times the disk mass,
where R is its radius in disk scale lengths — and proceeded to search
for the simplest shape-preserving warp as an eigenvalue problem involving
the drift rate, which proved to be a tiny -0.018 times $\Omega(3)$. The
corresponding warp shape is shown in Figure 2. I think it attests that
the idea has some merit — but that is all I can or ought to claim here.

This work was supported in part by grant MCS 78-04888 from the NSF,
and of course also by many old discussions with Chris Hunter.

REFERENCES

Bertin, G., and Casertano, S.: 1982, Astron.Astrophys. 106, 274.
Bertin, G., and Mark, J.W.-K.: 1980, Astron.Astrophys. 88, 289.
Binney, J.: 1978, M.N.R.A.S. 183, 779.
Binney, J.: 1981, M.N.R.A.S. 196, 455.
Hunter, C.: 1969, Studies in Appl.Math. 48, 55.
Hunter, C., and Toomre, A.: 1969, Astrophys.J. 155, 747. [= HT]
Kahn, F.D., and Woltjer, L.: 1959, Astrophys.J. 130, 705.
Lynden-Bell, D.: 1965, M.N.R.A.S. 129, 299.
Petrou, M.: 1980, M.N.R.A.S. 191, 767.
Rogstad, D.H., Lockhart, I.A., and Wright, M.C.H.: 1974, Astrophys.J.
     193, 309.
Saar, E.: 1979, in IAU Symposium No. 84, "The Large-Scale Characteristics
     of the Galaxy" (Dordrecht: Reidel), p.513.
Sancisi, R.: 1976, Astron.Astrophys. 53, 159.
Shu, F.H., Cuzzi, J.N., and Lissauer, J.J.: 1982, Icarus, in press.
Tubbs, A.D., and Sanders, R.H.: 1979, Astrophys.J. 230, 736.

DISCUSSION

BERTIN: From preliminary numerical work, Casertano and I have hints that
a D1 mode of the good kind needed for "flapping instability" may survive
in models that have a smooth disk density in the outer regions. We hope
to confirm this pretty soon.

TOOMRE: Bravo, if you can do it. And also don't forget those discrete,
slow m > 1 modes found long ago by Hunter.

SANDERS: I would like to stress that the large warps shown yesterday by
Dr. Sancisi exist mainly in the tenuous gas beyond the stellar disk, and
not in the self-gravitating stellar disk itself. This certainly suggests
to me that we may treat the warp as a collection of test particles; there
is no need to consider self-gravity in the warps. Moreover, we also see
in several cases — for instance, NGC 5907 and 628 — that the gaseous
warps begin precisely at the truncated edge of the exponential disk, or
where the gravitational potential becomes rapidly spherically symmetric
and the differential precession approaches zero. I submit that these
external galaxies where we actually observe the light distributions in
the disks and the morphology of their warps may be telling us more about
warp dynamics than our own galaxy where our picture is more restricted.

TOOMRE: You know I like to peek at observations myself, but I also think
any theoretical review worth its salt needs to challenge theorists mostly
on their own ground — and sometimes even on their own Galaxy.

SANDERS: I also have a question about the final model you discussed.
If there are two preferred planes — a galaxy disk and an outer plane
perhaps established by a tilted halo — would you expect the gaseous warp
to bend back toward the galaxy plane at larger radii, as it seems to in
one or two well-known cases?

TOOMRE: Indeed I wouldn't. If you are thinking of NGC 4762, that sure
has looked to me for years like some fairly mature but transient tidal
shape. My only real worry has been that its apparent neighbor, NGC 4754,
has that hefty excess redshift of around 500 km/s .

SANDAGE: How much mass must the tilted halo have, relative to the disk,
for this model of yours and of Dekel and Shlosman to produce the warp?

TOOMRE: Probably at least three times the disk mass. And that's counting
only within those ten or so disk scale lengths.

SANDAGE: Are you then not unhappy that the best cases where optical warps
are suggested — M33 and NGC 628 — are late Sc galaxies where the optical
halo is absent? For M33, I believe that no dark halo is suggested even
by the rotation curve.

TOOMRE: Yes, I am unhappy. But as you heard, I am even less happy about
most of the alternatives.

NELSON: All the models that you have described to us involve thin stellar
disks with no velocity in the z-direction. However, the warps are mainly
observed in the gas disks of galaxies, and there the vertical pressure
forces are comparable to the gravitational forces. Do you not think that
hydrodynamical models for warp dynamics should be considered?

TOOMRE: Sorry about neglecting those models here. Frankly, I could not
bring myself to believe that they are desperately relevant, and I had all
those other things to discuss.

NELSON: My point is that if you include the gas pressure and the finite
thickness, other wave modes appear for which the differential precession
can be much smaller (M.N.R.A.S. 177, 265, 1976; 196, 557, 1981).

TREMAINE: In view of the interest in hot thin disks, could you tell us
about work you have done on the firehose instability in such systems?

TOOMRE: The main thing I did in 1966 or thereabouts was to calm myself
down! I had first thought that such instabilities might set in even with
minor anisotropy, but then I spent several weeks calculating them (mostly
numerically) via the collisionless Boltzmann equation in a non-rotating
$sech^2z$ sheet of stars with different Gaussian distributions of random
velocities in the vertical and horizontal directions. And I was sorry
to find that this hose instability ceases already when the vertical rms
speed exceeds a mere 30 per cent of the horizontal rms speed. That is
where I lost interest, and why I never even published properly.

GALACTIC WARPS IN TILTED HALOS

Avishai Dekel
Caltech and Yale University

Izhak Shlosman
Tel-Aviv University

The current suggestions for the origin of warps in disk galaxies (see Toomre, this volume) find difficulties in explaining their frequent occurrence and an external driving mechanism seems to be required in order to maintain long-lived warps. Such a mechanism can be provided by an extended dark halo if a) it dominates the gravity at large radii while the inner disk is self-gravitating, b) it is slightly flattened and becomes flatter at larger radii, and c) it is tilted relative to the inner disk. Such a configuration may be formed as a result of tidal encounters, or of clouds infalling into halos.

Assume that the halo is oblate with its minor axis tilted relative to the disk, and consider test circular orbits. They precess about the local Laplace plane, where the external torque vanishes, which determines the mean shape of the system. It coincides with the disk plane at small radii while merging into the equitorial plane of the halo at large radii. An analogous problem is well known in celestial mechanics: the orbits of moons and rings deviate from the equitorial planes of their planets towards the ecliptic because of the torque exerted by the sun.

Toomre has considered the simple, but illustrative, case of a tilted massive ring. A detailed study that considers realistic halos is in progress from which results are briefly described here. The shape of edge-on warps in two disk-halo systems is shown in figure 1. The disk is thin and has an exponential surface density profile with a length scale r_d = 5, 3 kpc in a and b respectively, and a mass M_d = 15, 6 x 10^{10} $M_\odot$. The halo has a $1/r^2$ density profile with a core length scale r_h = 15, 18 kpc, and a mass of 10, 8 x 10^{11} $M_\odot$ inside 60 kpc. The eccentricity of the halo must grow with radius. Otherwise, there is no torque exerted by external shells on the inside. In figure 1 the eccentricity e = $[1 - (\text{axial ratio})^2]^{\frac{1}{2}}$ grows with the radius as a power law with a power 0.8, starting from e = 0.2 at 5 kpc. The relative tilt is 15°, which is comparable to the extreme observed cases (e.g. NGC 5907).

The gravitational potential is calculated analytically for discrete shells. Equipotential contours are plotted (dashed line). The warped

187

E. Athanassoula (ed.), Internal Kinematics and Dynamics of Galaxies, 187–188.

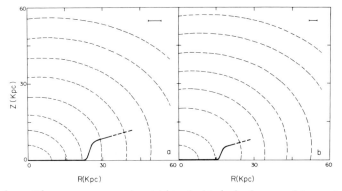

Figure 1. Edge on warps in tilted disk-halo configurations.

Laplace plane (thick line) is determined by the minima of the potential at
given radii. The deviation from the disk plane occurs quite abruptly, and
the Laplace plane coincides quickly with the halo equitorial plane, in
agreement with the geometry of observed warps (see Bosma and Sancisi in th
volume). The general shape of the warp is insensitive to the detailed
structure of the disk or the halo. The radius at which the warp occurs is
mostly determined by r_h/r_d.

The crucial point is that the <u>survival time</u> of the tilted disk-halo
configuration can be comparable to the Hubble time. Assume a simple
case in which the halo potential has only a monopole component and a
smaller quadropole component of the form $\phi_2(r)P_2(\cos\theta)$. A circular orbit
(r, v) initially in the disk plane (inclination i between halo and disk)
will precess about the halo equitorial plane with an angular frequency
$\omega_p = -3\phi_2(r)\cos i/2vr$. Except for the special case $\phi_2(r) \propto v(r)r$ the
precession tends to be differential, but in fact the disk would precess as
rigid body out to a few length scales because of its self-gravity: the
torque exerted by the disk on a particle slightly out of its plane is
dominant over the torque exerted by the halo out to > 5 r_d in the realisti
cases studied. The disk precession rate can be estimated at the giration
radius r_g (=$\sqrt{6}r_d$ for an exponential disk), and is slower than the rotation
rate there by a factor of $\sim \varepsilon_\phi(r_g)$, where $\varepsilon_\phi = 3\phi_2(r)/2v(r)^2$ is the
ellipticity of the equipotential surface at r. If $\varepsilon_\phi \leq 0.05$ at ~ 10 kpc th
precession rate is of the order of the Hubble time. In the case shown abc
$\varepsilon_\phi \cong 0.04$ at 10 kpc.

As a result of the precession of the disk inside the halo they
exchange angular momentum and suppress their mutual tilt. In a Hubble
time, the disk may affect the tilt of the inner halo only out to ~ 10 kpc.
The disk tends to settle down in the halo equitorial plane as a result of
dynamical friction. Our analytical study of the interaction between a
rigid precessing disk and a family of simple halo orbits shows that the ti
scale for the tilt dampening is longer than the precession time scale.
Preliminary results from an N-body experiment confirm the analytical
estimate, demonstrating the fact that the life time for a disk-halo tilt i
indeed very long (Shlosman, Gerhard and Dekel 1983, in preparation).

PERIODIC ORBITS AND WARPS

W.A. Mulder
Sterrewacht, Huygens Laboratorium, P.O. Box 9513,
2300 RA Leiden, The Netherlands

Orbit calculations were done in a rotating triaxial system with a
density distribution in accordance with recent observations of spiral
galaxies. A search was made for simple closed orbits which are tilted
with respect to the plane of the galaxy. A family of stable prograde
tilted orbits was found which can explain warps as stationary phenomena.

Tilted disks are common phenomena in many early type and elliptical
galaxies (Bertola et al., 1978; Hawarden et al., 1981). Spiral galaxies
often show warps. Attempts have been made to explain these configurations
in terms of stable closed orbits in rotating triaxial systems. Merritt
(Heisler, Merritt and Schwarzschild, 1982) found a family of retrograde
tilted orbits. Binney (1981) used Mathieu's equation to show that orbits
in the plane of a rotating triaxial potential can be unstable for perturbations perpendicular to that plane.

Here the triaxial density distribution is chosen to be $\rho = \rho_0\, m^p$
with $m^2 = x^2/a^2 + y^2/b^2 + z^2/c^2$ and a>b>c. For the bulge of our Galaxy
$p = -1.8$, c/a = 0.4 and $M = (9 \pm 2) \times 10^9\, M_\odot$ within 1 kpc (Sanders and
Lowinger, 1972; Isaacman, 1981). Burstein et al. (1982) found p=-1.7±0.1
from the rotation curves of 21 Sc galaxies. The density distribution
pertains this value even at large radii. Therefore, bulge and halo are
modelled with a single density distribution.

The axial ratio's were chosen to be b/a = 0.8 and c/a = 0.5 with
$p = -1.8$ and the rotation frequency of the triaxial system $\omega = 0.1$ around
the z-axis. Periodic orbits were calculated numerically with the method
described by Magnenat (1982). There are tilted orbits near every $(\omega_0-\omega):\omega_z$
resonance, where ω_0 is the rotation frequency in the non-rotating frame
and ω_z the oscillation frequency perpendicular to the plane. Most of
these orbits are self intersecting, have sharp turnings or are not symmetric with respect to the origin. The most important orbits for gas were
found at the ±1:1 resonance, occuring for the retrograde orbits within
CR and for the prograde orbits outside the OLR. The first correspond to
those found by Merritt (Heisler et al., 1982). The latter are shown in

189

fig. 1. They have an increasing tilt with increasing energy and start to
have loops at a specific value of the energy, implying that gas in this
type of orbit will have a maximum tilt.

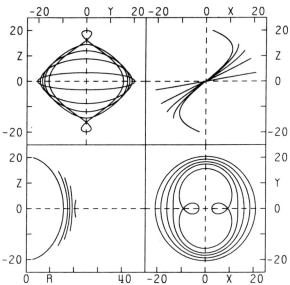

Fig.1. Stable prograde
orbits at the -1:1 resonance.
The plotted orbits have
energies E of 2.75, 3.00,
3.25 and 3.50. The orbit
with the loops is at E=4.40.
The tilt increases as E increases. CR is at r = 8.4.

The following conclusions can be drawn:
(1) Under the assumption that the motion of gas in the outer part of a
spiral galaxy can be described in terms of simple stable periodic orbits,
a warp can be explained as a stationary phenomenon in a rotating triaxial
system. As this assumption does not hold strongly, the presented calculations are only indicative.
(2) A model based upon these periodic orbits provides a length scale for
the galaxy, predicting the locations of the other resonances in the
assumed density distribution. It also defines the direction of the long
axis of the triaxial figure.
(3) It should be noted that highly non-circular motions can occur in a
rotating triaxial system. Interpretations of observations in terms of
circular motion might unnecessarily give rise to multi component models
for the mass distribution and notions as "expanding gas features".

References.

Bertola, R., Galleta, G.: 1978, Astrophys. J. 226, L115.
Binney, J.: 1981, Monthly Notices Roy. Astron. Soc. 196, 455.
Burstein, D., Rubin, V.C., Thonnard, N., Ford, W.K.: 1982, Astrophys. J.
     253, 70.
Hawarden, T.G., Elson, R.A.W., Longmore, A.J., Tritton, S.B., Corwin,
     H.G.: 1981, Monthly Notices Roy. Astron. Soc. 196, 747.
Heisler, J., Merritt, D., Schwarzschild, M.: 1982, Astrophys. J. 258, 490.
Isaacman, R.B.: 1981, Astron. Astrophys. 95, 46.
Magnenat, P.: 1982, Astron. Astrophys. 108, 84.
Sanders, R.H., Lowinger, T.: 1972, Astrophys. J. 77, 242.

IV

BARRED GALAXIES

MORPHOLOGY, STELLAR KINEMATICS AND DYNAMICS OF BARRED GALAXIES

John Kormendy
Dominion Astrophysical Observatory
Herzberg Institute of Astrophysics
Victoria, B.C., Canada

ABSTRACT

A brief review is given of the morphology of barred galaxies, following Kormendy (1981, 1982). The features illustrated include bulges, bars, disks, lenses, and inner and outer rings.

Most of the paper is devoted to a detailed discussion of the absorption-line velocity field of the prototypical SB0 galaxy NGC 936. The stars in the bar region show systematic non-circular streaming motions, with average orbits which are elongated parallel to the bar. Beyond the end of the bar, the data are consistent with circular orbits. The bar region also shows large random motions: the velocity dispersion at one-half of the radius of the bar is 1/2-2/3 as large as the maximum circular velocity. The observed kinematics are qualitatively and quantitatively similar to the behavior of n-body models by Miller and Smith (1979) and by Hohl and Zang (1979). The galaxy and the models show similar radial dependences of simple dimensionless parameters that characterize the dynamics. These include the local ratio of rotation velocity to velocity dispersion, which measures the relative importance of the ordered and random motions discussed above. Also similar are the residual streaming motions (relative to the circular velocity) in a frame of reference rotating with the bar. Circulation is in the same direction as rotation in all galaxies studied to date. Thus, except for the fact that NGC 936 has a slightly larger velocity dispersion, both n-body models are good first-order approximations to bars. Thus bars are different from elliptical galaxies, which in general are also triaxial, but which rotate slowly. This study of NGC 936 will be published in Kormendy (1983).

A brief discussion is given of the kinematics of lens components. In both barred and unbarred galaxies, the velocity dispersions in the inner parts of lenses are large. The ratio of rotational to random kinetic energy is ∿ 1/2 at 1/3-1/2 of the radius of the lens. This ratio then decreases to small values at the rim of the lens. Thus at least some kinds of disk components have large stellar velocity dispersions, even in unbarred galaxies.

E. Athanassoula (ed.), Internal Kinematics and Dynamics of Galaxies, 193–196.

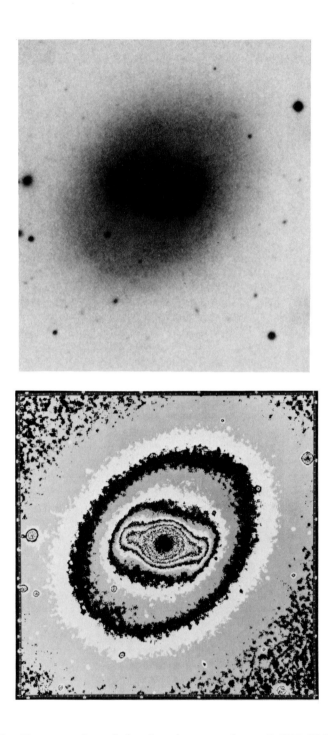

Fig. 1   Photograph and isodensity tracing of NGC 936.

REFERENCES

Hohl, F. and Zang, T.A. : 1979, Astron. J. 84, 585.
Kormendy, J. : 1981, in The Structure and Evolution of Normal Galaxies,
        ed. S.M. Fall and D. Lynden-Bell, p. 85, Cambridge : Cambridge
        University Press.
Kormendy, J. : 1982, in Morphology and Dynamics of Galaxies, Twelfth
        Advanced Course of the Swiss Society of Astronomy and Astrophysics
        ed. L. Martinet and M. Mayor, in press Sauverny : Geneva Obs.
Kormendy, J. : 1983, Astrophys.J., in preparation.
Miller, R.H. and Smith, B.F. : 1979, Astrophys.J. 227, 785.

DISCUSSION

RICHSTONE : You compare your observed dependence of V/σ along the bar
to that in N-body models. You could also compare V and σ separately with
a free scale factor. Have you tried that ?

KORMENDY : Not yet in detail. However, the rotation velocity $V(r)$ and
dispersion $\sigma(r)$ behave similarly in NGC936 and in the models beyond the
central region dominated by the bulge. $V(r)$ rises gradually to a V = constant rotation curve which is reached slightly interior to the end of the
bar. The dispersion drops by a factor $\sim$ 2-3 from near the center to the
end of the bar. At a more detailed level there are differences, some of
which are physically significant and some not. The galaxy has a bulge, so
its rotation curve is higher at small radii than in the models. The models
do not have dark halos, so $V(r)$ is not constant but is falling at large
radii. In addition the models are anisotropic in the equatorial plane,
with a larger dispersion parallel to the length of the bar than across it.
This is not seen in the data. More quantitative comparisons will be given
in a paper on NGC936, which will be published soon in Astrophys. J.

SELLWOOD : The Miller and Smith and Zang and Hohl models do not have
anything like the central concentration of NGC 936. Is it remarkable that
the velocity fields agree so well ?

KORMENDY : The central concentration of NGC 936 is contained in the
bulge. This produces a peak in $V(r)$ at small radii. In this region the
galaxy and models behave very differently. At larger radii the bulge is
apparently felt only as a fuzzy central mass, which is no longer very
important compared to the rest of the mass distribution. So it is perhaps
not too surprising that the galaxy and models are similar over the outer
two-thirds of the length of the bar. The agreement actually surprises
one more because the models are so much thicker than we assume the galaxy
to be.

SANDERS : If lenses are hot and flat, then the velocity distribution must
be highly anisotropic. Do you have any evidence for this ? Have you observed any face-on lenses and detected a small velocity dispersion ?

KORMENDY :   There is no direct evidence for the large velocity anisotropy.
The small axial dispersion is inferred from photometric studies of edge-on
galaxies with lenses. For example, Tsikoudi (1977, Ph. D. Thesis, University of Texas at Austin ; 1980, Astrophys. J. Suppl., 43, 365) and
Burstein (1979, Astrophys. J., 234, 829) conclude that the lens is part
of the thin disk in the edge-on, probably SBO galaxy NGC 4762. In general,
half of all edge-on galaxies are barred, but edge-on galaxies do not
obviously divide themselves into two groups with thin and thick disks.
All this suggests that lenses are flat, despite their large planar dispersion. Direct measurements of the axial dispersion are needed to test
this conclusion, but will be difficult to obtain.

CHRISTIAN :   You indicated that you chose galaxies which are likely to
have integrated spectra similar to the spectrum of a single template star.
Does the integrated spectrum of such a galaxy change at all with position
angle, that is, from along the bar to perpendicular to the bar ?

KORMENDY :   I have seen no sign of a change in spectral properties with
position angle. In SBO galaxies the bulge, bar and disk have similar
populations. Interestingly, at intermediate Hubble types ($\sim$ SBb ; e.g.,
NGC 2523), the bar has the same population as the bulge and not the disk
(which is blue and full of HII regions). This is true despite the fact
that the bar is part of the disk. The above preliminary remarks are
based on the morphology, on broad-band color measurements, and on a
qualitative examination of the present spectra. No detailed study has
been made of stellar populations in barred galaxies.

DISK STABILITY

J.A. Sellwood
Institute of Astronomy, Madingley Road, Cambridge, CB3 OHA.

What has prevented the formation of a strong bar in the majority of disc galaxies? No truly satisfactory answer has yet been given to this question and the difficulty remains a major obstacle to our understanding of the dynamics of these systems. In this review, I will discuss the implications of recent studies of this problem.

Should The Results Be Believed?

In order to determine whether a stellar dynamical model is globally stable one can either perform a linear stability analysis or run a computer simulation. Unfortunately, analytic studies are extremely difficult and have been completed in only a few simple cases (Kalnajs 1972, Zang 1976, Toomre 1981). Large n-body simulations are therefore necessary for most models of interest.

Since the conclusions are far reaching, it is important to demonstrate that the results can be trusted. One of the best tests is to show that identical results are obtained by the two methods for the same model; Sellwood (1982) finds discrepancies of only a few percent between the predicted eigen-frequencies and those observed in the simulations. Tests of the non-linear behaviour are more difficult to devise. Nishida et al (in preparation) are working with a completely different numerical technique in which they integrate the Vlasov equation. Comparison between their results and those from n-body codes in the non-linear regime will again provide a useful test of both methods.

Stability Criteria

The unavoidable conclusion from all studies of disc stability to date is that some fraction of the mass of an unbarred galaxy must reside in a spherically distributed component. The interesting question is what fraction? Despite some systematic searching, no criterion for stability has been found that is a reliable guide for any arbitrary model.

197

E. Athanassoula (ed.), Internal Kinematics and Dynamics of Galaxies, 197–202.
Copyright © 1983 by the IAU.

Ostriker and Peebles (1973) suggested that a single parameter, $t = KE_{rot}/|PE|$, was sufficient to discriminate stable from unstable stellar systems. They asserted that wherever t exceeded ∿0.14 the system should want to form a bar, and argued that only a small fraction of the mass of a galaxy could reside in a cool, rotationally supported disc. Their argument has often been cited as further evidence of the existence of massive halos around galaxies, yet it should be clear that this is evidence only for halo mass interior to the outer radius of the disc; any mass spherically distributed beyond that exerts zero gravitational force on the disc and therefore cannot affect its stability. Yet the t parameter can be made arbitrarily small by adding shells of matter at large radii. A revised parameter, t*, which avoids this nonsense was proposed by Lake and Ostriker (unpublished) and tested by Efstathiou et al (1982). They found that even the revised parameter was not an infallible guide to the stability of realistic models. Other counter-examples to the t* criterion can be found in Kalnajs' (1972) study of Maclaurin discs and in Zang's (1976) thesis work. Both parameters, therefore, are of little use.

Both Hohl (1976) and Sellwood (1980) concluded from n-body experiments of differentially rotating discs that the total spheroidal mass (well within the outer radius of the disc) had to be roughly twice the disc mass for the models to be stable. But, Hohl found less halo mass was needed if the disc rotates rigidly.

Efstathiou et al (1982) found that $Y = V_m(\alpha M_D G)^{-\frac{1}{2}}$ should be larger than ∿1.1 for a class of reasonably realistic model galaxies having exponential discs and a halo component chosen to give a predominantly flat rotation curve (Fall and Efstathiou 1980). V_m is the rotational speed of material on the flat part of the rotation curve, α^{-1} is the scale length of the exponential disc having a total mass M_D. This criterion can be applied only to galaxies with the assumed form of rotation curve.

A purely local criterion has been found by Toomre (1981). He finds that a parameter, X, should be larger than 3 to prevent swing amplification. X is defined as the ratio of the circumferential wavelength of the instability to the Jeans length in a cold rotating disc (Toomre 1964).

$$X = \frac{r \; \kappa^2}{2\pi m G \Sigma(r)}$$

where κ is the local epicyclic frequency, Σ is the surface density of matter in the disc and m is the number of arms: i.e. 2 for a bar instability. Unfortunately, X varies rapidly with radius for most galaxies, so this local criterion cannot easily be used to assess global stability.

While these parameters are rather diverse they all make similar demands on the distribution of mass, when applied to the Fall and Efstathiou models, for example. Figure 1(a) shows the variation with radius of the ratio of bulge+halo mass to the disc mass, both interior

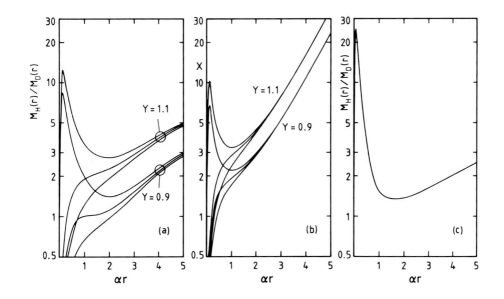

Figure 1. (a) Halo to disc ratio as a function of radius for some Fall
& Efstathiou models. Y has the indicated value for each group of three
curves which correspond to αr_m = .1, .8 & 1.3. (b) Toomre's X parameter
for the same models as in (a). (c) Halo to disc ratio for the model of
our Galaxy by Bahcall et al.

to r, for their models. Different values of Y determine the overall
halo to disc ratio while the parameter αr_m determines the prominence of
the bulge component. At two disc scale lengths from the centre, just
beyond the half-mass radius, the structure of the models is nearly independent of αr_m and the condition Y>1.1 is precisely that required to
ensure that the halo to disc ratio exceeds two at this radius. Figure
1(b) shows Toomre's X parameter for the same models. Again it is clear
that X reaches its critical value at two disc scale lengths for the
same value of Y. Thus for these models at least, three stability criteria agree that the bulge+halo mass interior to the half-mass radius
of the disc should be more than twice the disc mass in the same region.

Confrontation with Observations

The most serious attempt made to check stability criteria against
observational evidence is due to Efstathiou et al (1982). Taking all
existing rotation curve data that could be fitted to their models, they
concluded that the mass to light ratio of the discs of Sc galaxies could
not greatly exceed unity without violating their stability criterion.
This is rather uncomfortably low, but they claimed it to be consistent
with the Larson and Tinsley (1978) population models.

Mass models of our Galaxy are inconclusive: both the Bahcall et al (1982) model, shown in Figure 1(c), and the Caldwell and Ostriker (1981) model fail by a substantial margin to place sufficient mass in the bulge+ halo interior to the sun for the disc to be stable. However, it should be recognised that the disc mass in both these models is very dependent on Oort's measurement of the local surface density, a quantity which is rather badly determined by the existing observations.

We have also heard, presented at this conference, mass models of two edge-on disc galaxies. Van der Kruit, drawing together a number of strands of evidence, concluded that the bulge+halo mass interior to the outer edge of NGC 891 was roughly three times that of the disc. This galaxy should be well fitted by a model of the type shown in Figure 1 with $\alpha r = 4.5$ at the outer edge, and $Y = 1$, placing it marginally on the unstable side of the criterion. Cassertano's model of NGC 5907 falls well short of the stability criteria since he finds $M_H \sim 1.5 \ M_D$ at $\alpha r = 3.5$.

Fall and Efstathiou (1980, in their Figure 9) noticed that Y did not differentiate barred from unbarred galaxies when M was estimated from the disc luminosity, which led Efstathiou et al to conjecture that most galaxies lie on the borderline of stability. I feel that the hotch-potch of evidence reviewed here does not support their hypothesis, although it is too uncertain to rule it out: in every case it has to be stretched to find sufficient mass in the bulge+halo components for even marginal stability.

The Frequency of Barred Types Along the Hubble Sequence

One characteristic defining the Hubble sequence is the bulge to disc ratio; which decreases from type SO to Sc. If the fraction of luminous bulge is the principal determinant of disc stability one would expect the frequency of bars to increase towards later types. In fact, the opposite is the case.

Figure 2 shows the frequency of barred types taken from the Revised Shapley-Ames catalog of galaxies (Sandage and Tammann 1981). Amongst early type systems (SO to Sb) Sandage finds more than twice as many unbarred as barred galaxies, whereas slightly fewer bars occur in Sc types. The paucity of bars amongst bright Sc galaxies is made more significant when one takes into account a selection effect notices by Van den Bergh (1982) who showed that the barred Sc galaxies in the catalog are systematically less luminous than the unbarred. Leaving aside the smaller galaxies of types Sc and later, bars are actually much rarer amongst systems having little visible bulge.

Thus, if bulge+halo mass provides stability, one must conclude that on average Sc galaxies have relatively more mass in a dark halo than earlier type galaxies have in visible bulge and dark halo combined.

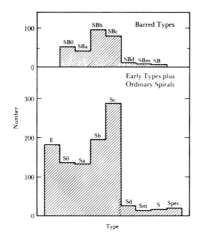

Figure 2. Histogram showing the frequency of Hubble types in the RSA catalogue. Intermediate types are combined into the main classes. Reproduced from Sandage and Tammann (1981).

Hot Discs?

All three stability criteria discussed above were determined on the basis that stellar discs are very largely rotationally supported, having sufficient random motion only to suppress local Jeans instabilities. However, global stability should be very substantially affected by a fair degree of pressure support among the stars of the disc near the centre. If we neglect this contribution we may overestimate the bulge+halo mass required.

Disc stars in the solar neighbourhood are believed to move on nearly circular orbits and to have a mean rotational velocity some five times the dispersion of the radial velocity components. But the sun lies far from the rotation axis and it is natural to expect that velocity dispersion will rise towards the centre. Observational evidence that this rise may be substantial is beginning to be found, although the measurements are still rather close to the limits of technical feasibility. (See reviews by Kormendy and Illingworth in this volume). In one case only is the observed random motion less than one third the mean rotation rate. Here, Illingworth and Schechter (1982) give an upper limit of 50 to 80 km/sec for the velocity dispersion of the disc stars in NGC 3115. However, as they clearly measured the tangential component, I would increase their limit by 50% since the radial component should be larger by roughly this amount.

Apart from reducing the bulge+halo mass required for global stability, hot discs have other advantages. For example, all globally unstable cool disc simulations give narrow, very strong bars, regardless of how marginally unstable they were. Unstable hot discs form fatter, weaker bars more consistent with intermediate SAB galaxies. Thus we could envisage a picture in which most galaxies have only moderate fractions of mass in the bulge+halo and the presence and strength of a bar reflects the extent of pressure support amongst the disc stars.

Much work needs to be done to put this hypothesis on a firmer basis. We need to determine the extent to which pressure support trades with bulge+halo mass at marginal stability. We also need to understand how galaxies can acquire large degrees of random motion in the plane when we expect disc stars to have formed on nearly circular orbits. A suggestion as to how this could be achieved is contained in the work of Carlberg and Sellwood (presented at this meeting). In their picture, the bar morphology could be related to the rate at which the disc is built up during galaxy formation.

References

Bahcall, J.N., Schmidt, M. & Soneira, R.M.: 1982, preprint.
Caldwell, J.A.R. & Ostriker, J.P.: 1981, Ap. J. 251, 61.
Efstathiou, G., Lake, G. & Negroponte, J.: 1982, M.N.R.A.S. 199, 1069.
Fall, S.M. & Efstathiou, G.: 1980, M.N.R.A.S. 193, 189.
Hohl, F.: 1976, Astroⸯ J. 81, 30.
Illingworth, G. & Schechter, P.L.: 1982, Ap. J. 251, 481.
Kalnajs, A.J.: 1972, Ap. J. 175, 63.
Larson, R.B. & Tinsley, B.M.: 1978, Ap. J. 219, 46.
Ostriker, J.P. & Peebles, P.J.E.: 1973, Ap. J. 186, 467.
Sandage, A. & Tammann, G.A.: 1981, "A Revised Shapley-Ames Catalog of
    Bright Galaxies" Carnegie Inst. Washington.
Sellwood, J.A.: 1980, Astron. & Ap. 89, 296.
Sellwood, J.A.: 1982, J. Comp. Phys. (in press).
Toomre, A.: 1964, Ap. J. 139, 1217.
Toomre, A.: 1981, "Normal Galaxies", Eds. Fall, S.M. & Lynden-Bell, D.
    Cambridge University Press.
Van den Bergh, S.: 1982, Astron. J. 87, 987.
Zang, T.A.: 1976, Ph.D. Thesis MIT.

INSTABILITIES OF HOT STELLAR DISCS

E. Athanassoula and J.A. Sellwood
Observatoire de Besancon, 25000 Besancon, France.
Institute of Astronomy, Madingley Road, Cambridge, England.

We use computer simulations to determine the dominant unstable modes
of stellar discs having the surface density distribution of Toomre's
(1963) model 1 and large random motions in the plane. Particles in the
models have their initial coordinates chosen from distribution functions
to ensure that the configurations would, in the limit of infinitely many
particles, be stationary solutions of the Vlasov and Poisson equations.
The 40,000 particles in our simulations are constrained to move on a
plane and we use a polar grid based Fourier method to determine the force
field (Sellwood 1982). The surface density is smoothly tapered to zero
at six scale lengths and the softening length is 1/30th of the outer
radius of the disc.

Kalnajs (1976) has given a family of distribution functions for this
disc which have a nearly constant Q at all radii when Q is near to 1, but
hotter members of the family have Qs which rise steadily outwards as
shown in Figure 1(a). The dominant modes of these discs were extracted
from simulations by a least squares fit to the logarithmic spiral Fourier
coefficients of the particle distribution. As expected, the growth rates
decrease substantially as pressure support is increased, but we still
find growth rates as great as one tenth the pattern speed when the mass
weighted average Q is as high as 2.04. (Pure pressure support corresponds
to a mean Q of 2.77). Composite models which contain two populations,
one having a large degree of pressure support, the other little random
motion, tend to be even more unstable for their mean Q than a single
population disc.

New distribution functions, which are generalisations of those given
by Kalnajs, enable us to study models where Q behaves differently with
radius (dashed curves, Figure 1(a)). These two models have a very similar
mean Q, but the growth rate of the model having the higher central Q is
one fifth that of the other.

All these results are plotted in Figure 1(b) which illustrates that
mean Q does not correlate well with growth rate. A somewhat better
correlation is found with the central value of Q (not shown here) but

E. Athanassoula (ed.), Internal Kinematics and Dynamics of Galaxies, 203–204.

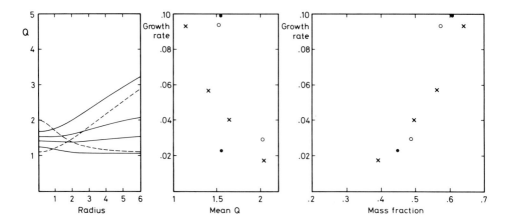

Figure 1. (a) The radial variation of Q for four Kalnajs functions (full curves) and two new functions (dashed curves). (b) Growth rates plotted against mean Q for Kalnajs functions (crosses), composite models (open circles) and new functions (filled circles). (c) The same growth rates plotted against fraction of mass on nearly circular orbits.

the tightest correlation we have found is shown in Figure 1(c). The absiccae are the fraction of mass in the models on "nearly circular" orbits, defined by the condition:

(a-p)/(a+p) < 0.5

where a and p are the apo- and peri-galactica of the unperturbed orbit of the star.

Three conclusions can be drawn from our results to date: (a) a large amount of random motion substantially reduces the growth rate of instabilities, (b) high Q near the centre is much more effective at reducing growth rates than increased random motion further out, and (c) the growth rate correlates with the fraction of mass on nearly circular orbits. It remains to be seen whether these conclusions can be generalised to models having different distributions of mass.

References

Kalnajs, A.J.: 1976, Ap. J. 205, 751.
Sellwood, J.A.: 1982, J. Comp. Phys. (in press).
Toomre, A.: 1963, Ap. J. 138, 385.

STABILIZING A COLD DISK WITH A 1/r FORCE LAW

Joel E. Tohline
Department of Physics and Astronomy
Louisiana State University

Massive "dark" halos are currently believed to surround individual galaxies and systems of galaxies primarily because: (a) individual galaxies exhibit "flat" rather than Keplerian rotation curves (see the review by Faber and Gallagher 1979); (b) the measured ratio of mass to light in systems of galaxies continues to increase roughly linearly with radius as one observes larger and larger systems (Rood 1982); and (c) rapidly spinning "cold" stellar disks are not dynamically stable if they are fully self-gravitating (Ostriker and Peebles 1973). It can be shown, however, that properties (a) and (b) of galaxies and systems of galaxies can be explained without invoking dark halos if one assumes that the force of gravity has the form

$$F = -\frac{GMm}{r^2}\left[1 + r/a\right] \qquad (1)$$

where r is the distance between the attracting bodies of masses M and m, and the scale "a" must assume a value $a \sim 1$ kpc. (See Milgrom 1982 and Bekenstein, these proceedings, for a discussion along similar lines.) Newton's Law is retrieved on scales $<<$ a but the force is proportional to 1/r on scales $>>$ a. Equation (1) should not be adopted in place of Newtonian gravity to explain the dynamical properties of galaxies unless it also explains property (c) without invoking the presence of dark halos.

Here a dimensionless form of Equation (1)

$$f = -\frac{1}{\eta^2}\left[1 + \eta\right] \quad , \qquad (2)$$

where $\eta \equiv r/a$, has been used to describe the gravitational attraction between individual particles in an "n-body" computer code and the stability of cold stellar disks has been tested on different scales. A softening length $\delta\eta = 0.03\ \eta_{max}$ (η_{max} is defined below) has also been added to each η in Equation (2) in order to soften direct 2-body encounters.

205

E. Athanassoula (ed.), Internal Kinematics and Dynamics of Galaxies, 205–206.

The following initial disk model has been chosen for each evolution:
    1.) 512 particles were placed in a disk of zero initial
thickness, in eleven concentric rings so that the disk of radius η_{max}
had a uniform surface density; 2.) Each particle was given a
circular velocity consistent with centrifugal balance in the
self-gravitating disk; 3.) Each particle was given an initial
z-velocity "vz" randomly chosen in the range $-vz_{max} < vz < +vz_{max}$
where vz_{max} was, for a given evolution, chosen to be a constant
fraction of the local initial circular velocity throughout the disk.
A larger vz_{max} produced a "hotter" initial disk.

A number of models, differing only in their initial values of η_{max}
and vz_{max}, were run for 2-3 rotation periods (measured in terms of the
disk's initial central rotation period) and the stability of each disk
was examined. The most obvious result was evident in plots of
particle positions in the equatorial plane of the rotating system:
Models having η_{max} = 0.01 developed strong nonaxisymmetric structure
in much less than one rotation period (this is exactly as was observed
by Ostriker and Peebles 1973) while models having η_{max} = 100.0
remained axisymmetric throughout each extended evolution. In addition
to this rather subjective measure of relative disk stability, the
quantity $\gamma \equiv$ (Random kinetic energy)/(Rotational kinetic energy)
was evaluated throughout each evolution. (In terms of the familiar
criterion proposed by Ostriker and Peebles [1973], a disk in virial
equilibrium should remain stable only if $\gamma \gtrsim 2.6$.) Disks having
η_{max} = 0.01 heated up monotonically with time during evolutions from
initial states of γ = 0.3 and γ = 0.05. In contrast to this, disks
having η_{max} = 100.0 did not heat up at all from initial states of
γ = 0.3 and γ = 0.05.

It appears, then that cold stellar disks are dynamically stable if the
force governing long range interactions between stars has a 1/r
dependence rather than a $1/r^2$ dependence. It is extremely curious
that the same force (Equation [1]) that has been empirically
formulated to explain properties (a) and (b) of galaxies as listed
above will also stabilize a cold axisymmetric disk if the disk's
radius is greater than, or on the order of, 100 times the scale "a".
Perhaps properties (a), (b), and (c) are not indicating the existence
of dark halos but are instead pieces of empirical data that are trying
to tell us: "Gravity has a 1/r dependence on the scale of galaxies."

REFERENCES

Faber, S. M., and Gallagher, J. S. 1979, Ann. Rev. Astron. Ap.,
    17, 135.
Milgrom, M. 1982, preprint.
Ostriker, J. P., and Peebles, P. J. E. 1973, Ap. J., 186, 467.
Rood, H. J. 1982, Ap. J. Suppl., 49, 111.

NUMERICAL EXPERIMENTS ON THE RESPONSE MECHANISM OF BARRED SPIRALS

K.O. Thielheim and H. Wolff
University of Kiel, F.R.G.

As a generating mechanism of spiral structure, we have recently studied the driving of density waves in the stellar component of disk galaxies by growing barlike perturbations or oval distortions. Numerical experiments (Thielheim and Wolff 1981, 1982) as well as analytical calculations using the first-order epicyclic approximation (Thielheim 1981; Thielheim and Wolff 1982) have been performed, demonstrating that this mechanism is capable of producing two-armed trailing spiral density waves in disks of noninteracting stars. These regular, global spiral structures are similar to those found in N-body experiments on self-consistent stellar disks that show bar instabilities which are weak enough to allow spiral patterns to persist (e.g., Hohl 1978; Berman and Mark 1979; Sellwood 1981). On account of this similarity, we take the view that the spiral structure observed in N-body experiments is primarily not an effect of the self-gravity of the stellar disk but a response phenomenon, caused by the formation of a weak central bar and its subsequent growth due to angular momentum extraction by interaction with the spiral as described by Lynden-Bell and Kalnajs (1972).

In the contributions mentioned above, we had to defer the proof of the secondary role of self-gravity since, to determine its influence on the spiral structure, it is necessary to compare the response of a disk of noninteracting stars to an oval distortion with the response of a self-consistent disk to the same time-dependent perturbation. We here report results of N-body experiments that have been performed to allow such a comparison. A method using biorthogonal pairs of potential and surface density functions described by Clutton-Brock (1972) is used to calculate the force field of the self-consistent stellar disk. Adopting this method, less than 10,000 stars are sufficient for the present two-dimensional disk galaxy simulations. To advance the particle coordinates and velocities, the usual time-centered leapfrog scheme proposed by Buneman (1967) is employed. The disk stars are given an initial velocity dispersion and are imbedded in a rigid bulge/halo potential. After constructing a stable axisymmetric configuration, an external oval distortion is slowly imposed and the resulting response is studied.

E. Athanassoula (ed.), Internal Kinematics and Dynamics of Galaxies, 207–208.
Copyright © 1983 by the IAU.

Inside corotation the response of the self-consistent disk is found to be barlike and in phase with the perturbation. Open two-armed trailing spiral density waves that emerge from the ends of the bar are observable for many rotation periods, although they dissolve as the perturbation asymptotically approaches its final strength.

Comparison runs equivalent to our previous response calculations are performed by excluding the self-gravity of the stellar disk. The response patterns of the disks with and without self-gravity are virtually identical in shape, particularly regarding the location of the spiral arms with respect to the principal resonances and their total azimuthal extension, though the spiral arms appear to be slightly narrower in the self-consistent disk. Differences are encountered in the amplitude of the bar and spiral response. Moreover, the self-consistent spirals are found to dissolve earlier. Both effects find an explanation in the fact that the amplitude of the self-consistent bar response does not strictly follow the time development of the imposed perturbation but increases faster in the beginning and slightly decreases at a late stage.

We come to the conclusion that the regular, global spiral patterns observed in N-body calculations are produced by a response mechanism invoked by the formation and subsequent growth of weak central bars. For this mechanism the self-gravity of the disk stars that participate in the spiral wave is not essential. Analysing the numerical experiments of Miller, Prendergast, and Quirk (1970), Quirk (1971) came to contradicting conclusions, however, he tried to drive spirals by nonevolving bars. Symmetry considerations (Thielheim 1980; Thielheim and Wolff 1981) point out the impossibility of exciting spirals in disks of noninteracting stars by nonevolving oval perturbations and this effect may also account for Quirk's results. N-body calculations performed by Sellwood (1981), which show regular spiral patterns associated with intermittent bar growth, support this interpretation.

REFERENCES

Berman, R.H. and Mark, J.W.-K. 1979, Astr. Ap., 77, 31.
Buneman, O. 1967, J. Comput. Phys., 1, 517.
Clutton-Brock, M. 1972, Ap. Space Sci., 16, 101.
Hohl, F. 1978, A.J., 83, 768.
Lynden-Bell, D. and Kalnajs, A.J. 1972, M.N.R.A.S., 157, 1.
Miller, R.H., Prendergast, K.H., and Quirk, W.J. 1970, Ap.J., 161, 903.
Quirk, W.J. 1971, Ap.J., 167, 7.
Sellwood, J.A. 1981, Astr. Ap., 99, 362.
Thielheim, K.O. 1980, Ap. Space Sci., 73, 499.
————. 1981, Ap. Space Sci., 76, 363.
Thielheim, K.O. and Wolff, H. 1981, Ap.J., 245, 39.
————. 1982, M.N.R.A.S., 199, 151.

ORDERED AND SEMI-ERGODIC MOTIONS IN BARRED GALAXIES

E. Athanassoula, O. Bienaymé
Observatoire de Besançon

L. Martinet, D. Pfenniger
Observatoire de Genève

A preliminary step towards the construction of self-consistent models for barred galaxies consists in understanding the stellar orbital behaviour in a given axisymmetrical + bar-like potential and, in particular, the influence of various parameters characterizing the bar on the different kinds of possible motions. To do this we undertook a systematic study of the main periodic and quasi-periodic orbits in a two-component mass model: An axisymmetrical part characterized by a rotation curve $V(r) = V_o (r/r_o)^{1-\delta}$ and a prolate bar-like perturbation whose density distribution is $\rho = \rho_o (1-(x^2+z^2)/b^2-y^2/a^2)^2$ (a>b). This rather realistic choice for the bar is in agreement with the available photometric data and has several advantages, i.e. all $\cos(m\theta)$ terms are included and the physical parameters such as the length of the bar a or its eccentricity e are explicitly included in the formulae. The ratio of bar to disk mass measured up to the outer Lindblad Resonance (ε) and the angular velocity (Ω_s) are also free parameters.

Fig. 1 shows the general aspect of the characteristic diagram representing the main families of symmetrical simple periodic orbits: every such orbit is represented by a point (H,x) (H is the Hamiltonian and x the distance to the centre where the orbit intersects the y=0 axis). Orbits A (elongated ellipses) and B' (rather round) are aligned perpendicularly to the bar. Orbits B (elongated) and A' (rather round) are aligned along the bar. We also find family L, starting from the Lagrangian point L_4 and continuing for larger H, and family R of retrograde orbits. A family of (1/1) resonant orbits (off-centered ellipses) exists outside the corotation radius r_{co}. The stability of these orbits as well as the influence of the different parameters mentioned above on the relative position of the various characteristics will be examined elsewhere in more detail[1].

The method of surface of section was used in order to get an insight into the quantity of trapped matter around the main stable periodic orbits inside and outside corotation. For different models and for given

E. Athanassoula (ed.), Internal Kinematics and Dynamics of Galaxies, 209–210.

values of H we estimated the percentage A of the accessible region in
the plane of section $(x,\dot{x})$ which is occupied by good invariant curves.
Fig. 2 shows A (in percentage) as a function of H, inside corotation,
for an axial ratio a/b = 4(1) and 7(2) of the bar. (The values of the
other parameters are ε = 0.1, a = r_{co}, Ω_s = 0.5, δ = 0.8, V_o = r_o = 1).
The sharp decreases of A correspond to the advent of semi-ergodic orbits
and become bigger with increasing axial ratio. Only direct orbits
are considered here (the large majority of retrograde orbits is regular
and trapped around family R). This and other similar curves show that
both ε and the eccentricity of the bar are essential in determining the
relative amount of ordered motions. The only orbits which strengthen the
bar inside corotation are orbits trapped around family B. Practically
only the Hamiltonian is a constraint for semi-ergodic orbits. In any
case the region they occupy is considerably less elongated than the bar.
We shall consider some consequences of these results for constructing
self-gravitating models of barred galaxies in another paper[1] as well as
the effect of increasing ε and Ω_s. Computations of surfaces of section
outside corotation (for ε = .1 and Ω_s = .5) give the following results:
1) Short bars (a = r_{co}) do not trigger semi-ergodicity, 2) long bars
(a = $2r_{co}$) induce dissolution of invariant curves between corotation and
the outer Lindblad resonance but large regions of phase space are still
occupied by ordered motions, in particular, there exists a non negligible
trapping of matter around the resonant stable periodic orbit 1/1 outside
OLR, 3) the presence of an extended stochastic region between corotation
and OLR implies a rather long and eccentric bar of mass greater than
roughly one tenth of that of the disk inside the outer Lindblad resonance.

[1] E. Athanassoula, O. Bienaymé, L. Martinet, D. Pfenniger, submitted to
Astronomy and Astrophysics.

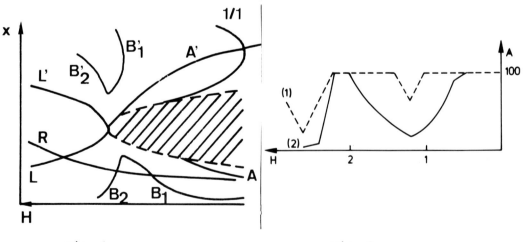

Fig. 1   Fig. 2

ONSET OF STOCHASTICITY IN BARRED SPIRALS

P.J.Teuben and R.H.Sanders
Kapteyn Astronomical Institute
University of Groningen, the Netherlands

Stellar orbits in the plane of a strong bar potential demonstrate the breakdown of the second integral of motion well inside corotation (Sanders(1980) and more recently Contopoulos(1982),Athanassoula et.al. (1983), Sanders and Teuben (1982)).
The onset of such stochastic motion may limit the bar strength (axial ratio), since stochastic orbits fill equipotential contours rather homogeneously.

The model adopted for the barred spiral is essentially the same as Sanders and Tubbs (1980), although the bar was modeled here by an inhomogeneous prolate ellipsoid with a density falling smoothly to zero at the edge of the bar.
The Surface-of-Section-method (Hénon & Heiles,1964) was used to estimate the percentage regularity in phase space by calculating the largest regular islands, belonging to the stable periodic orbits, in the $Y-\dot{Y}$-diagram (example fig.3). The onset of stochasticity turned out to be rather abrubt (although not as fast and complete as Hénon & Heiles) as a function of various parameters studied so far. For a particular model the percentage regularity as a function of Jacobi Integral (or radius) is shown in fig.1. In fig.2 we see a similar plot of percentage regularity vs. the deviation from axial symmetry, e.g. the axial ratio of the bar and the bar-disk mass ratio.
The pattern speed of the bar turned out to be not of importance for a realistic range of this parameter.

The onset of stochasticity is understood in terms of interaction of unstable periodic orbits. In our case it is the simple (e.g. not self-crossing) 3/1 (epicyclic) resonance interacting with various higher order resonances,such as the 8/3, 14/5 etc. Shortly after the bifurcation of this 3/1-family from the main simple periodic orbit (prograde), overlapping resonances leads to dissolution of invariant curves and to a high degree of stochasticity.

E. Athanassoula (ed.), Internal Kinematics and Dynamics of Galaxies, 211–212.
Copyright © 1983 by the IAU.

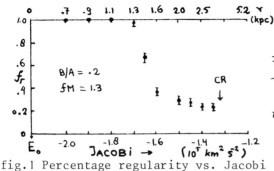

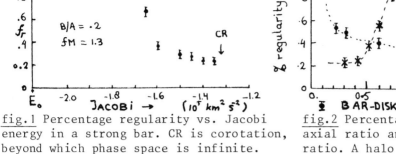

fig.1 Percentage regularity vs. Jacobi energy in a strong bar. CR is corotation, beyond which phase space is infinite.

fig.2 Percentage regularity vs. axial ratio and bar-disk mass ratio. A halo of 5 times the bar+disk mass was added.

We can conclude that for this model a limit on the bar strength can be derived, above which selfconsistent bars are probably not possible, although construction of bar and lens models maybe possible by populating both the regular and stochastic orbits.

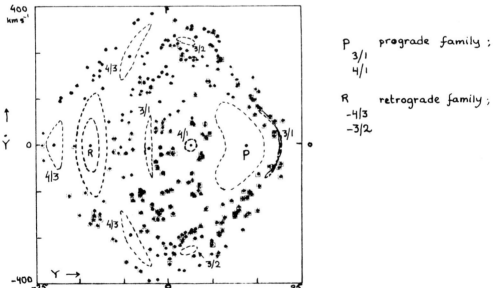

P   prograde family ;
3/1
4/1

R   retrograde family ;
-4/3
-3/2

fig.3 A surface of section in a strong bar model (fig.1, Jac = -1.4). The irregular points all stem from one orbit. Major periodic orbits appear as points, with their surrounding invariant curves.

REFERENCES

Athanassoula,E. ,Bienaymé,O. ,Martinet,L. and Pfeninger,D. , 1983,this boo
Contopoulos,G. , 1982,Astron. and Astrophys., in press.
Hénon,M. and Heiles,C. , 1964, Astr.J. 69,73.
Sanders,R.H. , 1980, Cambridge Workshop on Normal Galaxies.
Sanders,R.H. and Teuben,P.J. , 1982, in preparation.

ATTACKING THE PROBLEM OF A SELFCONSISTENT BAR

Maria Petrou
Astronomy Department, University of Athens,
Panepistimioupolis, Athens 621, Greece.

One way of approaching the problem of a selfconsistent bar, is
to examine the orbits of the stars which make up the bar. Given that
there are 10^{10} such stars and therefore 10^{10} such orbits, one has to
devise a way of studying and classifying them. Once one knows the most
important types of orbits which appear in a system and when they appear,
can proceed in constructing a selfconsistent bar.

The most important families of periodic orbits which appear in
a system are:
The x_1 family: When the potential is axially symmetric, this family
represents the circular orbits of various energies. If we add some bar
perturbation to the potential the x_1 family consists of distorted orbits
elongated along the bar.
The x_2, x_3 families: They are 2:1 resonant orbits which appear between
the two inner Lindblad resonances. They are elongated perpendicularly
to the bar and the x_2 are stable while the x_3 are unstable. The way these
families vary according to the strength of the bar is shown in figure 1
where we plot the radius of the orbit perpendicularly to the bar versus
the value of the Jacobi constant (assuming that we have a rotating
coordinate system).(Papayannopoulos & Petrou 1982).

The study of the periodic orbits is very important because the
majority of the non-periodic orbits in a galaxy are trapped around the
stable periodic ones. So, one periodic orbit is like a representative
of a large group of orbits.

The trapped orbits have, apart from the Jacobi constant, which
is an exact integral of the motion, other approximate constants too. The
orbits which have only the Jacobi constant as integral of the motion are
called ergodic and they fill up all the available space in the (x,y)
coordinates specified by their zero velocity curves. There are two
mechanisms which lead to ergodicity: 1)Many high order resonant families
of the form n:1 appear close to corotation. When n is even they are
separated from the x_1 family by a gap while when n is odd they bifurcate
from it. As the areas of importance of each resonance overlap the

E. Athanassoula (ed.), Internal Kinematics and Dynamics of Galaxies, 213–214.
Copyright © 1983 by the IAU.

approximate integrals dissolve. (Resonance interaction). This mechanism
appears in intermediate bars inside corotation (figure 2a).
2)An odd bifurcation of the x_1 family followed by a cascade of infinite
bifurcations. This is the Feigenbaum effect and has been found in the
case of strong bars immediately after the 3:1 resonant family bifurcates
from x_1 inside corotation and in intermediate bars immediately after the
1:1 family bifurcates from x_1 beyond corotation(fig.2b)(Contopoulos 1982).

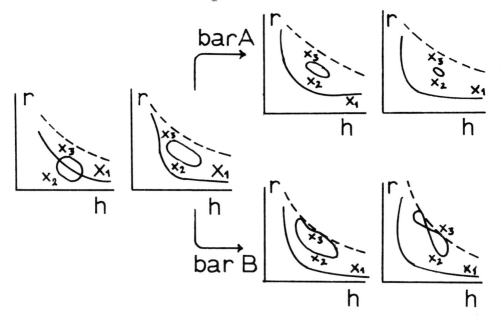

Fig.1:Evolution of the x_2,x_3 families as the bar becomes stronger from
left to right. The first diagram is without a bar. Bar A:a cos2θ bar.
Bar B:An inhomogenious prolate spheroid. The dashed line is the zero
velocity curve.

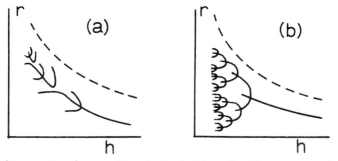

Fig.2:Ergodicity mechanisms. The dashed line is the zero velocity curve.

References
Contopoulos,G.:1982, Astron. Astrophys., in press.
Papayannopoulos,T. and Petrou,M.:1982, Astron. Astrophys., in press.

THEORETICAL STUDIES OF GAS FLOW IN BARRED SPIRALS GALAXIES

Kevin H. Prendergast
Department of Astronomy and Astrophysics
Columbia University, New York

We know far more about the velocity and density distributions of
gas in ordinary spirals than in barred spirals. There are several
reasons for this : the absence of a nearby barred spiral, the circumstance that several of the best objects lie rather far south, the necessity to map in two dimensions, and the surprisingly low intensity of the
emission lines, except in the nucleus and near the ends of the bar. The
older literature contains several investigations of radial velocities
measured along the bars of barred spirals (cf. Burbidge, Burbidge and
Prendergast, 1960 a,b for NGC 7479 and NGC 3504), and it has been known
for some time that rotation curves taken in different position angles
through the nuclei of barred spirals are not compatible with simple
circular motion. It was not until recently, however, that attempts were
made to map the velocity fields of barred spirals in two dimensions.
Chevalier and Furenlid (1978) studied NGC 7723, NGC 5383 has been mapped
by Peterson, Rubin, Ford and Thonnard (1978) in the optical and by
Sancisi, Allen and Sullivan (1979) in neutral hydrogen, and NGC 1300
has been mapped by Peterson and Huntley (1980). The most striking result
of these observations is that the isovelocity contours are elongated
in the direction of the bars, or, put differently, there is no straight
line through the center of a barred spiral along which the radial velocity is constant. This is clear proof that significant non-circular
motions are present in these galaxies. However, as it is impossible to
reconstruct a non-axisymmetric velocity field from radial velocity data
alone, there is no analog for barred spirals of the reduction models
which are used to estimate local mass densities for ordinary spirals
from observed rotation curves.

Theoretical studies have concentrated on the problem posed by non-
circular motions, and on certain morphological features of barred spirals,
particularly the smooth, rather straight dust lanes often found near
the leading edges of bars. The underlying mass distribution is assumed
to be independent of time when viewed in a coordinate system rotating
at constant angular velocity. The gravitational field is usually taken
to be invariant under reflection in each of three mutually perpendicular
planes, although some workers have adopted a two-armed spiral potential

215

E. Athanassoula (ed.), Internal Kinematics and Dynamics of Galaxies, 215–220.

in the outer parts of the system. In any case, several parameters are
required to specify the gravitational field, such as the degree of central concentration of the galaxy, and the strength, length, axis ratio
and angular velocity of the bar. The gas is presumed to lie in the equatorial plane, and to move under the influence of gravitational, centrifugal, Coriolis and pressure forces. The theoretical problem is to find
a flow pattern for the gas which is steady in the rotating coordinate
system.

Before proceeding to discuss detailed computational results, it may
be useful to consider certain qualitative aspects of the flow. The centrifugal and gravitational forces can be derived from an effective potential.
If we imagine this potential plotted (in three dimensions) as a function
of position in the equatorial plane, we would have, for an axisymmetric
galaxy, a surface resembling a volcano with a level rim at the corotation
radius, and all contour lines (equipotentials) would be circles. Adding
a strong bar deforms the circular equipotentials into ellipses oriented
along the bar inside corotation (i.e. inside the crater) and perpendicular to the bar outside. When there is a bar the rim of the volcano can
no longer be level, but will have two passes (X-type equilibrium points)
along the major axis of the bar, and two peaks (O-type equilibrium points)
at right angles to the bar. If we ignore for a moment the pressure and
inertia of the gas, the flow must be such that the Coriolis force balances
the forces derived from the potential, which implies that the streamlines
and equipotentials must coincide. The direction of flow must be in the
same sense as the rotation of the bar inside corotation, and in the opposite sense outside. The flow speed required to give the correct Coriolis
force depends on the local values of the gravitational and centrifugal
forces : where these are strong the velocity is high. Near the X and O
points, however, the applied forces are weak, and the inertia of the gas
cannot be neglected. In particular, a stream of gas moving at high speed
will overshoot the sharp bends where the equipotentials cross the major
axis of the bar ; the streamlines will be skewed with respect to the
equipotentials, such that the major axes of the streamlines lead the
major axis of the bar. In view of the formal analogy between compressible
gas dynamics and shallow-water theory, we can return to the picture of
the effective potential surface as a volcano with an elliptical crater,
and imagine a thin layer of water swirling at high speed within the crater. If the water does not wash over the lip of the crater (as it may
near the low parts of the rim) it sloshes up against the walls ; it is
not hard to imagine that a hydraulic jump forms when the flow stalls and
falls back on itself. The analog of this jump in gas dynamics is a shock,
and it will evidently be found near the leading edge of the bar.

The picture sketched above is probably appropriate only for galaxies
with strong, rapidly rotating bars. If the bar is weak or slowly rotating
the topology of the equipotentials is the same, but the features are less
pronounced. In this case the gas may be expected to respond to other dynamically significant singularities of the problem, such as the inner and
outer Lindblad resonances.

The dynamics of gas flow in a barred spiral is clearly a difficult problem, and one which can be approached on several levels. The simplest theories start from the assumption that the pressure is almost everywhere negligible, which is plausible, since the random velocities in the gas are an order of magnitude lower than the streaming velocities. If there were no pressure at all the streamlines of the flow would be identical to the orbits of non-interacting particles moving in the prevailing force field. The converse is <u>not</u> true : that is, one cannot take an arbitrary set of initial conditions for a collection of particles, solve for their motions and identify the resulting ensemble of trajectories with the streamlines, because the trajectories will cross one another. However, there are special sets of initial conditions which give stable periodic (loop) orbits, and there may be a family of such orbits which can be nested within one another without intersections. A trivial example is the family of circular orbits in an axisymmetric galaxy. If the bar contributes only a weak tangential component to the force field a slightly distorted version of this family may exist which can serve as an approximation to the streamlines. Unfortunately, the direction of elongation of the loop orbits changes by 90° at each Lindblad resonance, as well as at corotation, and the orbits intersect one another in these regions. If there were coupling between particles moving on adjacent orbits one might hope that crossings could be avoided. Hydrodynamic computations by Sanders and Huntley (1976), Sanders (1977), Berman, Pollard and Hockney (1979) and by Sorensen and Matsuda (1982) show that this hope can be realized. The loci of closest approach of orbits forms an open two-armed trailing spiral pattern extending as far as the outer Lindblad resonance. Some of the computations show large density gradients, indicating that shocks form where the streamlines are most closely crowded together.

The stronger the bar, the less likely it is that there exists an extensive family of nested stable loop orbits, and consequently the pressure must be considered from the outset. The next simplest thing to neglecting the pressure alltogether is to include only one component of the pressure gradient : the component perpendicular to the shock, say. This is the device adopted by Roberts, Huntley and van Albada (1979). These authors derive ordinary differential equations for an equivalent one-dimensional problem which can be solved with relative ease. They find that two kinds of streamlines are possible, both of which are usually crossed by shocks. For streamlines of the first kind the shock occurs after the streamline has reached its maximum distance from the nucleus, and post-shock flow is inwards. For streamlines of the second kind, the shock intervenes while the gas is still moving outwards ; the post-shock flow is directed towards the end of the bar, but eventually returns nearly parallel to itself after a sharp hairpin ben. Thus, the post-shock region is also one of high shear. In both cases a nested family of streamlines would show standing shock waves along the leading edges of the bar. Streamlines of the first kind are frequently found in two-dimensional simulations of gas flow in barred spirals. Flows with streamlines of the second kind are extremely difficult to model, partly because of resolution problems due to finite grid spacing, and partly because most codes are afflicted with a viscosity of purely numerical origin, which can be

devastating in regions of high shear. Van Albada and Roberts (1981) have
reported the results of a two-dimensional calculation on a fine grid
which shows a mild version of post-shock outflow. The real extent of the
phenomenon is difficult to estimate, because the one-dimensional calculation which exhibits the effect most clearly is only an approximation
to the two-dimensional problem that one would really like to solve.

The full hydrodynamic problem in two dimensions has been tackled by
many authors, using various mass models and employing a variety of numerical techniques. It is impossible to give an adequate summary in this
review of the numerous results obtained. One of the earliest investigations of this character is that of Sorensen, Matsuda and Fujimoto (1976),
who used a fluid-in-cell code. This method was also used by Berman,
Pollard and Hockney (1979) to investigate the influence of self-gravitation of the gas in a spiral driven by a weak oval distortion. The beam
scheme (Sanders and Prendergast 1974) has been extensively used for modelling flows in barred spirals (cf. Sanders and Huntley 1976, Huntley 1980,
Sanders and Tubbs 1980, Schempp 1982). The scheme is simple and rugged,
but the ruggedness is achieved at the price of large effective transport
coefficients : the code has roughly the same thermal conductivity, bulk
and shear viscosity a real gas would have if the mean free path were equal
to the grid spacing. Increasing the diffusivity of a numerical code blurs
the shock transitions, increases the apparent rate of gas flow into the
nucleus, decreases the sharpness of the bend between the spiral arms and
the bar, increases the angle by which the gas response leads the bar, and
increases the offset of the shocks within the bar. (Some of these effects
can also be produced by changing the mass distribution of the galaxy or
the rotation period of the bar). The influence of grid spacing has been
studied by van Albada and Roberts (1981), who also compared the beam
scheme and Godonov methods for the same problem and grid. The two results,
shown in their figures 5 an 13, are not identical, but the agreement is
rather more encouraging than otherwise. MacCormack's method has been used
by Sorensen and Matsuda (1981), a flux-corrected-transport method was
used by Jones, Nelson and Tosa (according to Matsuda 1981) and van Albada
has experimented with flux-splitting schemes. Several of the above methods
plus others, have been intercompared by van Albada, van Leer and Roberts
(1982) for a one-dimensional model of flow in a spiral gravitational field
Sanders (1977) has used a code due to Lucy which solves the hydrodynamic
equations by following the motion of particles having finite radii and
endowed with internal structure. N-body particle codes, supplemented with
a set of rules to govern the outcome of inelastic collisions between particles, have been used by Matsuda and Isaka (1980) and by Schwarz (1981).

The main conclusions suggested by these studies are :

1) A rigidly rotating bar or oval distortion is sufficient to drive
spiral structure in the gas, with or without self-gravitation ; the appearance of spirals is always accompanied by large departures from circular motion. Weak bars give open spiral patterns which extend throughout
the gas ; stronger bars give spirals which emerge at sharp angles to the
bar.

2) The gas response leads the bar, by an angle which is greater the weaker the bar.

3) Strong bars favor the appearance of strong shocks within the bar. When they occur, they lie near the leading edge, which is just where they should be if the identification of shocks and dust lanes is correct.

The conclusions above are relatively insensitive to the choice of code or grid spacing. Perhaps the most important aspect of the flow that is sensitive is the rate of infall of gas towards the nucleus. Simkin, Su and Schwarz (1980) have suggested that radial inflow induced by a rotating bar or oval distortion can feed gas to the nuclei of Seyfert galaxies, and Kormendy (1982) has proposed that the flat bulges of barred spirals are formed from gas that has drifted into the central regions over the lifetime of the bar. It would be premature to estimate rates of infall from present numerical computations, but it remains a challenging problem for the future.

REFERENCES

Burbidge, E.M., Burbidge, G.R. and Prendergast, K.H. 1960, Ap. J. 132, 654, 661.

Berman, R.H., Pollard, D.J. and Hockney, R.W. 1979, Astron. Astrophys. 78, 133.

Chevalier, R.A. and Furenlid, I. 1978, Ap. J. 225, 67.

Huntley, J.M. 1980, Ap. J. 238, 524.

Kormendy, J. 1982, Ap. J. 257, 75.

Matsuda, T. and Isaka, H. 1980, Prog. Theor. Phys. 64, 1265.

Matsuda, T. 1981, Prog. Theor. Phys. Supp. 70, 249.

Peterson, C.J., Rubin, V.C., Ford, W.K.Jr., and Thonnard, N. 1978, Ap. J. 219, 31.

Peterson, C.J. and Huntley, J.M. 1980, Ap. J. 242, 913.

Roberts, W.W.Jr., Huntley, J.M. and van Albada, G.D. 1979, Ap. J. 233, 67.

Sancisi, R., Allen, R.J. and Sullivan, W.T. 1979, Astron. Astrophys. 78, 217.

Sanders, R.H. and Prendergast, K.H. 1974, Ap. J. 188, 489.

Sanders, R.H. and Huntley, J.M. 1976, Ap. J. 209, 53.

Sanders, R.H. 1977, Ap. J. 217, 916.

Sanders, R.H. and Tubbs, A.D. 1980, Ap. J. 235, 803.

Schempp, W.V. 1982, Ap. J. 258, 96.

Schwarz, M.P. 1981, Ap. 247, 77.

Simkin, S.M., Su, H.J. and Schwarz, M.P. 1980, Ap. J. 237, 404.

Sorensen, S.A., Matsuda, T. and Fujimoto, M. 1976, Astrophys. and Space Sci. 43, 491.

Sorensen, S.A. and Matsuda, T. 1982, Mon. Not. R. Astr. Soc. 198, 865.

van Albada, G.D. and Roberts, W.W.Jr. 1981, Ap. J. 246, 740.

van Albada, G.D., van Leer, B. and Roberts, W.W.Jr. 1982, Astron. Astrophys. 108, 76.

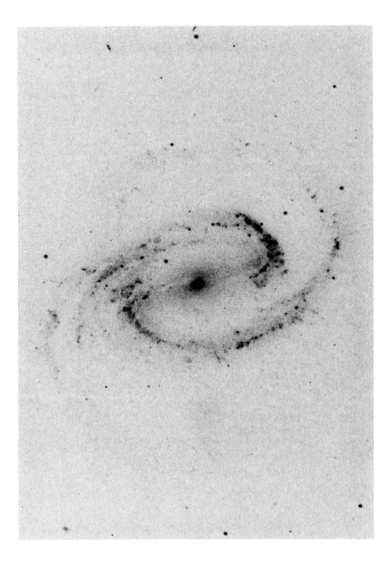

NGC 1300 (CTIO 4m,103aO+UG2, taken by A. Bosma)

PERIODIC ORBITS AND GAS FLOW IN BARRED SPIRAL GALAXIES

R.H. Sanders, P.J. Teuben, and G.D. van Albada
Kapteyn Astronomical Institute
University of Groningen, the Netherlands

One purpose for studying the gas flow in barred spiral galaxies is to use the observed distribution and kinematics of the gas as a tracer of the underlying gravitational field. By comparing model hydro-dynamical calculations with observations of actual systems, one would like to define three basic properties of barred galaxies:

1) The bar strength. How significant is the deviation from axial symmetry in the region of the bar, measured by some parameter such as q_t, maximum aximuthal force in terms of the mean radial force (Sanders and Tubbs, 1980).

2) The mean radial distribution of matter. Clearly in a system with large deviations from circular motion, the "rotation curve" gives no direct information on the radial mass distribution.

3) The angular velocity of the bar. Where is the co-rotation radius (or Lagrange points) with respect to the bar axes? Are other principal resonances present?

Deriving these properties through the use of numerical hydro-dynamical calulations is not unambiguous because the detailed results do depend upon the kind of numcerical technique used - specifically upon the magnitude of the unphysical numerical viscosity (G.D. van Albada, this volume). We can, however, place some definite constraints upon these properties by looking both at numerical hydrodynamics and the character of periodic orbits in non-axisymmetric potentials.

It is obvious that the hydrodynamical equation of motion written in Lagrangian form without the pressure term is the equation of motion of a particle. This means that in the absence of pressure forces (thermal, turbulent, viscous or magnetic) a gas streamline is an orbit, and steady state flow in some frame would correspond to simple non-looping periodic orbits. But, in fact, it is possible to make a much stronger statement. Suppose that we consider an ensemble of particles moving on a variety of trajectories essentially filling the volume of phase space allowed by the energy or Jacobi constant. And now suppose that we allow particles to be "sticky" in the sense that over some characteristic interaction distance, random velocities are reduced; that is, we

221

E. Athanassoula (ed.), Internal Kinematics and Dynamics of Galaxies, 221–224.
Copyright © 1983 by the IAU.

introduce a 'viscous' force which resists distortion of a fluid
element. Then such viscous dissipation forces particle motion toward
the simple periodic orbits, or, in the language of modern dynamics, the
periodic orbits appear as attractors in the phase space of the
Hamiltonian. An attractor, A, is a region of the phase space surrounded
by a neighbourhood U such that any particle within U approaches A as
t → ∞ (Treve, 1978). Attractors can only arise in dissipative systems;
in a conservative system a particle trajectory which is not periodic
obviously cannot become periodic. In a galactic potential (axisymmetric
or non-axisymmetric) there is one trivial attractor for a dissipative
medium - the center of the galaxy. But I suggest here that for all
practical purposes (t → 1/H) periodic orbits also arise as attractors
in the 4-dimensional phase space of the problem.

We do not provide here a general proof of this statement (see
Melnikov, 1963 and the discussion by Lake and Norman, 1982), but do
present two striking numerical examples.

The first involves an ensemble of particles distributed uniformly
through the phase space of the Henon-Heiles potential (Henon and
Heiles, 1964). If dissipation is added by an algorithm which reduces
the velocity dispersion over some interaction distance, then it is
found that within several orbit periods essentially all particle
trajectories penetrate a surface of section within one of the small
'islands' about a periodic orbit. The periodic orbit "attracts"
particle trajectories from a wide domain of the phase space (Sanders,
1982). The second example involves gas flow in the potential of a weak
bar. Here there is only one simple family of periodic orbits present
inside co-rotation, the parallel family, or X_1 in the notation of
Contopoulos and Papayannapoulos (1980). These orbits are shown in Fig.
1a. The gas flow in this potential calculated by a time-dependent
numerical hydrodynamical code (G.D. van Albada, this volume) is shown
in Fig. 1b. It is seen that gas streamlines inside co-rotation are
practically identical to family X_1. Moreover, because family X_1 is
simple (no self-crossing or looping) there are no shocks or gas inflow
toward the center.

a. **b.**

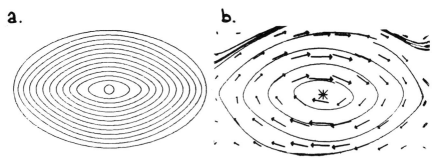

Fig. 1

But what if the lowest order periodic orbits are not so simple?
What if all families of periodic orbits are self-crossing or looping.
For example, if the bar becomes sufficiently strong, higher energy
orbits of family X_1 develop loops well inside co-rotation as is
illustrated in Fig. 2a. Clearly in this case the fully developed gas
flow cannot be along these orbits since streamlines cannot cross. The
flow may be attracted to these orbits but something else must happen,
and, of course, what happens is that shocks develop. In Fig. 2a we see
a number of particles moving in these periodic orbits of family X_1,
some of which loop. Pure particles, with no dissipation would stay on

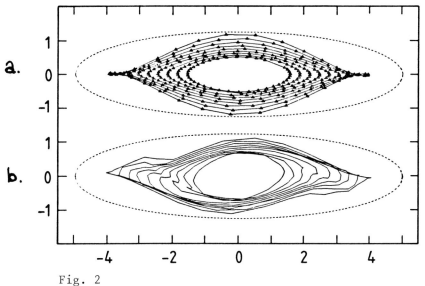

Fig. 2

such paths forever. But now if we add a bulk viscosity by means of an
algorithm described by Lucy (1977) we find, after a short time that the
rings of particles develop into the form shown in Fig. 2b (van Albada
and Sanders, 1982). Particles on looping orbits attempt to cross
through one another, lose energy and descend more abruptly toward the
center where they collide with particles on lower energy orbits. Or in
fluid dynamical terms a disturbance at the ends of the bar resulting
from the attempted looping, propagates downstream as shocks lying
roughly along characteristic curves; that is, linear shocks develop
along the leading edge of the bar which may be identified with the
linear off-set dust lanes seen in some SBb galaxies. This is
effectively the same physical idea proposed by Prendergast in
unpublished work more than 15 years ago.

Therefore the appearance of shocks in the gas flow in barred
spirals may be identified with the development of loops in the basic
parallel found of periodic orbits, X_1, and this only occurs in strong
bar potentials, $q_t \gtrsim 30\%$. This tells us that in at least some barred
galaxies, the bars are not a small density wave but a major deviation
from axial symmetry. Moreover, it is found that the shocks are stronger

and more off-set if the perpendicular family of orbits is present deep within the bar; that is, if the two inner resonances are present within the bar. This tells us that the mean radial distribution of matter in some barred galaxies must be strongly centrally condensed with respect to the bar. However, the bar cannot lie within two inner resonances because then the gas seems to be preferentially attracted to the perpendicular family of orbits (X_2); i.e., the gas distribution is elongated perpendicular to the stellar bar. This means that, for a reasonable radial mass distribution, co-rotation cannot be far from the ends of the bar (within one bar semi-major axis).

In summary, the clear relationship between the numerically calculated gas response and the character of the simple periodic orbits tells us that there are some barred galaxies which deviate strongly from axial symmetry, which are centrally condensed, and which have a rapidly tumbling bar figure.

REFERENCES

Albada, T.S. van, Sanders, R.H.: 1982, M.N.R.A.S. 200 (in press).
Contopoulos, G., Papayannapoulos, T.: 1980, Astron. Astrophys. 92, 33.
Henon, M., Heiles, C.: 1964, Astron. J. 69, 73.
Lake, G., Norman, C.: 1982, preprint.
Lucy, L.B.: 1977, Astron. J. 82, 1013.
Mel'nikov, V.K.: 1963, Trans. Moscow Math. Soc. 12, 1.
Sanders, R.H.: 1982, in preparation.
Sanders, R.H., Tubbs, A.D.: 1980, Astrophys. J. 235, 803.
Treve, Y.M.: 1978, in Topics in Nonlinear Dynamics A Tribute to Sir
    Edward Bullared, ed. S. Jorna (New York: American Institute of
    Physics), p. 147.

THE RESPONSE OF THE ENSEMBLE OF MOLECULAR CLOUDS TO BAR FORCING IN A
GALAXY DISK.

Combes, F. and Gerin, M.
Observatoire de Meudon - France

How a bar potential can force a spiral structure in the gas component
is now well understood. This response is due to the dissipative character
of the gas. Results of the computations of the response differ greatly
with the numerical code adopted, either particle (Schwarz 1981), or hydrodynamic (e.g. Sanders and Huntley, 1976) and within a given code, with the
spatial resolution used, since the artificial viscosity is thus varied
(see e.g. van Albada et al 1981). According to observations, the gas
component is mostly cloudy. Hence our aim is to compute the response to
a bar potential of the ensemble of molecular clouds for which collision
rate (and therefore dissipation rate) is relatively well-known.Using a
particle code and explicitly treating the collisions between clouds,
the viscosity parameter is more easily controlled and less artificial.
Also, since the mass transfer between clouds is taken into account, we
will obtain insight into the formation of large molecular clouds, which
are the preferential sites of active star formation.

The model = The bar potentials we use come from previous N-body computations (Combes and Sanders 1981), after transformation in analytic functions (polynomial fits). The collision model is described in Casoli and
Combes (1982). The gas component is described by molecular clouds from
$5 \ 10^2$ to $5 \ 10^5$ $M_\odot$. The result of a collision is 1,2 or 3 fragments according to the impact parameter. Giant Molecular Clouds ($M > 2 \ 10^5$ $M_\odot$) have
a finite lifetime of $\tau = 4 \ 10^7$ years, and are dispersed by star formation.
Their mass is recycled in small fragments with velocities of 15 Km/s.
Results = Figures 1 and 2 show the gas response to a strong bar (Qt the
ratio of total tangential force to radial one is 70 %). An open spiral
structure arises in the cloudy medium in less than one rotation and lasts
until the end of the simulation (7 rotations). GMC are concentrated in
the spiral arms. The spiral structure is even more contrasted in the
plot of the total energy dissipated in collisions (fig. 2) = most collisions occur in the arms. A test run has been computed with the same
potential but without collisions. A transient spiral structure also
arises, but disappears in less than one rotation. In the case of a weaker
bar (Qt ∿ 20 %) the spiral structure is less contrasted in the gas
component, but collisions also occur only in the arms. This case has

E. Athanassoula (ed.), Internal Kinematics and Dynamics of Galaxies, 225–226.
Copyright © 1983 by the IAU.

been run for 10 rotations. The spiral structure remains in the disk and
no ring is found to appear at outer Lindblad resonance.

strong bar (clouds)
                                        strong bar

                                    loci of dissipation

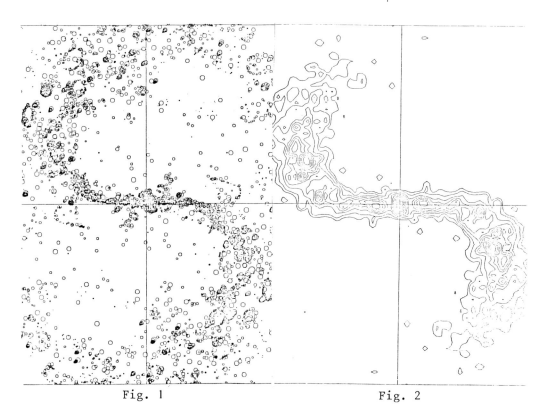

Fig. 1   Fig. 2

Fig. 1 : plot of molecular clouds of different sizes in a strong bar
(bar // Ox) at 2 rotations. Note the kinds of shell, consequences of
recycling of mass after star formation.

Fig. 2 : Plot of total energy dissipated in collisions for the same run.

References

Casoli, F., Combes, F., (1982) Astron. Astrophys. 110, 287
Combes, F., Sanders, R.H. (1981) Astron. Astrophys. 96, 164
Sanders, R.H., Huntley, J.M. (1976)Astrophys.J, 209, 53
Schwarz, M.P. (1981) Astrophys. J 247, 77
Van Albada, G.D., van Leer B., Roberts W.W. (1981) Astrophys. J preprint

GAS FLOW MODELS FOR BARRED SPIRALS

G.D. van Albada
Kapteyn Laboratorium
University of Groningen, the Netherlands

ABSTRACT

This paper is a progress report on a research project studying the gas flow in barred spirals. The parametrization of the potential and the gasdynamics code being used are described in section 1, some preliminary conclusions and a sample result are presented in section 2.

1. PARAMETRIZATION OF THE POTENTIAL; THE GASDYNAMICS CODE

The gas flow is computed in a series of potentials derived from a three component model of a barred spiral galaxy. The model is somewhat ad hoc, designed for computational simplicity as much as to give a reasonable description of the type of potential actually expected. The components are:

1) A prolate inhomogeneous spheroidal bar with

$$\rho_{BAR} = \rho_{0,BAR} \cdot \left[1 - x^2/a^2 - (y^2 + z^2)/b^2\right] \tag{1.1}$$

The free parameters for the bar are the axial ratio a/b and the quadrupole moment, while $a = 5$ kpc is fixed.
2) A Hubble density law bulge. The free parameter in this case is the sum of the central densities of the bar and the bulge, while the core radius of the bulge is derived from the constraint that the sum of the masses of the bar and the bulge within 10 kpc is fixed.
3) A $n=1$ Toomre-Kuzmin disk leading to an essentially flat rotation curve outside 10 kpc. This disk has no adjustable parameters. A fourth adjustable parameter is the corotation radius, or more precisely the radius of the Lagrangian point at the end of the bar.

The gasdynamics code is based on the FS2 code in Van Albada, Van Leer, and Roberts (1982), which is a second order accurate upwide centered method. It has been adapted for two-dimensional computations by using time-splitting, also to second order accuracy. In extensive tests

227

E. Athanassoula (ed.), Internal Kinematics and Dynamics of Galaxies, 227–228.

it was found that the properties of the particular code used can
strongly affect the derived gasflow, thus at the very least affecting
the quantitative results of a parameter study. The code effects for the
FS2 code are much smaller than for previously used codes (assuming a
perfect gas), but still non-neglibible. They were further reduced by
using a finer grid than usual.

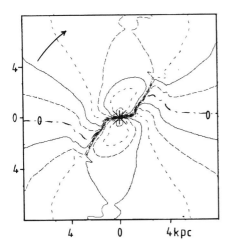

Fig. 1.
Model velocity field obtained for
a bar with $a/b = 3.0$, corotation
at 6 kpc, $\rho_{CEN} = 30$ M$_\odot$ pc^{-3} and
$QT_{max} = 33\%$. The bar is at $45°$
with the horizontal axis; the
galaxy rotates clockwise. Shown
are 50 km s^{-1} contours of the
velocity in the x-direction.

2. RESULTS

The parameter space allowed by the four free parameters in the model is quite large; thus far the emphasis has been on finding that part
of the parameter space where models similar to the classical SBb galaxies like NGC 1300 are obtained. This implies that we are looking for
long, straight shocks, more or less parallel to the axis of the bar and
offset to the leading edge. In fact it has proved to be quite easy to
produce such models for strong bars ($QT_{max} > 40\%$, as defined in Sanders
and Tubbs, 1980), as long as the central mass density is sufficiently
high (over 18 M$_\odot$ pc^{-3}). However, N-body models of bars so far have not
resulted in QT_{max} exceeding about 30% for more than a dynamical time
scale (Sanders, private communation). Therefore, the present research
is primarily directed at finding models with weak bars. Some of such
models have already produced quite satisfactory results, the velocity
field shown in Fig. 1 is an example of a model with $QT_{max} = 33\%$.

REFERENCES

Albada, G.D. van, Leer, B. van, Roberts, W.W.Jr., 1982, Astron. Astro-
     phys. 108, 76
Sanders, R.H., Tubbs, A.D., 1980, Astrophys. J. 235, 803

PHOTOMETRY, KINEMATICS AND DYNAMICS OF THE BARRED GALAXY NGC 5383

E. Athanassoula
Observatoire de Besançon and European Southern Observatory
M.F. Duval
Observatoire de Marseille

NGC 5383 has often been used as a prototype of SB(s)b type galaxies, particulary in studies of gas flow. We have therefore set out to complement the existing data with surface photometry, velocity measurements in the bar and a dynamical model.

The surface photometry of this galaxy shows the existence of four components: a bulge, a disk, a bar and a lens (Fig. 1). Their geometrical parameters are given below:

	position of the major axis	$<b/a> = q$
bulge	$90° \pm 5°E$	0.71 ± 0.04
bar	$135° \pm 2°E$	0.38 ± 0.02
lens	$107° \pm 2°E$	0.88 ± 0.02
disk	$85° \pm 3°E$	0.66 ± 0.03

Their effective luminosity distributions are described for the bulge and the disk by the classical formulas:

bulge $\log I_b = 0.08 - 3.33 [(r*/0.16)^{1/4} - 1]$ $I_{o_b} = 3.2 \ 10^5 \ L_\Theta \ pc^{-2}$

disk $\log I_d = -0.9 - 0.456 \ r*$ $I_{o_d} = 15.5 \ L_\Theta \ pc^{-2}$

We obtain: $L_b/L_T = 0.32$ and $L_d/L_T = 0.33$

From the luminosity profiles of the bar and the lens, we deduce the best representative simple laws :

bar $I_B = 0.42(1-(r*/0.60)^2)^{1/2}$ $I_{o_B} = 51.9 \ L_\Theta \ pc^{-2}$

lens $I_\ell = 0.18(1-(r*/1.35)^2)$ for $r* > 0!6$ and

$I_\ell = 0.$ for $r* < 0!6$

(r* in arcmin)
So that: $L_B/L_T = L_\ell/L_T = 0.15$

E. Athanassoula (ed.), Internal Kinematics and Dynamics of Galaxies, 229–230.
Copyright © 1983 by the IAU.

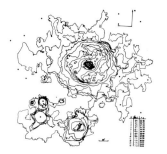

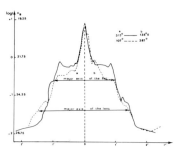

Figure 1. Isophotal map in blue   Figure 2. Luminosity profiles
         light

Several spectra have been taken with slit positions complementing those
of Duval (1977) and Peterson et al. (1978), thus providing good coverage
of the bar region despite the fact that the bar does not contain much
ionized gas.

    Using the photometric properties of the four components, a set of
models has been built and the velocity fields calculated with the help
of the beam scheme, and compared to the observations. From the best
fitting model we deduce the M/L ratios of each component after correction
for galactic and internal absorption :

bulge	bar	lens	disk + halo	
$1.4h^{-1}$	$6.5h^{-1}$	$9.3h^{-1}$	$27h^{-1}$	in M_θ/L_θ

(distance : 23.5h Mpc and h = 100/H km s^{-1} Mpc^{-1})

We note the similarity of the M/L ratios of the bar and the lens. Various
"rotation curves" (Fig. 3) issued from the adopted model have been
compared to the axisymmetric curve derived from HI data
(Sancisi et al., 1979).

Figure 3.

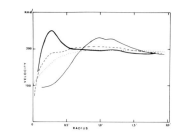

    ─ cut along the bar
    ── cut across the bar
    ─── rotation velocities
        meaned in annuli
    ... Axisymmetric rotation
        curve from HI data

REFERENCES

Duval, M.F. : 1977, Astrophys. Space Sci. 48, 103.
Peterson, C.J., Rubin, V.C., Ford, W.K., Thonnard, N.: 1978, Astrophys.
    J. 219, 31.
Sancisi, R., Allen, R.J., Sullivan, W.T.: 1979, Astron. Astrophys. 78, 217.

KINEMATICS OF THE BARRED GALAXY NGC 1365

Steven Jörsäter, Per Olof Lindblad and Charles J. Peterson
Stockholm Observatory   University of Missouri-Columbia

NGC 1365 is a galaxy which has lately received a lot of attention from people studying the structure and dynamics of barred galaxies. This is not surprising since it is one of the best suited objects in the sky. We have obtained a number of long-slit spectra in the red region ($H\alpha$, [NII], [SII]) with the ESO 3.6 m (Lindblad and Jörsäter) and with the CTIO 4 m (Peterson) telescopes. In addition, a couple of Fabry-Perot $H\alpha$ interferograms have kindly been given to us by G. Comte and Y. Georgelin. Some preliminary results are presented here. Fig. 1 shows the positions of measured velocity points. The digits along the vertical axis indicate distance from the nucleus in seconds of arc. The dashed line at P.A. 48 deg indicates the line of nodes as determined from photometry of the outer features of the galaxy (Lindblad 1978). An arbitrary isophote has been sketched to aid the orientation. The emission lines in the bar are surprisingly weak which is the reason for the scarcity of velocity points there. Fig. 2 shows a rotation curve based on the P.A. of the line of nodes of 48 deg and an inclination of 55 deg (Lindblad 1978). Only velocity measurements within 50 deg of the line of nodes have been used in this diagram in order to avoid large projection errors. The distance used is 20 Mpc. The spread is quite large indicating a significant amount of non-circular motion.

REFERENCES

Lindblad, P.O. 1978 in : Astronomical papers dedicated to Bengt Strömgren, ed. A. Reiz and T. Andersen, p. 403.

E. Athanassoula (ed.), Internal Kinematics and Dynamics of Galaxies, 231–232.

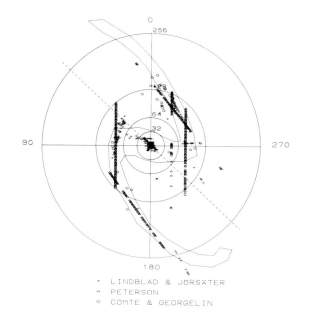

+ LINDBLAD & JÖRSÄTER
△ PETERSON
○ COMTE & GEORGELIN

Fig. 1

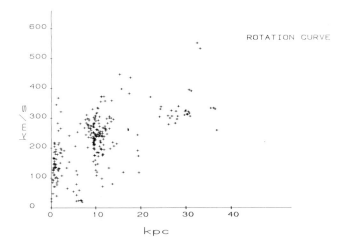

Fig. 2

HI IN THE BARRED SPIRAL GALAXIES NGC 1365 AND NGC 1097.

J.M. van der Hulst[1],M.P. Ondrechen[2],J.H. van Gorkom[3],E.Hummel[4]

1.Netherlands Foundation for Radio Astronomy, The Netherlands
2.University of Minnesota, Minneapolis, MN, USA.
3.National Radio Astronomy Observatory, VLA, Socorro,NM,USA.
4.University of New Mexico,Albuquerque, NM,USA.

In this paper we present preliminary results from 21-cm line observations with the Very Large Array (VLA) of the southern barred spiral galaxies NGC 1365 and NGC 1097. Despite a wealth of theoretical models describing the gas flow in a non-axisymmetric bar potential (see Prendergast this volume), few observations of the HI distribution and motions in barred spiral galaxies exist. A notable exception is NGC 5383 (Sancisi et al. 1979). The observations we performed with the VLA are described below. The velocity resolution is 25 km sec^{-1}. The angular resolution is 28"x20", p.a. 20° for NGC 1365 and 30"x25", p.a. 20° for NGC 1097. Velocities are heliocentric.

Figure 1 shows the HI column density distribution and velocity field of NGC 1365 superposed on a print from the SRC sky survey. We find a very good correspondence between the bright optical spiral structure and the neutral gas, with the largest concentrations of gas close

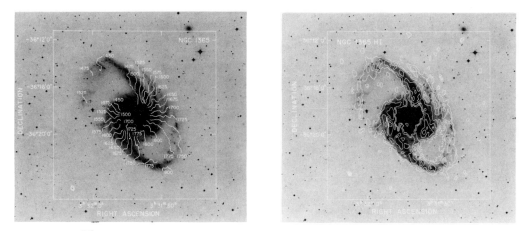

Figure 1. HI column density distribution (left) and velocity field (right) of NGC 1365. Column density contours are 1.7, 5.1, 8.5 and 11.9x10^{20} cm^{-2}. Velocity contours are labelled in km sec^{-1}. The beam is indicated by the ellipse in the lower left corner of each diagram,

E. Athanassoula (ed.), Internal Kinematics and Dynamics of Galaxies, 233–234.

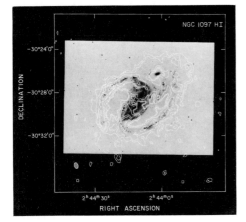 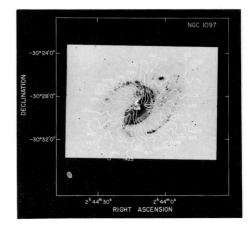

Figure 2. HI column density distribution (left) and velocity field (right) of NGC 1097. Column density contours are 4, 6, 10, 12, 15 and 17×10^{20} cm^{-2}. Velocity contours are labelled in km sec^{-1}. The beam is indicated by the ellipse in the lower left corner of each diagram.

to the edges of the bar. Peak column densities are 1.0–1.5×10^{20} cm^{-2}. In between the prominent spiral arms and in the bar region we do, however, not detect HI emission above our present sensitivity of 10^{20} cm^{-2}. The latter is particularly surprising because most models predict an accumulation of gas in the central part of the bar due to loss of angular momentum of the orbiting gas clouds. On the other hand, the HI column density is greatest at the leading edges of the bar, consistent with theoretical gasflow calculations. The velocity field is fairly symmetric and shows the gradual S-shape of the line of nodes characteristic for galaxies with a bar or an oval distortion (Bosma 1981).

Figure 2 shows the HI distribution and velocity field of NGC 1097 superposed on a photograph of Arp (see Lorre 1978). As in NGC 1365 there is a good correspondence between the dense HI and prominent optical features. Also the faint outer structures to the northwest, around the companion galaxy, coincide with an optical extension which is best seen in deep photographs of Arp processed by Lorre (1978). This faint asymmetric outer HI is possibly due to gravitational interaction of NGC 1097 and its companion galaxy. In NGC 1097 we do detect HI in the bar with a slight enhancement in the central region as expected from the various model calculations (e.g. Roberts et al. 1979). The velocity field of NGC 1097 shows the same characteristics of that of NGC 1365: a curving line of nodes and streaming motions along the spiral arms.
In the central region of the bar we find a large velocity gradient. With the presence of gas in the bar it becomes possible to probe the gas response in the central regions and a detailed comparison with theoretical models will be undertaken.

Bosma, A.: 1981, A.J. 86 1791, 1825.
Lorre, J.J.:1978, Ap.J. (Letters) 222, L99.
Roberts, W.W., Huntley, J.M. and van Albada, D.G.: 1979, Ap.J. 223, 67.
Sancisi, R., Allen, R.J. and Sullivan, W.T.: 1979, A. & A. 78, 217.

OBSERVATIONS OF THE NEUTRAL HYDROGEN IN THE BARRED SPIRAL GALAXIES NGC
3992 AND NGC 4731

S. T. Gottesman, J. R. Ball and J. H. Hunter
Department of Astronomy, University of Florida, Gainesville,
Florida, U. S. A.

We report observations of the atomic hydrogen properties of the
barred spiral galaxies NGC 3992 and NGC 4731. These systems were observed in 1980 and 1981 with the VLA telescope of the National Radio
Astronomy Observatory. In Table 1 we list the systemic parameters of
interest.

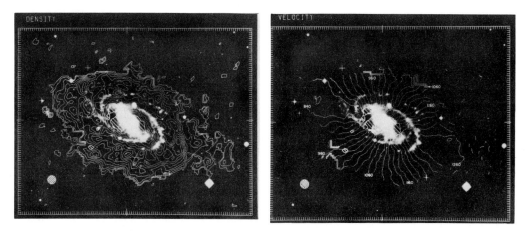

(a)   (b)

Figures la and lb. Show the column density and velocity field of the
HI in NGC 3992. The central region is devoid of gas. The (α, δ) scales
are 6 arcseconds per interval. The synthesized beam is shown

In figures la and lb we show the neutral hydrogen column density
and the observed, temperature-weighted, mean velocity field of NGC 3992
superimposed on optical photographs of the galaxy. The disk and spiral arm
structure of this system emit strongly and we observe a low density tail
at the preceding edge of the galaxy. Unfortunately, the inner zone that
is dominated by the bar is deficient in HI. The velocity field shows
perturbations of the order of 10 km/sec, where the line-of-sight crosses
the spiral arms. Also the velocities drop sharply in the low density

235

E. Athanassoula (ed.), Internal Kinematics and Dynamics of Galaxies, 235–236.

tail indicating that this material may not be rotating in the fundamental
plane. Owing to the poor signal-to-noise ratio in the central region,
not very much can be said about the kinematics of the barred zone.

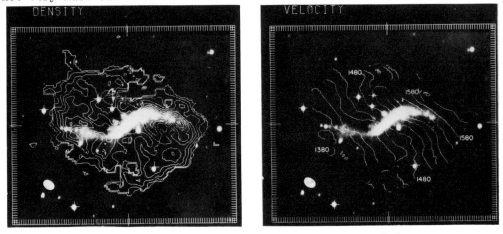

(a)   (b)

Figures 2a and 2b. Show the column density and velocity field of the
HI in NGC 4731. The (α, δ) scales are 6 arcseconds per interval and
synthesized beam is shown.

In figures 2a and 2b density and velocity data for the galaxy NGC
4731 are displayed. Both the bar and spiral arms are bright and well
resolved by these observations. The effect of the bar on the kinematics
of the system are strong, especially if contrasted with NGC 3992. As
the line-of-sight crosses the bar, velocity perturbations of about ±10
km/sec are seen. Also, the apparent line-of-nodes, as defined by the
velocity field, is rotated substantially with respect to the HI distribution. This research is partially supported by the NSF.

Table 1

Systemic and Observational Parameters for the
Galaxies NGC 3992 and NGC 4731

	NGC 3992	NGC 4731
Optical Diameter	7.6 arc. min.	6.5 arc. min.
Type	SBT 4	SBS 6
Systemic Velocity	1046 $\pm$ 5 km/sec	1488 $\pm$ 5 km/sec
Distance	14.2 Mpc	10.5 Mpc
Angular Resolution	22''9 x 20''7	33''2 x 23''2
Linear Resolution	1.58 x 1.43 kpc	1.69 x 1.18 kpc
Velocity Resolution	25.2 km/sec	25.3 km/sec
RMS sensitivity: Janskies/Beam	0.96 x 10^{-3}	2.46 x 10^{-3}
Kelvins	1.3	2.0
Detected HI mass (solar units)	2.6 x 10^9	1.8 x 10^9
Maximum HI surface Density	1.4 x 10^{21} Cm^{-2}	2.5 x 10^{21} Cm^{-2}

THE BARRED GALAXY NGC 7741

M.F. Duval
Observatoire de Marseille
2 Place Le Verrier
13248 Marseille Cedex 4 (France)

The type of this galaxy (SB(s)cd) is intermediate between symmetrical early type galaxies and asymmetric late-type barred spirals. Three different components are clearly seen: a bar (B), a central disk (c) and an extended disk (d).

The centre of the central disk is approximately coincident with the centre of the stellar continuum detected in the bar but does not correspond with the geometrical centre of the bar (Fig. 1).

The interior of the central disk is bordered by an oval distribution of HII regions (Fig. 4).

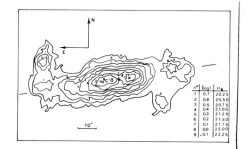

Fig. 1. +: stellar continuum centre
        C: geometrical centre

Fig. 2. isophotal map in blue light

From a surface photometry study in blue light (Fig. 2), we derive the geometrical parameters of the different components:

	position of the major axis	$< b/a > = q$
bar	$98° \pm 2°E$	0.30 ± 0.02
central disk	$160° \pm 3°E$	0.71 ± 0.03
extended disk	$160° \pm 5°E$	0.68 ± 0.03

and their intrinsic luminosity distribution:

237

E. Athanassoula (ed.), Internal Kinematics and Dynamics of Galaxies, 237–238.

bar $\qquad$ $I_B = 4.3(1-(r*/0.32)^2)^{72}$ $\quad I_{oB} = 422 \ L_\odot pc^{-2}$

central disk: $\quad I_c = 0.45(1-(r*/1.8)^2)^{32}$ $\quad I_{oc} = 44 \ L_\odot pc^{-2}$

extended disk: $\quad \log I_d = -0.5 \ -0.611 \ r*$ $\quad I_{od} = 31 \ L_\odot pc^{-2}$

These laws give: $L_B/L_T = 0.09$, $L_c/L_T = 0.56$, $L_d/L_T = 0.31$; we obtain

for the bright inner arm: $L_a/L_T = 0.04$. The luminosity of the bar in

this galaxy is principally dominated by strong HII regions.
The velocity field has been obtained from spectra and interferograms.
It shows strong deviations with respect to the velocities expected from
an axially symmetric mass distribution as it can be seen on the "rotation curves" drawn along the major axis of the galaxy and along the major axis of the bar (Fig. 3). Otherwise, two zones of constant radial velocity are observed at 20" to the north and south of the centre of the bar.

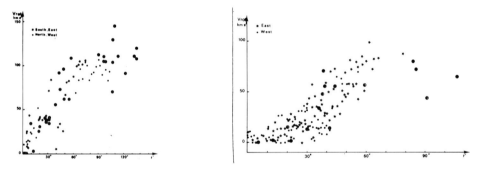

Figure 3. "rotation curves"
major axis of the galaxy $\qquad$ major axis of the bar
◯circled points are obtained from 2 interferograms,✶and ● from 6 spectra

Using Toomre's models for the determination of the central disk and
extended disk masses we obtain, after correction for galactic and
internal absorption, the following M/L ratios:
bar + central disk $\quad$ M/L $\simeq 1h^{-1} \ M_\odot/L_\odot$

extended disk + halo $\quad$ M/L $\simeq 30h^{-1} \ M_\odot/L_\odot$

(distance adopted: 10 Mh pc with h = 100/H km s^{-1} Mpc^{-1})

Fig. 4

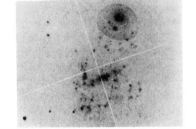

Hα photograph taken by
J.Boulesteix, G.Courtès,
H.Petit at the prime focus of
the 6m telescope at Zelenchus-
kaya equiped with a focal reducer and a image tube. 15min
exposure. FWHM of the filter
6 Å.

HYDRODYNAMICAL MODELS OF OFFCENTERED BARRED SPIRALS

J. Colin and E. Athanassoula
Observatoire de Besançon   ERA 07904
25000 Besançon - France

We have studied the·flow of gas in barred spirals in which the centers of the bar and the disk do not coincide. This is often observed in late type galaxies like the L.M.C., NGC 1313, NGC 4618 etc...

Starting from the centered barred galaxy model presented by Sanders and Tubbs (1980), we introduced an offset and studied its effects on the gas distribution and velocity field.

When the orbital period of the bar around the disk center is equal to that of rotation around its own center, it is possible to

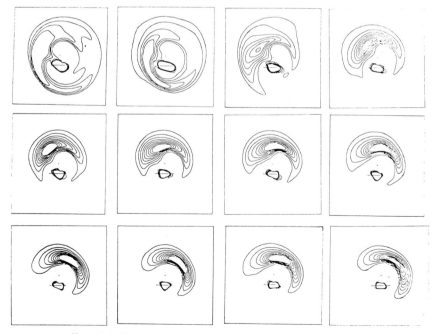

Fig. 1 Evolution of the isodensities of a gas flow in an offcentered bar galaxy.

239

E. Athanassoula (ed.), Internal Kinematics and Dynamics of Galaxies, 239–240.

obtain a stationary gas flow. An example of this is given in figure 1 which shows the lopsided response to this type of offcentered forcing. The bar is a homogeneous oblate spheroïd of major axis 1 kpc and minor axis 0.4 kpc displaced by 0.8 kpc in the sense perpendicular to its major axis. The corresponding potential is shown in figure 2 a.

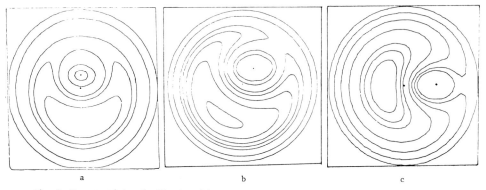

Fig. 2 Isopotentials of offcentered bars
         a. displacement along the minor axis of the bar.
         b. Displacement in a random direction.
         c. displacement along the major axis of the bar.
         In all cases, bar is horizontal.

The stationarity in the gas flow was obtained after roughly $3.5 \ 10^8$ years. Several other examples of displacements parallel (figure 2c) or perpendicular (figure 2a) to the bar major axis as well as more random orientations (figure 2b) have been studied. For cases where the orbital period of the bar around the disk is not equal to that of rotation around its own center, we find a very complicated gas flow pattern and no stationarity although the short-lived configurations obtained are good representations of late type asymetrical galaxies.

Reference
Sanders, R.H., Tubbs, A.D. : 1980, Astrophys. J., 235,803.

STRUCTURE AND DYNAMICS OF MAGELLANIC TYPE GALAXIES

J. V. Feitzinger
Astronomical Institut, Ruhr-University Bochum, FRG

Morphology

In continuation of former work (Feitzinger 1980) the morphological structure of 68 SB(s)m systems, greater than 2!5 on ESO/SRC survey plates,
was investigated. Three morphological features should be mentioned:
1. Stellar bars located asymmetrically with respect to the disk are
   found in 7o% of the cases; the displacement relative to the disk diameter is Λ = o.1 - o.35.
2. The disk is populated in 45 % of the cases by chains of HII regions
   (spiral arm filaments).
3. Though not very frequent nearly 25 % of the SBm systems have a dominating HII region near the outer periphery and predominantly near one
   end of the bar.

Model Calculations

To obtain a first understanding of the structure of SBm systems and the
above mentioned features gas dynamical calculations were performed to
clear the gas-bar interaction (Feitzinger, Schmidt-Kaler, Weiss 1983).
In the case of a symmetrical embedded prolate bar in an oblate disk two
stable and two unstable Lagrange points exist apart from the center. If
we shift the bar out of the center of the disk one of the stable points
(N_4 or N_5) is lost (de Vaucouleurs, Freeman 1973). For an inertial observer the rotation curve of such a galaxy, measured perpendicular to
the symmetry line of the bar, shows two disturbances (Fig. 1). The stable
circulation around the N_4 point diminishes, if we shift the bar out of
the disk center and the rotation curve becomes asymmetric.
Generally near the end of the bar a density increase develops and also
short density enhancements (spiral arm filaments) are set up in the disk.
The stability and life time of these structures depends mainly on the
bar rotation. In the disturbed gas disk stochastic selfpropagating star
formation processes have been run (Feitzinger et al. 1981). We see a pre-
ponderance of star formation in the denser parts of the disk. The ends
of the bar are very active sites, and star formation is also going on in
the density enhancements of the disk. Observationally such structures
may be identified with the spiral arm filaments and the giant and supergiant HII regions at the ends of the bars of Magellanic type galaxies.

241

E. Athanassoula (ed.), Internal Kinematics and Dynamics of Galaxies, 241–242.
Copyright © 1983 by the IAU.

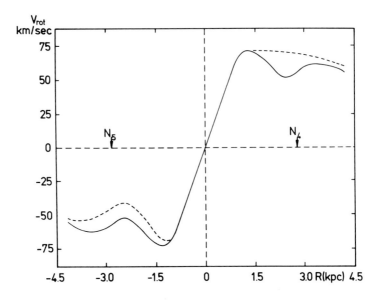

Fig. 1 Rotation curve, measured perpendicular to the bar.
       Large Magellanic Cloud parameters are used (compare also Christi-
       ansen and Jefferys1976). Full line: Bar and disk center coincide.
       Broken line: Bar and disk center are shifted by 5oo pc; the rota-
       tion curve becomes asymmetric.

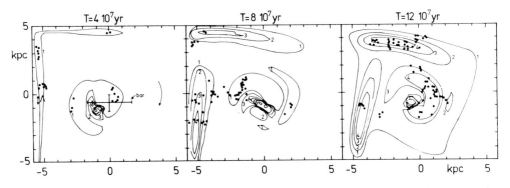

Fig. 2 Density distribution in percent of the undisturbed disk at three
       time steps with the same bar position. Dots indicate star forming
       regions. The time step of star formation is 10^7 yr and only the
       most recent sites are indicated; no aging is shown.

References:
Christiansen, J.H., Jefferys, W.H., 1976, ApJ 2o5 , 52
de Vaucouleurs, G., Freeman, K.C., 1973, Vistas 14, 163
Feitzinger, J.V., 198o, Space Science Review 27, 35
Feitzinger, J.V., Glassgold, A.E., Gerola, H., Seiden, P.E., 1981, Astro-
       nomy Astrophys. 98, 371
Feitzinger, J.V., Schmidt-Kaler, Th., Weiss, G., 1983, in preparation

FORMATION OF RINGS AND LENSES

E. Athanassoula
Observatoire de Besançon

1. RINGS

Both inner rings and outer rings are frequently observed in disk galaxies (de Vaucouleurs, 1963 and 1975). As an example, Fig. 1 shows NGC 2217 in which both occur. Further examples, e.g. NGC 2859 and NGC 3081, can be found in the Hubble Atlas (Sandage, 1961), the Revised Shapley Ames Catalogue (Sandage and Tammann, 1981), and in the morphological study of nearby barred spirals by Kormendy (1979). In this talk I will concentrate on some theoretical questions concerning rings, and in particular on their formation.

Schwarz (1979, 1981) studied the response of a gaseous disk to a rotating bar potential. He models the gas by "clouds" which can collide inelastically, thereby losing an important fraction of their relative motions. A spiral is formed which extends roughly from corotation (hereafter CR) to the outer Lindblad resonance (OLR). Due to the torque exerted by the bar, this spiral evolves into a ring at the OLR. Schwarz associates this kind of ring with the outer rings frequently observed in external galaxies. Depending on the value of the pattern speed used in the calculations, a second ring was sometimes formed. This occured near the ultra harmonic resonance (UHR) where $\Omega_p = \Omega - \kappa/4$, and/or the inner Lindblad resonance (ILR). Schwarz associated these rings with the observed inner rings.

The problem with Schwarz's calculations is that the time scale for ring formation is very short. Indeed, the rings form in just a few bar rotations, yet a large percentage of observed galaxies do not have rings. Schwarz proposed three possible ways out, favouring the third one :
a. The mass distribution in his model bar differs from that in real galaxies
b. The cloud collisions have not been correctly modelled
c. The gas between CR and OLR is replenished by mass loss from stars or by infall.

E. Athanassoula (ed.), Internal Kinematics and Dynamics of Galaxies, 243–252.
Copyright © 1983 by the IAU.

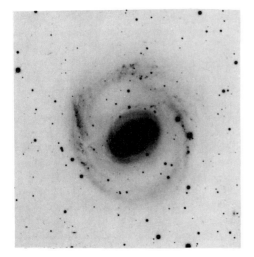

Figure 1. NGC 2217 (Prime
focus, CFHT 3.6 m by
Athanassoula and Bosma)

   One of the virtues of Schwarz's model is the definite prediction
that rings will form at resonances. Athanassoula et al. (1982, hereafter
ABCS) tested this prediction by analyzing data on the sizes of rings and
lenses as given in the catalogue of de Vaucouleurs and Buta (1980) for
532 galaxies. For galaxies having both outer and inner rings, the ratio
of the two sizes, R/r, was formed and histograms were made of the R/r
frequency distribution for both barred and non-barred galaxies, as shown
in Fig. 2. Clearly the two histograms are very different. The barred
galaxies have a peak around R/r = 2.2, while the non-barred galaxies
have a rather flat distribution. The form of the barred spiral histogram
can be explained (cf ABCS) if outer rings form near the OLR and inner
rings near the CR or the UHR. Thus, this statistical analysis agrees
nicely with Schwarz's prediction.

   Two further points can be made here. First, rings in barred spirals
seem to avoid the ILR (at least the regular inner and outer rings ; nuclear
rings are not discussed here). Since Schwarz claims he found rings at the
ILR in his models, maybe the existence of this resonance should be

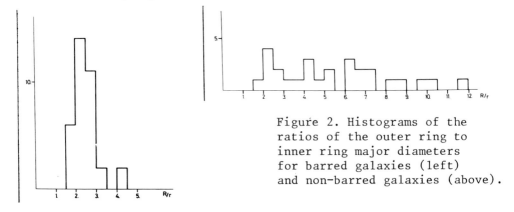

Figure 2. Histograms of the
ratios of the outer ring to
inner ring major diameters
for barred galaxies (left)
and non-barred galaxies (above).

questioned (see also Van Albada and Sanders, 1982). Second, rough knowledge of the rotation curve at the distance of the outer ring and of its
diameter allows calculation of the pattern speed in barred galaxies with
outer rings. Inner rings are less useful in this case since at that distance the velocity field can be highly perturbed by the bar.

The form of the histogram for SA galaxies is more difficult to
explain. One possibility is that rings in SA's are not associated with
resonances. However this is hard to accept in view of the continuity of
some of the properties along the "families" SB-SAB-SA. Furthermore,
Schwarz showed that a non-barred, but spiral forcing was as efficient
in forming rings, which were again associated with resonances. An alternative explanation is that in SA galaxies more resonances come into play.
In the histogram of R/r we would then have a superposition of several
peaks which, together with the small numbers involved, leads to the observed flat distribution. The question remains then, why we don't observe
SA galaxies with three large rings.

In Schwarz's models, the orbits between CR and OLR are unstable and
their eccentricity increases with time. A similar result, for a different
form of the potential, was stressed by Contopoulos and Papayannopoulos
(1980). They found that the main families of periodic orbits for strong
bars are unstable between CR and OLR, but stable around the OLR. This
behaviour, they conclude, explains the appearance of rings ; the instability of the main periodic orbits between CR and OLR leads to a depletion
of that region while the stable orbits around OLR trap quasi-periodic
orbits around them (e.g. Arnold and Avez 1967). Nevertheless they correctly note that only knowledge of the distribution function i.e. of the number of particles following a given orbit can definitely settle the question.

However, the main families of periodic orbits between CR and OLR
are not unstable in every model of a barred galaxy. A counterexample has
been discussed by Athanassoula et al. (this volume and 1982), who used
a nonhomogeneous prolate spheroid to model the bar, and a power law
rotation curve to model the axisymmetric background. They found that if
the length of the bar equals the corotation radius (CR as defined by the
axisymmetric background), the main families of periodic orbits between
CR and OLR are stable for reasonable values of bar eccentricity and of
ratio of bar to disk mass. They are followed by regular invariant curves
on all surfaces of section. The way to introduce stochasticity (i.e. unstable orbits) in this model is to allow the bar to extend beyond corotation.

Thus, depending on how mass is distributed between the axisymmetric
background and the bar, the models may have stable or unstable periodic
orbits between CR and OLR. It then becomes crucial to know what the relevant mass distributions in real barred galaxies are, and whether they
differ in galaxies with rings and without rings. Rather than exploring
all the theoretical possibilities, it seems more sensible to turn to observations. With this aim in mind, Athanassoula and Bosma (1982) have taken a number of plates of galaxies with bars or ovals, with and without

rings. From the light distribution and various assumptions about the M/L ratios, they plan to derive the potential function of the bar in the galaxy plane. A calculation of the main families of periodic orbits or, better still, of the gas response to these potentials should shed more light on the ring formation process.

2. LENSES

A lens is a shelf in the luminosity distribution of a galaxy occuring between the bulge and the disk. A typical example is NGC 1553 (Freeman, 1975). Further examples (e.g. NGC 3245, NGC 4150, NGC 4262, etc) can be found in the Hubble Atlas (Sandage, 1961) and the Revised Shapley-Ames Catalogue (Sandage and Tammann, 1981).

Kormendy (1981) would like to further subdivide objects fitting the above description and distinguish between those located in early type galaxies (SO to Sa), which he calls lenses, and those located in later types (Sb to Sm), which he calls ovals. This distinction would, according to Kormendy , be based on their different kinematical properties. I believe that further work is needed to show whether such a distinction is real and necessary. I will thus, for the purposes of this review, not distinguish between the two, but use the terms lenses and ovals alternatively.

The ABCS statistical study showed that lenses have axial ratios between 0.5 and 1. Hence an important fraction of them are eccentric enough to have sizeable nonaxisymmetric forces, thus qualitatively resembling bars. There are further links or similarities between bars and lenses. Kormendy (1979) has noted that for galaxies with both a bar and a lens, the major axes of the two components have the same length and position angle (hereafter PA) in 17 out of 21 cases he studied. This he terms, "the bar fills the lens in one dimension". Furthermore, for NGC 5383, the only galaxy with both a bar and a lens for which a M/L estimate is available for both components (Duval and Athanassoula, 1982 and this volume),the values were found to be the same within 30 %. This argues for similar stellar populations in the two components and a common origin or link in their formation.

The following scenarios have been put forward for the formation of lenses : Bosma (this volume) thinks that lenses are primary components, formed as part of the disk formation process. The outer edge formed where the initial gas density dropped low enough so that star formation stopped abruptly. This process would then account for all sharp outer edges in lenses and disks.

According to Kormendy (1979, 1981) lenses are secondary components formed by secular evolution of a bar to a more axisymmetric state. This could be due to the interaction of the stars in the bar with a very rapidly rotation bulge.

Let me put forward here yet a third scenario. According to it, lenses like bars, are due to instabilities of galactic disks. The difference in eccentricity between the two components is due to the different amounts of random motions initially present in the disk, i.e. the different initial distribution functions.

This hypothesis is based on the results of numerical simulations (Athanassoula and Sellwood, this volume). We studied the effects of velocity dispersions on the stability of galactic disks. The aim was to find out whether, and if so by how much, velocity dispersions reduce the growth rate of the bar instability and whether disk stability could be reached with little or no help from a halo. The initial mass distribution was identical for all runs, namely a Toomre $n = 1$ disk (Toomre, 1963). The initial velocities of the 40.000 particles used in the simulations were taken according to distribution functions calculated analytically. Starting therefore from initial conditions that are stationary solutions of the Boltzmann equation, i.e. much closer to equilibrium than is usually done with other starting conditions, we have less reason to fear that imprecise initial conditions provoke bar instability. We let the runs evolve and the stellar density distribution change shape due to the bar instability. A bar or oval was invariably the end product of the evolution. Its shape however, differed systematically from one simulation to the next depending on the velocity dispersion in the initial distribution function.

Let me first discuss the results of four simulations whose radial velocity dispersions, $\sigma_u(r)$ are given by full curves in Fig.3. The particular set of distribution functions were taken from Kalnajs (1976). A quantitative measure of the resulting bars or ovals was obtained by fitting ellipses to the isodensity contours in the bar using a software package by M. Cawson (1982). The ellipses approximate the isodensity contours very well in the bar region in nearly all cases, while their eccentricity changed little with radius. Only when we used the coolest of the four distribution functions were the isodensity contours somewhat box shaped and the eccentricity of the best fitting ellipse less representative. In Fig. 4. we plot the mean eccentricity of the bar isodensities, $1 - b/a$ with b and a the minor and major axes of the bar or oval, as function of the mean mass averaged Q (i.e. the ratio of actual dispersion to what is required to prevent axisymmetric instabilities, cf. Toomre, 1964) in the initial disk. A very clear trend shows up in the sense that initially colder disks evolve into thin bars, and initially hotter disks evolve into fatter ovals. Other series of runs confirm this trend. We make here the comparisons in terms of the dimensionless parameter mean Q rather than in σ_u, although the latter might be more appropriate in case initially different mass distributions are to be compared. We also used the dispersion of velocities in the initial distribution functions, since, after the bars or ovals have formed, all quantities depend on the angle as well as the radius, thus making comparisons more difficult and less meaningful. Note that the differences between the dispersions of velocities of the four models are largely reduced as the runs evolve, since the initially cooler models heat up more than do the

initially hotter ones. However a detailed comparison is too lengthy to
be included here.

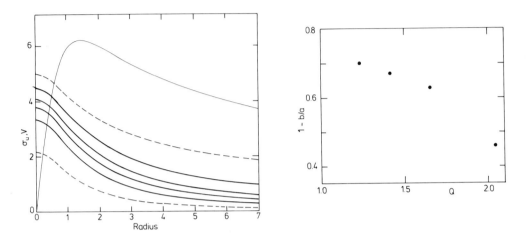

Figure 3. Velocity dispersions
(thick and dotted lines) and
circular velocity (thin lines)

Figure 4. Mean eccentricity of
the bar isodensities as a function
of the mean mass averaged Q.

One of the previous runs was analysed by dividing stars into two
groups : the hottest half and the coldest half. At the beginning of the
simulation all stars at a given energy were divided into two halves i.e.
those with the largest angular momentum and those with the smallest. Since
we tagged each star we could follow the evolution of each group separately, and found that the above trend still applied very well ; that is,
the coldest half had isodensities well approximated by eccentric ellipses,
and the hottest half had near circular isodensity contours. Furthermore
the PA of the major axes coincided to within the (small) measuring errors.

We also studied the evolution of disks in which two populations were
initially present, each with its own distribution function f(E,J). Several
such cases were run, all leading to similar conclusions, so we will only
present one example here. In this particular run, 50 % of the stars followed a cool distribution function and 50 % followed a hot one. Their corresponding radial velocity dispersions are given by the dotted curves in
Fig. 3. As expected the hotter population formed a near circular oval,
while the cooler one evolved into a much narrower bar. The ellipses fitted
to the isodensity contours are given in Figure 5. The PA and, to the extent they could be defined, the lengths of the major axes of the two components are the same.

The basic trend which can be observed from these runs is that the

final ellipticity of the bar or oval depends on how hot the initial distribution function was. Cool systems make thin bars while hot systems
make fat ovals.

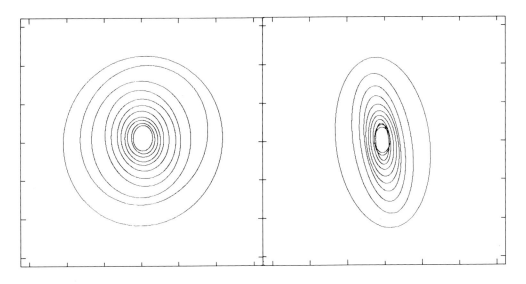

Figure 5

These results may well be relevant to the formation of lenses.
Indeed if lenses are so similar to bars, the main difference being their
shape, it may well be that their formation process is the same, i.e.,
that they are both due to an instability of the galactic disk. The differences in shape would then be due to the different amount of random
motions initially present in the disk. In cases where, for whatever
reason, two distinct populations were present, the coolest one would
form a bar and the hottest one a lens, the major axes of both components
having the same PA and approximately the same length in agreement with
Kormendy's maxim. Furthermore the velocity dispersions in our models are
in the same range as those borne out by the observations of Kormendy
(1981) and Bosma and Freeman (1982).

Several interesting questions arise if lenses are indeed as hot as
suggested by the observations (Kormendy, 1981) and our scenario. In particular, the coexistence of a relatively sharp edge and large velocity
dispersions (which do not necessarily extend out to the edge) has to be
investigated further. Some lenses seem to be devoid of structure while
disks with sharp edges might be prone to unstable edge modes (Toomre,
1981). However, a detailed study of the influence of sharpness and velocity dispersion at the edge on the occurence of edge instabilities has not
been done yet. Finally, lenses, and bars also, are thin in the z direction
(Tsikoudi, 1980, Burstein, 1979), while being hot in the plane. Such large
anisotropies in the velocity distribution pose another interesting theoretical problem.

REFERENCES

van Albada, T.S. and Sanders, R.H. : 1982, Monthly Notices Roy. Astron. Soc. 201,303.

Athanassoula,E., Bienaymé, O., Martinet, L. and Pfenninger D. : 1982, submitted to Astron. and Astrophys.

Athanassoula, E. and Bosma, A. : 1982, in preparation

Athanassoula, E., Bosma, A., Crézé, M. and Schwarz, M.P. : 1982, Astron. Astrophys. 106,101.

Arnold, V. and Avez, A. : 1967, Problèmes ergodiques de la Mécanique classique, Gauthier-Villars, Paris.

Bosma, A. and Freeman, K.C. : 1982, in preparation.

Burstein, D. : 1979, Astrophys. J. 234,829.

Cawson M. : 1982, in preparation.

Contopoulos, G. and Papayannopoulos, T. : 1980, Astron. 92,33.

Duval, M.F. and Athanassoula, E. : 1982, Astron. and Astrophys. in press.

Freeman, K.C. : 1975, in IAU Symposium, No. 69, Dynamics of Stellar Systems, ed. A. Hayli, Reidel, p. 367.

Kalnajs, A. : 1976, Astrophys. J. 205,751.

Kormendy, J. : 1979, Astrophys. J. 227,714.

Kormendy, J. : 1981, in Structure and Evolution of Normal Galaxies, ed. S.M. Fall and D. Lynden-Bell, Cambridge University Press, p. 85.

Sandage, A. : 1961, Hubble Atlas of Galaxies, Carnegie Institution of Washington.

Sandage A. and Tammann, G.A. : 1981, A Revised Shapley Ames Catalog, Carnegie Institution of Washington.

Schwarz M.P. : 1979, Ph. D. Thesis, Australian National University.

Schwarz M.P. : 1981, Astrophys. J. 247,77.

Toomre, A. : 1963, Astrophys. J. 138,385.

Toomre, A. : 1964, Astrophys. J. 139,1217.

Toomre, A. : 1981, in Structure and Evolution of Normal Galaxies, ed. S.M. Fall and D. Lynden-Bell, Cambridge University Press, p.111.

Tsikoudi, V. : 1980, Astrophys, J. Suppl. 43,365.

de Vaucouleurs, G. : 1963, Astrophys. J. Suppl. 8,31.

de Vaucouleurs, G. : 1975, Astrophys. J. Suppl. 29,193.

de Vaucouleurs, G. and Buta R. : 1980, Astron. J. 85,637.

DISCUSSION

SANDAGE : It is true that the galaxies NGC 2217, NGC 5101, NGC 5566, and other similar early type Sa galaxies that you showed at the beginning have very large external "near" rings of star formation. Yet these structures are not true rings. They are segments of spiral arms (m = 2) very tightly wound, but clearly separated from each other after each winding of $\sim 180°$. The same happens in the well-known Sa galaxy NGC 3185 where, on low resolution plates a completed ring is strongly suggested, yet the structure is, in fact, two spiral segments. This feature, present in a subset of ~ 30 Shapley-Ames early Sa galaxies resembles Schwarz's Figure in the lowe right of this diagram which you showed. (See Ap.J. 247,77, Figure 10).

The question is : what happens much later in the Schwarz sequence as time proceeds ? Can a true <u>closed</u> ring be formed of old stars such as in NGC 2859 or NGC 4736 ? Invariably, when these true closed rings are seen, and there are some even in Shapley-Ames galaxies, they have red colors and a very smooth intensity distribution, indicating stellar ages of at least 10^9 years.
If, then, your answer is yes, are there later "Schwarz true ring" forms stable for 5 to 10 orbital periods ?

KENNICUTT :   The models discussed here can account for the narrow gaseous rings seen in the outer parts of galaxies, but many galaxies seem to possess very broad stellar rings, extending over many kpc in radius, and occuring in both barred (NGC 3945) and non-barred systems (NGC 4736). Do you think that this latter type of ring is a distinct phenomenon dynamically, or do you think that these models can provide insights into them as well ?

ATHANASSOULA (to both SANDAGE and KENNICUTT) :   In the evolutionary sequence I showed (Figure 10 of Schwarz 1981), only the gaseous component feels the bar and responds to it. The response is initially a two-armed spiral. During the evolution there is an outward flux of particles from the region between CR and OLR to the OLR. Thus this region is depleted and the response becomes more ring-like and less spiral-like with time. Exactly where one should draw the line between the two, I don't know. The gas particles will follow more and more closely a small range of periodic orbits around the OLR. Thus collisions will be minimized and further evolution in the model stopped. However, the subsequent evolution in real galaxies is less obvious. Schwarz's calculations do not include all physical processes which can be of importance in the formation of stellar rings. One plausible hypothesis is that, because of the high gas concentration, rings are regions of enhanced star formation and thus a stellar ring will be formed at the same position as the gaseous one. Whether and for how long the gaseous ring will be maintained will depend on several, not very well understood, parameters like the rate of star formation in the ring and whether new gas will be added to it from the region between CR and OLR. The maintenance of stellar rings can also be questioned since in all N-body (i.e. stellar) simulations, rings seem to be transients. If however this structure is maintained while the stellar population ages and acquires higher dispersion of velocities by mechanisms analogous to those discussed by Schwarzschild and Spitzer (Astrophys. J. 114, 385 and 118, 106), then this can account for the redder and relatively broad stellar rings.

VAN WOERDEN :   The distinction between rings and completely wrapped-up spirals is semantic. Schwarz's time sequence of evolution of a spiral pattern shows that structures may develop which morphologically cannot be distinguished from rings.

MILLER :   Rings in N-body simulations are very unstable if they contain much mass. Stellar rings, especially those in outmost regions of galaxy images, are likely to contain enough mass to cause trouble even if they have low surface brightness.

ATHANASSOULA : Rings have often formed in the numerical N-body simulations described above as well as in previous ones (e.g. Sellwood, Astron.
Astrophys. 99, 362), but they were always transient. I have not looked
into how the mass in the ring may influence its lifetime. However, I do
not know to what extent these results can be extrapolated to real galaxies,
since there can be a number of differences. In particular the dispersion
of velocities and the amount of softening present in the models can be
of great importance.

HUNTER : What resonances do you have in mind that might be responsible
for the formation of rings in non-barred spirals ? In the absence of a
bar, there is no obvious pattern speed to use in calculating the location
of the outer Lindblad resonance.

ATHANASSOULA : Anything rotating and nonaxisymmetric can provide a pattern speed. Bars, ovals or spirals seem the obvious candidates. In fact
Schwarz (Astrophys. J. 247, 77) pointed out that a spiral potential can
produce rings in time scales comparable to those of bar forcing.

JORSATER : Contopoulos (Astron. Astrophys. 81, 198) suggested that bars
end at corotation. However in your discussion of rings you presented
results of orbit calculations in a model with a substantially longer
bar. Could you comment on this ?

ATHANASSOULA : Because of the rather centrally condensed density distribution in the bar, only 10-15 % of the bar mass was actually outside corotation in the examples shown. Also note that the bar introduces a sizeable nonaxisymmetric force so that the very definition of corotation is
not unique. Athanassoula et al used, for simplicity, the background axisymmetric rotation curve and did not include the axisymmetric contribution
of the bar in their definition of CR.

KENNICUTT : Lenses often exhibit very sharp outer cutoffs in their
radial distribution. How can such a sharp transition be produced from a
protogalaxy that must have possessed a much smoother radial distribution ?

ATHANASSOULA : The question holds true for both bars and lenses since
they both have rather sharp cutoffs. Neither the scenario presented here
nor that of Kormendy claims to answer it. According to Bosma's scenario,
the lens stops where the density of the initial gas dropped low enough
so that star formation stopped abruptly.

LENSES AND LOW SURFACE BRIGHTNESS DISKS

A. Bosma
Sterrewacht
Leiden

The 21.65-"law" for disk galaxies has been debated ever since Freeman's (1970) paper in which he found that for 28 out of 36 galaxies the extrapolated central surface brightness of the exponential disk component I_0, follows this rule with little intrinsic scatter. Some people think it significant, while others invoke selection effects. Bosma and Freeman (1982) made a new attempt to clarify this problem by studying ratios of diameters of disk galaxies on the various Sky Surveys in a region of overlap. The limiting surface brightness levels were calibrated to be 24.6 and 25.6 magn/arcsec^2 for the Palomar blue prints and the SRC J films, resp. The distribution of the ratio Γ = diameter (SRC) / diameter (PAL) gives a measure of the true distribution of I_0 if the galaxy has an exponential disk in the brightness interval 24.6 to 25.6; e.g. I_0 = 21.6 corresponds to Γ = 1.32, I_0 = 22.6 to Γ = 1.50 and I_0 = 23.6 to Γ = 1.90, etc.

From micrometric measurements of several hundreds of galaxies the distribution of Γ turned out to be quite different from the one expected on the basis of the 21.65 law, even allowing for measuring errors, plate variations, etc. From a detailed analysis we conclude:
- 55% of the galaxies can indeed follow the 21.65 law with some scatter.
- 25% of the galaxies have Γ's only slightly larger than 1, and have apparently a steep outer edge reminiscent of those found in some edge-on's by Van der Kruit and Searle (1982).
- 20% of the galaxies have Γ's much larger than 1.3 and therefore have a disk of much lower central surface brightness than 21.65. Most of these galaxies are probably giants (they do not look at all like magellanic irregulars, and for several of them we measure high rotational velocities), and most of these giants have a bright inner zone and a faint outer disk.

This variety in photometric properties of disk galaxies is also suggested by e.g. different Γ-distributions for galaxies with sharp edges (i.e. size defined by outer spiral or ring structure) or with smooth edges (size defined by the old stellar disk), or between ordinary and barred spirals. A similar result can be inferred from Boroson (1981), who could decompose only 15 out of 27 galaxies into a standard $r^{\frac{1}{4}}$-law bulge and an exponential disk with I_0 = 21.7 $\pm$ 0.6. The other 12 have outer cutoffs, and/or bright inner zones, and/or faint outer disks.

253

E. Athanassoula (ed.), Internal Kinematics and Dynamics of Galaxies, 253–254.
Copyright © 1983 by the IAU.

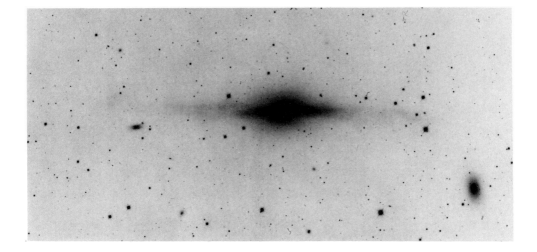

Figure 1. NGC 5084 (Prime focus, CFHT 3.6m by Bosma/Athanassoula)

We have made several follow up observations on the galaxies with the two zone structure. Of particular interest are galaxies like NGC 2090 and NGC 5084. For NGC 2090, a Sc galaxy, comparison of plates in the UV and the red shows that the inner zone is quite a bit redder than the outer zone, similar to the situation in NGC 5963 (cf. Romanishin et al., 1982). In NGC 5084 the outer disk is warped (Figure 1) and tilted slightly with respect to the inner lens. This suggests that the lens is dynamically distinct from the outer disk. The maximum rotational velocity of the stars in the lens of NGC 5084 is $\sim$200 km sec^{-1}, while that of the HI in the outer disk is $\sim$300 km sec^{-1}.

These data suggest the hypothesis that lenses are formed as a primary component relatively early on in the formation of a disk galaxy. Only in some, unknown, circumstances an inner flat component is formed which quickly forms stars out to a radius where the gas density becomes too low. The remaining gas in the protogalaxy settles into an outer disk much later. The formation of an edge in the lens is presumably similar to the formation of a cutoff in the disk à la Van der Kruit and Searle. The connection between bars and lenses as reviewed e.g. by Athanassoula (this volume) arises when the inner disk happens to be unstable to a bar mode.

Further work on these galaxies is in progress.

REFERENCES:

Boroson, T.A., 1981, *Astrophys. J. Suppl. 46*, 177
Bosma, A., and Freeman, K.C., 1982, in preparation
Freeman, K.C., 1970, *Astrophys. J. 160*, 811
Romanishin, W., Strom, K.M., and Strom, S.E., 1982, *Astrophys. J. 258*, 77
Van der Kruit, P.C., and Searle, L., 1932, *Astron. Astrophys. 110*, 61

V

SPHEROIDAL SYSTEMS

DYNAMICS OF EARLY-TYPE GALAXIES

Garth Illingworth
Kitt Peak National Observatory

The form of the velocity and velocity dispersion profiles in
elliptical galaxies, and their implications for anisotropy and mass-to-
light ratio changes in ellipticals are discussed. Recent results on the
luminosity dependence of the rotational properties of ellipticals are
summarized, as are questions concerning the shapes of ellipticals. The
relation of bulges to ellipticals, in the light of these new data, is
considered.

I. ELLIPTICALS: $V(r)$ and $\sigma(r)$ PROFILES

A recent paper (Davies et. al. 1983) summarized available kinematic
data for some 43 elliptical galaxies. A significant fraction of these
galaxies have dispersion (σ) and velocity (V) profiles that extend out
to near a de Vaucouleur's radius, r_e. I have selected a sample of these
ellipticals, covering a range of luminosities, and plotted both rotation
and dispersion profiles in dimensionless coordinates. These are by no
means an unbiased sample of V and σ profiles. They were chosen to
illustrate the range of behavior of such profiles. Velocity dispersion
profiles are of particular interest since we still do not know if
ellipticals have massive halos, i.e., if the mass-to-light ratio, M/L,
is approximately constant or increases steadily outward. Unlike disk
systems in which M/L variations can, in principle, be readily derived
from the rotation curve of the gaseous component, the derivation of the
run of M/L with radius in ellipticals is complicated by anisotropy in
the stellar velocity distribution.

$V(r)$. The sample rotation curves have been normalized by the mean
maximum rotation velocity $\langle V_{max} \rangle$, and plotted against r/r_e in Figures 1
and 2. The galaxies in this sample have a range of ellipticities (E2-
E5) and are not, in general, rotationally flattened. All have $V_{max} \gtrsim 50$
km s^{-1}, so that observational uncertainties do not significantly affect
the rotation curves. Typically, $15'' \lesssim r_e \lesssim 70''$ and $50 \lesssim V_{max} \lesssim 150$ km
s^{-1}.

257

E. Athanassoula (ed.), Internal Kinematics and Dynamics of Galaxies, 257–270.
Copyright © 1983 by the IAU.

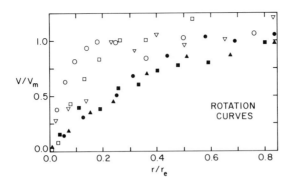

Figure 1. Absorption line rotation curves are shown for six
ellipticals. The rotation velocity V has been normalized by the mean
maximum rotation V_m, and plotted against radius r normalized by the de
Vaucouleurs radius r_e. This sample was selected to have relatively
shallow gradients. Compare these with the ellipticals in Figure 2.

        Several examples of slowly rising rotation curves are shown in
Figure 1. Interestingly, the three with the shallowest gradients (the
filled symbols - NGC 3605, 4387, and 4478) are all low luminosity
ellipticals from Davies et. al. (1983). Some ellipticals, on the other
hand, show a rapid rise to $V \sim V_{max}$ and a roughly flat rotation curve out
to $r \sim r_e$. Examples are shown in Figure 2. The turnover radii r_m of the
three most rapidly rising rotation curves (for NGC 1052, 3377 and 4697)
could be strongly affected by seeing. For example, in NGC 1052 the
apparent turnover radius is ~2", making the actual turnover
radius $<$ 300pc. It will be interesting to see if the ellipticity
remains large at such small radii as well, suggesting that anisotropy is
significant even close to the nucleus (see Jedrzejewski et al. 1983,
Davies and Illingworth 1983c).

        While the sample is small most normal ellipticals show no
significant rotation along the projected figure minor axis. To date it
is only those ellipticals whose apparent principal axes show gross
position angle changes (e.g., NGC 596 - Schechter and Gunn 1979, and
Williams 1981), or those showing obvious signs of a recent interaction
(e.g. NGC 4125, Bertola 1981, 1983), that show clear differences between
the position angles of the kinematic and photometric major axes. More
minor axis rotation data and dispersion profiles would be valuable.

$\sigma(r)$. Velocity dispersion profiles are shown in Figures 3 and 4,
normalized to the central dispersion σ_0 and to r_e as before, but plotted
on logarithmic scales. Again the galaxies plotted cover a range of
ellipticities and luminosities. Error bars have not been plotted in the
interest of clarity; the point-to-point scatter for each galaxy gives
some indication of the uncertainty involved. Most elliptical galaxies
show $\sigma(r)$ decreasing with radius, as exemplified in Figure 3. This is
the behavior expected for a constant M/L galaxy (see e.g., model fits
for NGC 3379 in Davies and Illingworth 1983a) with an isotropic velocity

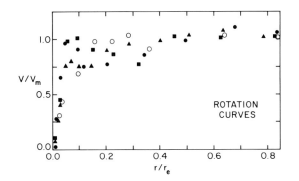

Figure 2. As for Figure 1, but for a sample of ellipticals with rapidly rising rotation curves that are roughly flat beyond the turnover.

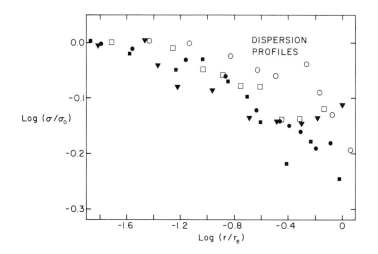

Figure 3. Velocity dispersion profiles for a sample of five ellipticals, normalized by the central dispersion σ_0, and by r_e as in Figure 1, and plotted on logarithmic scales. This sample was selected to have decreasing $\sigma(r)$ profiles.

dispersion tensor. Alternatively, if M/L increases outwards as doctrine would have it, then the radial component σ_r of the velocity dispersion must become increasingly important at the expense of the azimuthal component σ_t such that the observed (line-of-sight projection) dispersion decreases. In terms of the usual anisotropy parameter, $\beta = 1 - (\sigma_t/\sigma_r)^2$, $\beta \to 1$ as the radius increases ($\beta = 0$ for isotropy and $\beta \to -\infty$ in azimuthally anisotropic systems). Such radial anisotropy is not unreasonable given that collapse models of galaxy formation are likely to result in velocity distributions with radially elongated velocity ellipsoids.

While the majority of ellipticals show $\sigma(r)$ profiles that decrease

with radius, a significant fraction have $\sigma(r)$ constant or even
increasing. Several examples of constant $\sigma(r)$ profiles are shown in
Figure 4. The cD galaxy in the center of the cluster A2029 has a $\sigma(r)$
profile that increases significantly with radius (Dressler 1979). The
comparison of his observed dispersion profile with that expected for a
constant M/L model is striking (his Figure 5); at $r\sim 100$ kpc,
$\sigma_{obs} \sim 2\sigma_{model}$.

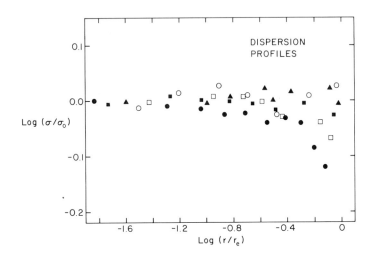

Figure 4. As in Figure 3, but for a sample of ellipticals whose
dispersion profiles are approximately constant with radius. An
interesting feature of this log-log plot is that one galaxy is seen to
have a very rapid change in slope (filled circles). This is unlikely to
be real and probably indicates a problem with sky subtraction at the
very low surface brightness levels involved. On the usual linear $\sigma(r)$
plots this problem was not as apparent.

These flat or slowly rising dispersion profiles are inconsistent
with constant M/L, isotropic-dispersion models. However, as before,
anisotropy changes can mimic M/L changes. In these cases, the azimuthal
component of the dispersion would need to increase at the expense of the
radial component to be consistent with a constant M/L galaxy; i.e., $\beta < 0$
(see e.g., Tonry 1983). Is this physically reasonable? Not if
ellipticals form by the usual dissipational or dissipationless collapse
picture. It is not at all apparent how such a galaxy could come to have
an azimuthally elongated velocity ellipsoid. However, if mergers of
disk galaxies contribute significantly to the elliptical population,
such anisotropy may be quite reasonable, as Tonry (1983) has argued.
There is no theoretical support, though, for this situation. It would
be valuable if N-body models of disk-system mergers that result in
elliptical-like objects with appropriate values of σ and V/σ, could be
investigated to see if $\sigma_t > \sigma_r$ in the body of the model.

In summary, it appears possible for ellipticals to be constant M/L objects. We would require some galaxies to have radially-anisotropic envelopes and some to have azimuthal anisotropy. Whether such a situation stays tenable as individual galaxies are investigated in more detail remains to be answered. For it is clear from the models of flat $\sigma(r)$ profile galaxies (Efstathiou 1982, Tonry 1983) that anisotropy can only compensate for the decrease in σ_{total} out to $r \sim r_e$; beyond that radius σ_{obs} decreases because essentially all the dispersion is in the azimuthal component. Measurement of σ for $r > r_e$ in ellipticals with increasing $\sigma(r)$ would be valuable. Tonry's (1983) fit to Dressler's (1979) A2029 data (his Figure 4) exemplifies the problem. With $r_e \sim 100$ kpc, Tonry managed to fit the dispersion profile with a constant M/L but highly anisotropic model (with $\beta \sim -15$). However, $r_e < 100$ kpc in A2029, being probably more like 70 kpc. In this case, it is likely that the M/L will need to increase with radius, regardless of the amount of anisotropy present. Of course, the actual change in M/L will depend greatly upon the degree of anistropy, being greatest in the isotropic models suggested by Dressler and least in a low β ($\beta \ll 0$) model.

A sufficiently large sample (e.g., Davies et. al. 1983) of elliptical galaxy dispersion profiles now exist that we can investigate the distribution of σ trends with radius (Fried and Illingworth 1983b). For example, by fitting power laws to $\sigma(r)$ we find that $d(\ln \sigma)/d(\ln r)$ ranges from ~ 0.1 to -0.3, i.e., slightly increasing as in A2029 to decreasing as in NGC 3379. The median value lies between 0.0 and -0.1, larger than the ~ -0.15 typical of constant M/L isotropic models. There appear to be no significant trends with ellipticity and luminosity. This, as well as the detailed $\sigma(r)$ and $V(r)$ profiles, will be discussed more extensively in Fried and Illingworth (1983b).

II.   LUMINOSITY DEPENDENCE OF ROTATION PROPERTIES

The usual $V/\bar{\sigma}$ vs ε plot is shown in Figure 5, where V is the maximum rotation velocity, $\bar{\sigma}$ is the mean velocity dispersion within $r_e/2$ ($\bar{\sigma} \sim \sigma_0$, the central dispersion; see Davies et. al. 1983), and ε is the ellipticity. The oblate line defines where rotationally-flattened oblate galaxies fall irrespective of inclination, whereas the prolate line shows the median of the distribution expected for rotationally-flattened prolate galaxies (Binney 1978, Illingworth 1981). The filled symbols are for ellipticals brighter than $M_B = -20.5$. Note that there are very few oblate rotators ($\lesssim$ 15-20%), and that these bright ellipticals are not consistent with being prolate rotators.

The open symbols are for ellipticals fainter than $M_B = -20.5$ (and one likely S0 at $V/\sigma > 1.0$ and $\varepsilon = 0.33$). All these galaxies are consistent with being oblate rotators. This striking result is taken from Davies et. al. (1983). This variation of rotational properties is better demonstrated in Figure 6. Here the ε dependence of V/σ has been removed by normalizing each observed V/σ by that expected for an oblate rotator at the appropriate ε. This parameter,

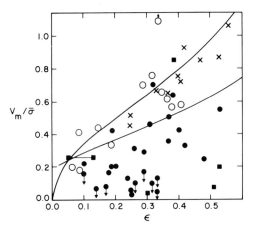

Figure 5. The maximum rotation velocity V_m, normalized by the mean dispersion $\bar{\sigma}$ within $r_e/2$ is plotted against ellipticity ε for 32 ellipticals (filled symbols) brighter than $M_B = -20.5$, and for 11 ellipticals (open circles) fainter than $M_B = -20.5$ (and one probable S0 with $V_m/\bar{\sigma} > 1$). Like data for the bulges of disk galaxies are indicated by crosses (see § IV). The oblate (upper) and prolate (lower) model lines are explained in the text. Arrows indicate upper limits. The two filled connected squares indicate the range of ε seen in NGC 3379 for $r \lesssim r_e$ (Davies and Illingworth 1983a).

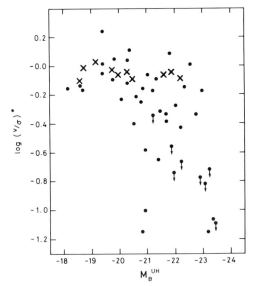

Figure 6. The observed $V_m/\bar{\sigma}$ for each elliptical in Figure 5 has been normalized by V/σ from the oblate line in that figure at the appropriate ellipticity, and plotted, for all M_B, as filled symbols against absolute magnitude derived assuming a uniform Hubble flow model with a Hubble constant $H_0 = 50$ km s^{-1} Mpc^{-1}. Use of Virgocentric flow magnitudes makes no significant difference to the distribution. The crosses are bulges (see § IV).

$(V/\sigma)* = (V/\sigma)_{obs}/(V/\sigma)_{Model} = 1$ for oblate rotators. Log $(V/\sigma)*$ is plotted against absolute magnitude M_B for all the ellipticals of Figure 5. In the mean, a trend is seen in that fainter ellipticals rotate more rapidly, in a dimensionless sense. However, it must be remembered that the scatter for $M_B < -21$ is real, and that the data tell us that slowly-rotating low-luminosity ellipticals are rare. Any formation theory for elliptical galaxies must reproduce the distribution in Figure 6. A first attempt (see Davies et. al. 1983) shows that this constraint is a severe one, particularly for hierarchical clustering collapse models, both with and without dissipation.

III. SHAPES OF ELLIPTICALS

A variety of statistical tests aimed at establishing the intrinsic shapes of ellipticals have been applied to photometric and kinematic data of these galaxies (e.g., Marchant and Olsen 1979, Richstone 1979, Lake 1979, Merritt 1982). Generally, such tests have shown that ellipticals are more likely to be oblate, but with very marginal significance. In addition, a potentially serious problem arises with these tests if there exist correlations between the physical properties of galaxies (e.g., M/L) and their intrinsic flattening, as has been suggested by Terlevich et al. (1981). They found, for example, that the velocity dispersion at a given absolute magnitude appears to be correlated with flattening. More and better data, both photometric and kinematic, are needed before these tests can place statistically significant constraints on the shapes of ellipticals.

For the moment only the distribution and kinematics of gas and dust in ellipticals (see e.g., Bertola and Galletta 1978, Tohline et al. 1982), and the V(r) and $\sigma(r)$ data (see e.g., Davies et. al. 1983) can be used to constrain the shapes of ellipticals. From these latter data we know that ellipticals are not rotationally-flattened isotropic-dispersion prolate objects. For the more luminous ellipticals, at least, little more can be established from their distribution in Figure 5. Their figures could range from prolate through triaxial to oblate, flattened, in all cases, as a result of an anisotropic velocity dispersion tensor. However, we have seen that low luminosity ellipticals are consistent with being oblate rotators. Since the photometric and kinematic properties of ellipticals appear to change smoothly with luminosity, one could argue that there is unlikely to be a dichotomy in the shapes of ellipticals, and that the more luminous galaxies are also oblate, or if triaxial are closer to being oblate than prolate. This argument is not very compelling, being based mainly on Occam's Razor and so should not be given too much weight. Clearly, observational programs on ellipticals should include galaxies with a range of luminosities and should compare, in particular, the properties of those brighter than $M_B = -20.5$ with those of fainter galaxies.

Ellipticals with gas and dust may provide clues as to their structure. Bertola and Galletta (1978) claimed that galaxies with gas

and dust along their minor axes are prolate objects, initiating a lively
and ongoing controversy. These early-type galaxies, generically known
as polar ring galaxies, are characterised by an extensive ring of gas,
dust, and sometimes an obvious young stellar population, whose rotation
axis appears to be orthogonal, or nearly so, to that of the central
galaxy. Examples are NGC 2685, the "Spindle", and Centaurus A, NGC
5128.

While it is clear that stable orbits about the long axis do exist
in prolata with slow or zero figure rotation, such orbits are not unique
to prolate figures. Stable polar orbits do exist in triaxial forms, and
in very mildly triaxial forms at that. A recent paper by Steiman-
Cameron and Durisen (1983) shows that for a nearly oblate figure
flattened by 2:1, i.e., an E5 galaxy, in which the axes in the plane
differ by only 2% (i.e., a:b:c = 1.0:0.98:0.5), fully 15% of any
infalling material will ultimately reside along the minor axis. That
is, gas will dissipate onto the short axis plane from a very significant
fraction (15%) of configuration space for the orbital angular momentum
vector of the infalling material, even though the triaxiality in this
case is very small. Thus, these 'prolate' figures could in reality be
nearly oblate figures. Those in which the infalling material ultimately
dissipates onto the long axis plane become merely 'normal' S0 and Sa-Sb
galaxies.

We cannot establish, of course, that any individual galaxy is a
mildly triaxial, nearly oblate figure. However, we can check the
prolate hypothesis for consistency with the data. There are already
indications that two objects of this class, NGC 2685 (Schechter and Gunn
1978) and NGC 4650A (Ulrich and Schechter 1983), are not prolate. Both
galaxies show large rotation velocities in a dimensionless V/σ sense,
and, in one case, there are indications that the rotation curve turns
over and that the object rotates differentially. Such rapidly rotating
objects, especially if they are rotating differentially, appear to be
quite difficult to construct as stationary prolata. Yet even a quite
slow figure rotation moves the stable orbits away from being polar, and
ultimately, as the figure rotation increases, no stable polar or near
polar orbits exist. Thus, while these polar ring or minor axis gas and
dust-lane galaxies do not provide unambiguous evidence for a particular
shape, they are still a valuable shape diagnostic and further study will
be interesting (see e.g., Fried and Illingworth 1983a, Illingworth and
Davies 1983).

In summary, no data exist that uniquely require any ellipticals to
be prolate or nearly so. Assuming ellipticals to be oblate, or oblate-
triaxial is consistent with available data.

IV. BULGES OF DISK GALAXIES

Illingworth and Schechter (1982) and Kormendy and Illingworth
(1982) obtained velocity and velocity dispersion profiles for 10 bulges

of both S0 galaxies and later-type spirals. All are consistent with
being rotationally-flattened oblate systems (see Figure 5), particularly
when the flattening due to the disk potential is taken into account.
Dressler and Sandage (1983) have further increased the sample of bulges
with a similar result. The dynamical difference between bulges and
ellipticals found by the above authors is now seen to be primarily a
luminosity effect. As the above authors had noted, rapidly rotating
bulges with absolute magnitudes $M_B \gtrsim -20$ were being compared to slowly
rotating ellipticals with $-23 < M_B < -21$. The low luminosity ellipticals
($M_B \gtrsim -20$) studied recently by Davies et. al. (1983) show dynamical
properties similar to bulges (Figure 5 and §II). Are bulges to be
viewed once again as little ellipticals with a disk? Probably, but with
some interesting differences in detail, as we shall see.

Before discussing these differences, I want to emphasize again an
observational problem that can seriously bias velocity and velocity
dispersion data in mixed population systems, e.g., measuring bulge
properties where there is some disk contamination. Whitmore (1980)
showed that velocity dispersion measurements in two component averages
give higher weight to the narrow line component (i.e., the low
dispersion component). Illingworth and Schechter (1982) confirmed this
and found an even stronger effect for velocity measurements (see also
McElroy 1983). They found that velocity measurements in two component
averages typically weighted the narrow line component by a factor 3 or
more, whenever it contributed $> 10\%$ to the average. However, the
weight factor depends upon the relative contribution of the two
components, their velocity dispersions, and also their velocity
difference. Since as little as 10% of the narrow line component can
significantly affect the measurement, one should be very careful, for
example, when determining bulge rotations in the presence of disk
components. If disks are typically 21-22 mag arcsecond^{-2} in the bulge,
then errors can be introduced even at bulge surface brightnesses of 19
mag arcsecond^{-2}, i.e., one will derive dispersions that are too low and
rotation velocities that are too high. While such problems have been
found to occur with the Fourier quotient method, similar effects will
probably be found with the Tonry-Davis (1980) CCF method and others.

For velocity measurements, at least, it should be possible to test
for the effect of the narrow line component by restricting the frequency
range of the fit in the Fourier plane, or equivalently by convolving the
data by an appropriate (e.g., Gaussian) broadening function. Likewise
for velocity dispersions, through considerable care will be required
since one is attempting to derive line widths in this case.

As we mentioned above, are we now to assume that bulges are but
little ellipticals with a disk, or are there still significant
differences? While more data are really needed to answer this question
it does appear that some interesting differences still exist. First we
can try and establish the luminosity dependence of the rotational
properties of bulges, i.e., how does V/σ vary with M_B. Is rotation
dynamically less important for more luminous bulges, and do they show

the scatter in (V/σ)* seen for bright ellipticals ? The answer is a
tentative no. Tentative because luminous bulges are rare. Kormendy and
Illingworth's (1982) bulge sample contained only one, NGC 4594, the
Sombrero. Davies and Illingworth (1983a) discuss a further two examples.
All three have -22.5 < M_B < 21.5, and all show significant rotation.
This can be seen in Figure 8 where the bulges (crosses), unlike the
ellipticals, appear to show no trend with luminosity. However, this is
a small sample. Dynamical data for more luminous bulges would be valuable (see also Dressler and Sandage 1983).

There are a further three areas where bulges may be compared with
ellipticals, but where the data are lacking or ambiguous, and so further
observations are needed. First, as mentioned above, luminous bulges are
rare, suggesting that the luminosity function for bulges differs from
that for ellipticals. This needs to be quantified. Second, more bulge
surface brightness distributions are needed for comparison with those of
ellipticals. Both the direct contribution of the disk to the observed
luminosity distribution and the dynamical effect of the disk's potential
(see e.g. Monet, Richstone and Schechter 1981) must be considered when
making such comparisons. Third, are bulges intrisically more flattened
than ellipticals ? Kormendy and Illingworth (1982) suggested that they
are, as have Dressler and Sandage (1983). However, Boroson's (1981)
surface photometry suggests that they are less flattened. More decomposition studies are required.

There remains one outstanding difference between ellipticals and
bulges. "Box" - or "peanut" -shaped bulges (e.g., NGC 128, the prototypical "peanut" bulge) are common. An example, NGC 1381, an edge-
on SO in Fornax with a "box" -shaped bulge is shown in Figure 9 (see also

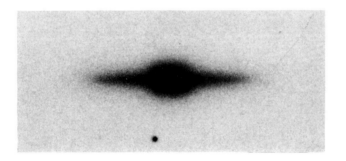

Figure 9. Printed from a IIa-0 + GG13 plate of NGC 1381, an edge-on
SO galaxy in Fornax. The "box" -shape of the bulge is apparent in the
flattening of the bulge isophotes at Z ∿ 10" above the plane. I am
grateful to K. Freeman for this print.

Davies and Illingworth 1983b). However, ellipticals with such a characteristic shape are unknown (to me at least!), while a significant fraction of bulges are "box" -shaped (10-20%? - it would be valuable to
quantify this). This structure is unlikely to be due to any direct dyna-

mical effect of the disk upon the bulge. Such "box" -shaped structure occurs at small radii where the potential would appear to be dominated by the bulge. It has been suggested that the "box" -shaped bulge of NGC 4565 is a bar seen end-on (Jensen and Thuan 1983). Simple morphological arguments regarding the frequency of occurence of bars, their size, and the lack of thick (in z) components of that size in edge-on galaxies indicate that bars, and their associated lenses, are a stellar component that has no appreciable z thickness, i.e., they are "cold" in z, like disks and are therefore highly triaxial (see, e.g., Kormendy 1982).

Another feature of "box" -shaped bulges that distinguishes them from other bulges and ellipticals is that there appears to be a 1:1 correspondence between the existence of "box" - or "peanut" -shaped structure and differential cylindrical rotation, i.e., rotation that is constant, or nearly so, with height z above the plane (Kormendy and Illingworth 1982). Again, the sample is small, namely 3 galaxies but indicative (see Davies and Illingworth 1983a for more discussion). "Normal" bulges and ellipticals show $V(z)$ profiles that decreases with z (Illingworth and Schechter 1982, Kormendy and Illingworth 1982, Davies and Illingworth 1983c). Observational and theoretical work on these "box" -shaped bulges could be very rewarding.

While ellipticals and bulges are similar dynamically, it appears that significant differences in detail do exist.

Much of this discussion is qualitative. This is of necessity in many areas, since little data exist, and our theoretical understanding is even less. However, we have learned much in the last few years. Hopefully we will consolidate these observational gains and improve our understanding of the structure of these galaxies, and the proceses by which they formed.

ACKNOWLEDGMENTS

    I am grateful to numerous colleagues for stimulating discussions. In particular, I would like to thank J. Binney, R. Davies, A. Dressler, G. Efstathiou, S. Faber, M. Fall, K. Freeman, G. Lake, C. Norman, D. Richstone and P. Schechter.

REFERENCES

Bertola, F. : 1981, in S.M. Fall and D. Lynden-Bell (eds.), The
        Structure and Evolution of Normal Galaxies, Cambridge University
        Press, Cambridge, p. 13.
Bertola, F. : 1983, this volume.
Bertola, F. and Galletta, G. : 1978, Astrophys. J. Letters, 226, L115.
Binney, J. : 1978, Monthly Notices Roy. Astron. Soc., 183, 779.
Boroson, T. : 1981, Astrophys. J. Suppl., 46, 177.

Davies, R.L., Efstathiou, G., Fall, S.M., Illingworth, G. and Schechter, P.L. : 1983, Astrophys. J., 266, p. 000.

Davies, R.L. and Illingworth, G. : 1983a, Astrophys. J., 266, p.000.

Davies, R.L. and Illingworth, G. : 1983b, Astron. J., in preparation.

Davies, R.L. and Illingworth, G. : 1983c, Astrophys. J., in preparation.

Dressler, A. : 1979, Astrophys. J., 231, 659.

Dressler, A. and Sandage, A. : 1983, Astrophys. J., in press.

Efstathiou, G. : 1982, private communication.

Fried, J. and Illingworth, G. : 1983a, Astrophys. J. Letters, in preparation.

Fried, J. and Illingworth, G. : 1983b, Astrophys. J., in preparation.

Illingworth, G. : 1981, in S.M. Fall and D. Lynden-Bell (cds.), The Structure and Evolution of Normal Galaxies, Cambridge University Press, Cambridge, p. 27.

Illingworth, G. and Davies, R.L. : 1983, Astrophys. J. Letters, in preparation.

Illingworth, G. and Schechter, P.L. : 1982, Astrophys. J., 256, 481.

Jedrzejewski, R., Davies, R.L. and Illingworth, G. : 1983, Monthly Notices Roy. Astron. Soc., in preparation.

Jensen, E.B. and Thuan, T.X. : 1983, Astrophys. J., in press.

Kormendy, J. : 1982, in L. Martinet and M. Major (eds.), Morphology and Dynamics of Galaxies, Twelth in the Saas-Fee Lecture Series, Geneva Observatory, in press.

Kormendy, J. and Illingworth, G. : 1982, Astrophys. J., 256, 460.

Lake, G. : 1979, in D.S. Evans (ed.), Photometry, Kinematics and Dynamics of Galaxies, Department of Astronomy, University of Texas at Austin, p. 381.

McElroy, D.B. : 1983, Astrophys. J., in press.

Monet, D.G., Richstone, D.O. and Schechter, P.L. : 1981, Astrophys. J. 245, 454.

Marchant, A.B. and Olson, D.W. : 1979, Astrophys. J. Letters, 230, L157.

Merritt, D. : 1982, Astron. J., 87, 1279.

Richstone, D.O. : 1979, Astrophys. J., 234, 825.

Schechter, P.L. and Gunn, J.E. : 1978, Astron. J., 83, 1360.

Schechter, P.L. and Gunn, J.E. : 1979, Astrophys. J., 229, 472.

Steiman-Cameron, T.Y. and Durisen, R.H. : 1983, Astrophys. J. Letters, in press.

Terlevich, R., Davies, R.L., Faber, S.M. and Burstein, D. : 1981, Monthly Notices Roy. Astron. Soc., 196, 381.

Tohline, J.E., Simonson, G.F. and Caldwell, N. : 1982, Astrophys. J., 252, 92.

Tonry, J.L. : 1983, Astrophys. J., 264, p. 000.

Tonry, J.L. and Davis, M. : 1980, Astron. J., 84, 1511.

Ulrich, M.H. and Schechter, P.L. : 1983, Astrophys. J., in preparation.

Whitmore, B.C. : 1980, Astrophys. J., 242, 53.

Williams, T.B. : 1981, Astrophys. J., 244, 458.

DISCUSSION

THUAN :   What is the light distribution along the minor axis in the "bulge" component of the box-shaped galaxies ? Do they follow the $r^{1/4}$ law like "normal" spheroids do ? I know that at least in the case of NGC 4565, the light drop-off along the minor axis is exponential, which is much steeper than the $r^{1/4}$ law and which suggests that the "bulge" component in NGC 4565 is not a "normal" spheroid.

ILLINGWORTH :   The evidence for the light distribution in spheroids being well-fit by an $r^{1/4}$ law is minimal. More data are needed. The light distributions of the "box-shaped" bulges do not appear to differ in a consistent sense from the $r^{1/4}$ law. In the outer parts, the light distributions in NGC 4565 and NGC 4111 differ in opposite senses from an $r^{1/4}$ law, one showing a deficit, the other an excess. The "box-shaped" spheroids are clearly not "normal". However, they are very common, and we need to understand why such shaped objects are only seen in disk galaxies. Why are there no "box-shaped" elliptical galaxies ?

INAGAKI   :   In your diagrams some galaxies have velocity dispersions decreasing inwards. Do you think that these observations are reliable ? It seems to me that it is difficult to construct a dynamical model which has such a velocity dispersion distribution profile.

ILLINGWORTH :   I do not think that there is any statistically significant example of a velocity dispersion profile that decreases inwards near the center of a galaxy. A constant M/L $r^{1/4}$ -law galaxy should show such a decrease for $r \lesssim r_e/20$ (1"-3" for typical ellipticals). However, this type of effect would be very difficult to observe because seeing becomes a problem at such small radii. Changes on larger scales in the sense described would imply either that the M/L is increasing outwards or that the azimuthal component of the velocity dispersion is preferentially increasing at the expense of the radial component.

SELLWOOD :   What type of stellar population are you measuring in the disk of NGC 488 ?

ILLINGWORTH :   This is very difficult to say because the data have quite low signal-to-noise. Disks are low surface brightness objects ! The hydrogen lines do not appear to be strong, yet the metal lines appear to be weaker than in the bulges and ellipticals that we have studied. Our solutions did however give quite acceptable values for the line strength parameter.

RICHSTONE :   Is there any believable evidence for rotation along the minor axis or for velocity dispersion rising with height above the major axis ?

ILLINGWORTH :   The best example is NGC 596, studied extensively by Ted Williams, (published in Astrophys. J. 1981, 244, 458). Others exist but usually they are ellipticals that appear to be disturbed or have some photometrically striking features. "Normal" ellipticals do not show minor axis rotation, in general. However, the sample is small. The few cases that have been studied show that the velocity dispersion is either constant or decreases with distance from the major axis.

SIMKIN :   Although you have not included them in your discussion, I would like to note that the cD radio galaxies for which I have measured absorption line velocities show very large rotations comparable to that found for the disks of Sa galaxies (peak V sini velocities of 150-350 km/s).

ILLINGWORTH :   These objects are quite an enigma. Such large values of V/σ are more consistent with disks than spheroidal systems. If true, they would suggest that the highest radio luminosity cD galaxies are SO's. Further data on these objects would be valuable.

ROTATIONAL VELOCITIES AND CENTRAL VELOCITY DISPERSIONS FOR A SAMPLE
OF S0 GALAXIES

Alan Dressler
Mount Wilson and Las Campanas Observatories of the
Carnegie Institution of Washington

The following discussion is based on a paper of the same title co-
authored with Allan Sandage in the January 1, 1983 Astrophysical Journal.

Central velocity dispersions and rotation curves to radii of ∿5 kpc
have been measured for 32 galaxies, mostly field S0s. Our rotation
curves confirm the result of Kormendy and Illingworth (1982) that the
bulges of S0 galaxies are in rapid rotation, with enough rotational
kinetic energy to account for their flattenings. The V/σ-ellipticity
relation we find for S0 bulges is compared with similar data for

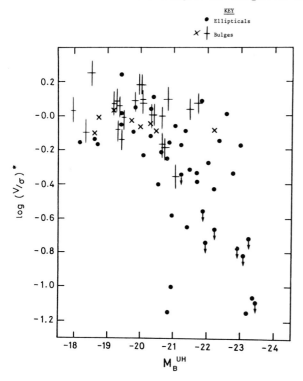

Fig. 1. Davis et
al. Fig. 5 showing
ratio of rotational-
to-pressure support
for ellipticals and
bulges. A value of
$(V/\sigma)* \approx 1$ indicates
an isotropic oblate
rotator. Bulges
from our sample are
shown as crosses.
All bulges, even
luminous ones, are
consistent with
oblate models.

271

E. Athanassoula (ed.), Internal Kinematics and Dynamics of Galaxies, 271–274.

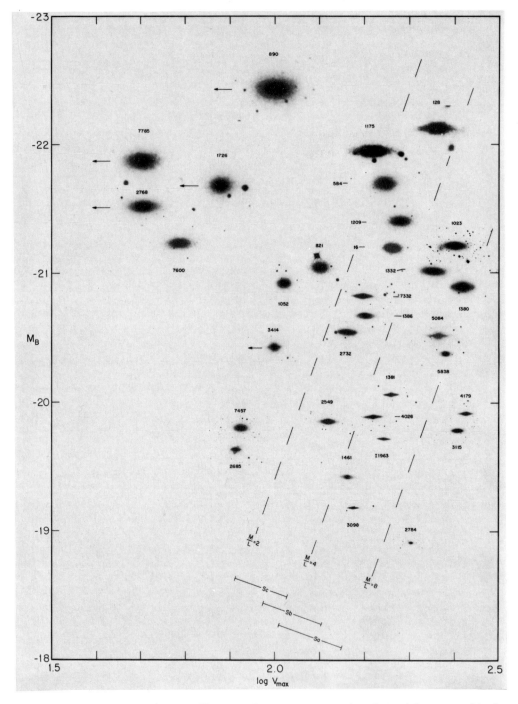

Fig. 2. The Tully-Fisher diagram log V_{max} vs. absolute blue magnitude for field S0s. The position of each galaxy is marked by a photo which indicates its relative size. Representative lines of M/L = constant are shown as well as the region of the diagram occupied by spirals of different Hubble type.

elliptical galaxies from Davies et al. We conclude that (1) faint S0 bulges and elliptical galaxies (M_B fainter than -20.5) are both consistent with oblate rotators with isotropic velocity dispersions (although the S0 bulges in our sample are flatter than ellipticals, on the average) and (2) bright S0 bulges, $-22.0 < M_B < -20.5$ are also consistent with oblate models but bright ellipticals are not. This is demonstrated in a plot of M_B vs. $(V/\sigma)^* \equiv (V/\sigma)_{obs}/(V/\sigma)_{oblate\ model}$ (Fig. 1) from Davies et al. (1982) to which we have added our data for S0 bulges (the crosses). It is clear that all bulges even the luminous ones, have $(V/\sigma)^* \sim 1$ and are thus consistent with oblate, isotropic models. Hence a significant kinematic difference persists between S0 bulges and ellipticals. This difference suggests that disk galaxies and ellipticals did not share a common history of spheroidal formation.

After correcting the measured rotational velocity for the inclination of the S0 disks to the line of sight and for the integration through them we produce the Tully-Fisher velocity-absolute magnitude diagram for our sample. In the Tully-Fisher diagram, Fig. 2, the positions of the galaxies in the diagram are represented by photographs from the POSS that have been enlarged in proportion to the galaxy's redshift (i.e., all are brought to the same distance). Prepared in this way, the diagram shows that galaxy radius has a small spread at a given absolute magnitude, and that surface brightness is relatively constant over this range in luminosity. It is also apparent that bulge luminosity covers a wider range than disk luminosity, therefore bright S0s are bulge dominated while faint S0s are usually disk dominated.

In this diagram, the S0s scatter over the entire region inhabited by spirals of all types. The absence of a tight Tully-Fisher relationship could be due to a combination of (1) M/L variations, (2) contamination of disk light by bulge light, (3) variations from galaxy to galaxy in the run of velocity with radius, i.e., in the dynamical structure, and (4) inclusion of galaxies without true disks in which the internal kinetic energy is then due more to velocity dispersion than to rotation. The large scatter in the Tully-Fisher diagram indicates that, in terms of disk kinematics, field S0s are a heterogeneous class, and are therefore less promising than spirals for mapping perturbations in the local Hubble flow. The S0 class seems to include transition objects from ellipticals to true disk systems without arms. These "diskless S0s" have luminosity profiles that mimic the presence of a disk, but their kinematics are indicative of a rotationally flattened bulge without a component flat enough to be considered a disk.

Finally, we compare the bulge luminosities of S0s with those of elliptical and spiral galaxies, by using the central velocity dispersion as an indicator of the luminosity (and mass) of the spheroidal component. The central velocity dispersions of a representative sample of different Hubble types are presented in histogram form in Fig. 3. The histograms show that the average central velocity dispersions of S0 bulges are intermediate between those of E galaxies and spirals of all Hubble types, indicating that S0 bulges are more massive, on the average, than the bulges of spiral galaxies. This is further evidence that the S0 phenomenon is intrinsic to the Hubble sequence rather than a result of simple stripping of the materials required for star formation from present-day spirals.

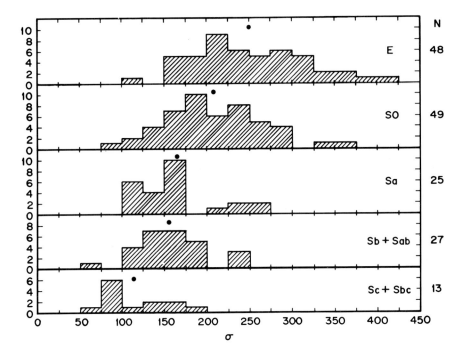

Fig. 3. Histograms showing the central velocity dispersions for a representative sample of different Hubble types.

REFERENCES

Davies, R. L., Efstation, G., Fall, S. M., Illingworth, G., and
    Schechter, P. L.: 1982, preprint The Kinematical Properties of
    Faint Elliptical Galaxies.
Kormendy, J. and Illingworth, G.: 1982, Astrophys. J., 256, 460.

MODELS OF ELLIPTICALS AND BULGES

James Binney
Department of Theoretical Physics,
Oxford, OX1 3NP, England

1. INTRODUCTION

During the last seven years we have become much more aware of the importance of velocity anisotropy in spheroidal components. There were never any sound arguments for assuming that the velocity ellipsoids in spheroidal components would be spherical, but the mathematical convenience of this assumption is such that velocity anisotropy was either absent from or unimportant in the models that seemed so promising at the last Besancon meeting in 1974. With the advent of accurate velocity information from absorption-line studies of early-type systems, it became apparent that the real world is a good deal more complex than it might have been, and the theoretical situation is now less satisfactory than it seemed in 1974. All I can do here is to report on our somewhat painful efforts to pick ourselves up from the floor to which the observers knocked us in 1975-7.

2. SPHERICAL SYSTEMS

As is well known, the most general distribution function for a system that is spherically symmetric in all its properties, is a function $f(E,L)$ of the specific stellar energy $E=\frac{1}{2}v^2+\Phi$ and angular momentum $L=|\mathbf{r}\mathbf{x}\mathbf{v}|$, and the velocity ellipsoids in the galaxy are everywhere spherical if and only if $(\partial f/\partial L)\equiv 0$. Numerical simulations of the relaxation of spherical star clusters from Gott (1973) to van Albada (1982) have tended to show that while the velocity ellipsoids at the centres of the final equilibrium systems are spherical, those near the periphery are elongated along the local radial direction. Thus in these systems $(\partial f/\partial L)\neq 0$, and we should be wary of assuming that the distribution functions of spherical galaxies have $f(E)$.

E. Athanassoula (ed.), Internal Kinematics and Dynamics of Galaxies, 275–283.

However, until recently we have tended to think in terms of
models whose distribution functions depend on E only. A nice
illustration of how misleading this can be is furnished by M87.
Young et al (1978) and Sargent et al (1978) obtained CCD
photometry and long-slit spectroscopy of this galaxy and
interpreted their observations on the assumption that the
velocity dispersion is everywhere isotropic. They concluded that
the ratio M(r)/L(r) of the mass contained interior to radius r to
the light inside r, rises steeply from $M/L_V < 9$ at r>600 pc to
$M/L_V > 60$ at r<200 pc. However, if one drops the assumption of
velocity isotropy, these same observations are consistent with a
constant mass-to-light ratio $M/L_V = 7.6$ (Binney and Mamon 1982).
Furthermore, it can be argued that the radial variation of the
anisotropy parameter $\beta \equiv (1 - \sigma_\theta^2 / \sigma_r^2)$ that is implied by the
assumption of constant mass-to-light ratio in M87, is of the same
type as one would have predicted from the theory of Tremaine et
al (1975) that galactic nuclei are formed as a result of massive
globular clusters becoming trapped in galactic centres through
the action of dynamical friction.

Mamon and I only showed that the first moment of the Vlasov
equation can be satisfied by a constant M/L model of M87. We did
not prove that the Vlasov equation can itself be satisfied.
Unfortunately we do not yet know how to find a distribution
function f(E,L) that generates given surface density and velocity
dispersion profiles $\Sigma(R)$ and $\sigma_v(R)$. The different information
contents of one function of two variables f(E,J), and two
functions $\Sigma(R)$ and $\sigma_v(R)$ of one variable, suggests that if any
non-negative f(E,L) generates the given profiles, many other
distribution functions will also be possible. But it is not clear
under what circumstances no non-negative distribution function is
compatible with a set of data. Duncan and Wheeler (1980) and
Tremaine and Ostriker (1982) have tackled the problem of
interpreting the brightness and velocity dispersion profiles of
M87 and M31 from this more demanding point of view.

3. AXISYMMETRIC SYSTEMS

3.1 Systems with $f(E,L_z)$

The classical model of an axisymmetric galaxy (e.g. Wilson 1975)
has a distribution function $f(E,L_z)$ that depends on energy and
the component L_z of angular momentum along the symmetry axis.
These models are often referred to as "isotropic" because the
velocity ellipsoids cut meridional planes in circles (see Fig.

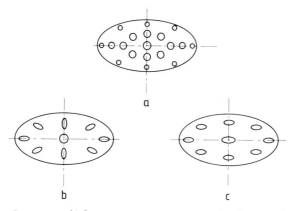

Figure 1. Possible arrangements of the velocity ellipsoids in the meridional plane of an axisymmetric galaxy.

1a). Models of this type have been at a discount since Bertola and Capaccioli (1975) and Illingworth (1977) demonstrated that most giant elliptical galaxies are not flattened by rotation. Actually there has never been any hard evidence that giant elliptical galaxies cannot be modelled in this way: Any model based on $f(E,L_z)$ immediately gives rise to a family of models in which the distribution functions $f(E,L_z)$ differ from each other only in the parts that are odd in L_z. Since the rotation speed of a model is proportional to the part of the distribution function that is odd in L_z and does not contribute to the density distribution, all these models have the same density, but among them are models that have very small or even zero rotation rates.

Personally I have always been strongly prejudiced against models of this type since I can see no obvious way of ensuring that $f=f(E,L_z)$, while flattened but non-rotating ellipticals with $f{\neq}f(E,L_z)$ are a natural consequence of either the Zel'dovich (1970, 1978) pancake theory of galaxy formation (Binney 1976), or the merger picture of the formation of ellipticals (White, this symposium). However two recent developments give pause for thought:

    (1) Frenk and White (1980) have shown that the velocities of the galactic globular clusters are nearly isotropically distributed, rather than being strongly biased around the radial direction as are the velocities of the RR-Lyrae stars (Woolley 1978). Freeman (this symposium) cautions us against assuming blindly that globular clusters are necessarily typical of the spheroidal component, but the conclusion of Frenk and White shows that velocity anisotropy does not always play an important role in spheroidal systems.

(2) The observations of early-type disk galaxies and of low luminosity ellipticals that have been reviewed by Illingworth (this symposium), show that if the velocity ellipsoids in these galaxies are not spherical, they must be elongated along the radial directions (see Fig. 1b) so that anisotropy does not make an appreciable contribution to the flattenings of these system.

Jarvis (1981) has recently fitted models based on $f(E,L_z)$ to observations of the disk galaxies NGCs 4594, 7123 and 7814. He models the bulges of these galaxies as systems with $f=f_0[\exp(-E/\sigma^2)-1]\exp(\Omega L_z/\sigma^2)$ that are placed in the disk-like gravitational potential $\Phi_d(R,z)=-GM_d\{R^2+[a+(z^2+b^2)^{1/2}]^2\}^{-1/2}$ that was introduced by Miyamoto and Nagai (1975). Figure 2 illustrates the effect of a disk with a quarter of the spheroid's mass on the position of the spheroid in the usual v_m/σ_0 diagram.

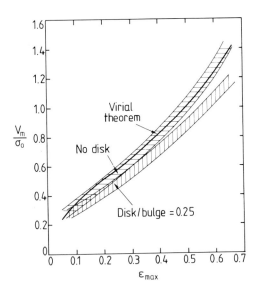

Figure 2. Maximum rotation speed over central velocity dispersion for Jarvis' models. The models with and without a disk lie in the shaded areas, while the full curve is the ratio of global parameters that one obtains from the virial theorem.

By adjusting the five parameters of his models, Jarvis is able to fit extensive photometric and kinematic data for the earliest of his galaxies, NGC 7814, extremely well. Frankly I find the quality of Jarvis' fits disconcerting since, as Hunter (1977) has pointed out, any model with a distribution function such as that of Jarvis, of the form $f(E,L_z)=g(E)h(L_z)$, will tend to become spherical near the centre. Furthermore, it is straightforward to show that near the centre of Jarvis' models, the spheroid must rotate as a solid body as regards the variation of $\langle v_\phi \rangle$ with both radius and height above the disk. My guess is that neither the ellipticity profiles nor the velocity fields of bulges have these characteristics. Unfortunately, the disks of Jarvis' galaxies effectively obliterate the central regions. It will be interesting to see whether rotationally-flattened low-luminosity ellipticals are consistent with distribution functions of the form $g(E)h(L_z)$.

3.2 Systems with $f(E,L_z,I_3)$

The great majority of orbits in potentials like those of disk galaxies admit a third integral of motion in addition to L_z and E (eg. Martinet and Mayer 1975). There is no unique form for the third integral since, given any third integral I_3, any non-trivial function $I_3'(E,L_z,I_3)$ yields a new third integral I_3'. However, the potentials of spheroidal components are usually fairly spherical, and in this case it is natural to consider I_3 to be a generalization of the magnitude $L \equiv |\mathbf{r} \mathbf{x} \mathbf{v}|$ of the angular momentum vector (Saaf 1968, Innanen and Papp 1977, Richstone 1982). Adopting the convention that I_3 is the natural generalization of L, consider the general structure of models based on the following simple distribution functions.

(i) $\quad f = f_K(E)\exp[\frac{\Omega L_z}{\sigma^2} - (\frac{I_3}{r_a\sigma})^2]$

(ii) $\quad f = f_K(E)\exp[-(I_3{}^2-L_z{}^2)/(r_a\sigma)^2]$

(iii) $\quad f = f_K(E)\exp[\frac{\Omega L_z}{\sigma^2} - (I_3-L_z{}^2)/(r_a\sigma)^2]$

where $f_K=f_o[\exp(-E/\sigma^2)-1]$ is King's (1966) distribution function, and Ω and r_a are parameters.

A model built around the first of these distribution functions will rotate with central angular speed Ω like the models of Prendergast and Tomer (1970), Wilson (1975) and Jarvis (1981), while having radially elongated velocity ellipsoids as in the models of Michie and Bodenheimer (1963). Figure 1b is a caricature of a model of this type. Such models may account for observations of rotationally-flattened spheroidal components and globular clusters. Lupton and Gunn (in preparation) have constructed models of this type and fitted them to observations of globular clusters, by assuming $I_3=L$. Petrou (1982 and this symposium) has constructed models in which a more sophisticated approximation to I_3 is employed.

Since the distribution function (ii) is even in L_z, it leads to models which do not rotate. However, these models will be flattened because the argument of the exponential is approximately equal to $[-(L_x^2+L_y^2)]$, and so orbits that carry stars far from the equatorial plane will be depopulated. Clearly we can set a model of this type rotating by adding a component to f that is odd in L_z as in (iii). Petrou (1982) has used a distribution function akin to (iii) to build models which are flattened by anisotropy rather than rotation.

Two problems that should not be difficult to solve, but have been outstanding for some years, are (a) to find the distribution function that generates a realistic spheroidal system with a flat rotation curve, and (b) to find the distribution function that generates a box shaped bulge like that of NGC 128 (p.7 of the Hubble Atlas).

3.3 Schwarzschild's Method

Schwarzschild (1979) introduced a technique into galactic dynamics which enables one to construct a model with a predetermined density distribution, without assuming anything about non-classical integrals such as I_3. In Schwarzschild's technique, one chooses a convenient potential and then uses the computer to calculate a library of orbits in this potential. Linear programming techniques are then used to determine whether these orbits can be populated in such a way as to generate the initially assumed potential.

This technique has so far only been applied by Schwarzschild to the construction of triaxial systems, and by Richstone (1980 and this symposium) and Meys et al (this symposium) to the construction of rather special scale-free models. It is a pity that nobody has yet used Schwarzschild's method to construct a realistic axisymmetric model, since the technique is very well suited to this problem, and the labour involved, though considerable, is much less than that involved in the construction of Schwarzschild's triaxial models.

4. TRIAXIAL SYSTEMS

Over the last four years a wide range of triaxial equilibrium stellar models have been published. Aarseth and Binney (1978) and

Wilkinson and James (1982) have described n-body models in which the triaxial figure is stationary in space, while Schwarzschild (1979) has used his technique to construct a model of this type around a predetermined Hubble-like density profile. Wilkinson and James (1982) and Schwarzschild (1982) have shown that anisotropy-supported bars of this type can be generalized to include figure rotation. Hohl and Zang (1979) and Miller and Smith (1979) have shown that rapidly rotating clouds of stars invariably relax to tumbling bars. The product of a galaxy merger (White, this symposium) is oblate if the galaxies spiral together from a large impact parameter encounter, and prolate if the galaxies collide head-on and their spins are dynamically unimportant. A head-on collision between galaxies with suitably aligned spins can generate a system which is part prolate and part oblate (Gerhard 1982a) and may not even settle to a state that is steady in a suitable rotating frame of reference (Gerhard 1982b).

Thus numerical work indicates that triaxial equilibria are the outcome of a wide variety of initial conditions. Attempts to determine whether elliptical galaxies are more often prolate or oblate (Marchant and Olson 1979, Richstone 1979, Lake 1979, Merritt 1982) have yet to produce a definite result for want of sufficient photometric and spectroscopic data. At the moment the best hope of pinning down the shapes and figure rotation speeds of ellipticals seems to lie in gas in and around these galaxies. Westerbork observations (Knapp, this symposium) indicate that the velocity fields of many of these gas features are remarkably regular. This suggests that each element of gas is moving on a closed orbit. The spatial and velocity structure of these orbits should betray the figure and rotation speed of the underlying potential (Binney 1978, 1981, van Albada et al 1981, Heisler et al 1982, Magnenat 1982, Tohline and Durisen 1982).

REFERENCES

Aarseth, S.J. & Binney, J.J., 1978. M.N.R.A.S., 185, 227.
Binney, J.J., 1976. M.N.R.A.S., 177, 19.
_____, 1978. M.N.R.A.S., 183, 799.
_____, 1981. M.N.R.A.S., 196, 455.
Binney, J.J. & Mamon, G.A., 1982. M.N.R.A.S., 200, 361.
Bertola, F. & Capaccioli, M., 1975. Ap. J., 200, 439.
Duncan, M.J. & Wheeler, J.C., 1980. Ap. J. Lett., 237, L27.
Frenk, C.S. & White, S.D.M., 1980. M.N.R.A.S., 193, 295.
Gerhard, O.E., 1982a. M.N.R.A.S. in press.
_____, 1982b. Ph.D. Thesis University of Cambridge.
Gott, J.R., 1973. Ap. J., 186, 481.

Heisler, J., Merritt, D. & Schwarzschild, M., 1982. Ap. J. 258, 490..

Hohl, F. & Zang, T.A., 1979. Astron. J., 84, 585.

Hunter, C., 1977. Astron. J., 82, 271.

Illingworth, G., 1977. Ap. J. Lett., 218, L43.

Innanen, K.A. & Papp, K.A., 1977. Astron. J., 82, 322.

Jarvis, B.J. 1981. Ph.D. thesis, Australian National University.

King, I.R., 1966. Astron. J., 76, 64.

Lake, G., 1979. Photometry, Kinematics and Dynamics of Galaxies, ed D.S. Evans, pp 381 (Austin: Department of Astronomy, University of Texas at Austin).

Magnenat, P., 1982. Astron. Astrophys., 108, 89.

Marchant, A.B. & Olson, D.W., 1979. Ap. J. Lett., 230, L157.

Martinet, L. & Mayer, F., 1975. Astron. Astrophys., 44, 45.

Michie, R.W. & Bodenheimer, P.H., 1963. M.N.R.A.S., 126, 269.

Miller, R.H. & Smith, B., 1979. Ap. J., 227, 407.

Miyamoto, M. & Nagai, R., 1975. Publ. Astron. Soc. Japan, 27, 533.

Petrou, M., 1982. M.N.R.A.S., in press.

Prendergast, K.H. & Tomer, E., 1970. Astron. J., 75, 674.

Richstone, D.O., 1979. Ap. J., 234, 825.

_____, 1980. Ap. J., 238, 103.

_____, 1982. Ap. J., 252, 496.

Saaf, A.F., 1968. Ap. J., 154, 483.

Sargent, W.L.W., Young, P.J., Boksenberg, A., Shortridge, K., Lynds, C.R. & Hartwick, F.D.A., 1978. Ap. J. 221, 731.

Schwarzschild, M., 1979. Ap. J., 232, 236.

_____, 1982. Ap. J. in press.

Tohline, J.E. & Durisen, R.H., 1982. Ap. J., 257, 94.

Tremaine, S.D. & Ostriker, J.P., 1982. Ap. J., 256, 435.

Tremaine, S.D., Ostriker, J.P. & Spitzer, L., 1975. Ap. J., 196, 407.

van Albada, T.S., Kotanyi, C.G. & Schwarzschild, M., 1981. M.N.R.A.S., 198, 303.

Wilkinson, A. & James, R., 1982. M.N.R.A.S., 199, 171.

Wilson, C.P., 1975. Astron. J., 80, 175.

Woolley, R., 1978. M.N.R.A.S., 184, 311.

Young, P.J., Westphal, J.A., Kristian, J., Wilson, C.P. & Landauer, F.P., 1978. Ap. J., 221, 721.

Zel'dovich, Ya.B., 1970. Astron. Astrophys., 5, 84.

_____, 1978. The Large-Scale Structure of the Universe, ed M.S. Longair & J. Einasto, pp 409. (D. Reidel Publ. Co., Dordrecht, Holland)

DISCUSSION

RICHSTONE : In the work with Mamon on M87 you used a hydrostatic
approach. Don't you think considerations of the collisionless Boltzmann
equation make the large jumps in your anisotropy parameter (as a function
of radius) somewhat unreasonable ?

BINNEY : At present Jean's moment equations provide the only flexible
framework within which to analyse observational data. Of course, one
would prefer to think about observations in terms of an algorithm that
generates distribution functions $f(E,L)$ that are compatible with given
observations, but we don't have such a treasure. So Mamon and I thought
we would probe the limits of what can be achieved with Jean's equations,
and emphasize the danger of assuming with Sargent et al that β can be
simply set to zero.
The following simple argument shows that very rapid changes in β are
possible in principle : in the portion of an unconfined Michie model in
which $\beta \simeq 1$, the density $\rho(r) \simeq \rho_1 (r_1/r)^{3.5}$. If we set this system in
the low-density core of a giant elliptical we will have

$$\beta(r) \simeq 1 - \rho_2 \sigma_\theta^2 \left[\rho_1 s^2 \left(\frac{r_1}{r}\right)^{3.5} + \rho_2 \sigma_r^2 \right]^{-1}$$

where ρ_2, σ_θ and σ_r are the radius-independent parameters of the elliptical envelope, and s is the radially-directed velocity dispersion
towards the outside of the Michie model. If one now sets $\rho=\rho, s= \sigma_r$
and $\beta(\infty) = 0.4$, one finds that $\beta(0.75r_1) = 0.84$ and $\beta(1.5r_1) = 0.52$.
When account is taken of the potential generated by the elliptical
envelope, the density of the anisotropic core will fall more steeply
than $\rho \sim r^{-3.5}$, and so β will decrease even more rapidly than in this
naive model. In our model of M87, β changes from 0.85 at 100 pc to 0.4
at 200 pc.

INAGAKI : Is it possible,or even easy to construct dynamically stable
models with velocity dispersion decreasing inwards ?

BINNEY : An example of a system of this type is the spherical galaxy
with $f(E)$ that obeys the $R^{1/4}$ law in projection (Mon. Not. R. Astr. Soc.
200, 951). Antonov's work (Vestnik Leningrad Univ. No 19 : 96) shows
that this system is stable to all types of perturbation, and σ decreases
interior to $0.07R_e$. It is easy to construct systems of this sort by
inserting cold quasi-isothermal models into hot models of the same type
(Mon. Not. R. Astr. Soc. 190, 873).

THE STABILITY OF AXISYMMETRIC GALAXY MODELS WITH ANISOTROPIC DISPERSIONS

Tim de Zeeuw, Marijn Franx, Jacques Meys, Karel Brink and
Harm Habing
Sterrewacht Leiden

The possible equilibrium configurations for elliptical galaxies and the bulges of spiral galaxies are no longer thought to be confined to the small class of axisymmetric systems with isotropic velocity dispersions. Many of these systems are not supported by rotation but instead by anisotropic velocity dispersions which are maintained by nonclassical integrals of motion (e.g., Binney 1978), and may be triaxial. Few theoretical models for such systems exist (Schwarzschild 1981). Here we discuss some axisymmetric models that we have constructed by means of Schwarzschild's (1979) selfconsistent method, and in particular their stability.

We consider oblate spheroidal models in which the density is constant on spheroids of fixed eccentricity e, and is given by a power-law

$$\rho(\tilde{\omega}, z) = \rho_n s^n = \rho_n \{\tilde{\omega}^2 + z^2/(1 - e^2)\}^{n/2}, \qquad 0 \le s \le a, \qquad (1)$$
$$= 0, \qquad a < s.$$

Here $(\tilde{\omega}, \phi, z)$ are cylindrical coordinates, with the z-axis along the symmetry axis. The corresponding gravitational potential can easily be derived. Within the boundary of the model the equations of motion for this potential are invariant under the scaling transformation

$$\tilde{\omega} \to \alpha\tilde{\omega}, \qquad\qquad z \to \alpha z, \qquad\qquad (2)$$
$$\phi \to \phi, \qquad\qquad t \to \alpha^{-n/2} t.$$

This scaling property greatly reduces the numerical work necessary to obtain a complete catalog of orbits within the model. It is sufficient to compute orbits for only one energy; orbits at all other energies can be obtained by simple scaling. This property of power-law density distributions was used by Richstone (1980) for the special case of the logarithmic potential ($n = -2$).

The results presented below are for the case $n = -1.8$, $e = 0.91$. The corresponding mass model is thought to be representative of the bulge of our own galaxy, up to about 300 pc from the center (e.g., Isaacman 1980).

285

E. Athanassoula (ed.), Internal Kinematics and Dynamics of Galaxies, 285–286.

A complete catalog of orbits was used to construct a number of equilibrium models with Schwarzschild's linear program. A model consists of a
set of orbits with positive occupation numbers which reproduces the density distribution in which the orbits were calculated. For each model we
have the freedom to choose the fraction of retrograde stars in each orbit
since in an axisymmetric model there is no distinction between retrograde
and direct motion around the symmetry-axis. All orbits turn out to have
three isolating integrals of motion, so that the resulting models have
anisotropic velocity dispersions. The primary difference from Richstone's
(1980) models is the existence of a definite boundary, so that the problem
becomes inherently two-dimensional.

We have used van Albada's axisymmetric N-body program (van Albada
and van Gorkum 1977) to test the stability of the equilibrium models
against axisymmetric perturbations. We find that models which have a
local radial velocity dispersion below a certain critical value are unstable to such perturbations. We suspect that a generalized stability
criterion for the local, anisotropic, velocity dispersion may exist,
analogous to the result for infinitely thin disks (e.g., Toomre 1964).
This stability of course depends only on the orbits that constitute the
model, not on the fraction of retrograde stars in each orbit.

The models that are stable against axisymmetric perturbations may
still be unstable to bar-like perturbations. This instability depends on
the total angular momentum of the model (Ostriker and Peebles 1973), so
now the fraction of retrograde stars is important. Using Vandervoort's
(1982) refinement of the Ostriker-Peebles criterion, we find that some of
our models should be stable even if all stars are direct, but that other
models can be made stable only when a fraction of the stars is retrograde.
We are currently testing Vandervoort's criterion with three-dimensional
N-body calculations.

The stable models still show a wide variety in observable properties.
The combination of linear programming and N-body methods used here will
allow us to investigate the full range of stable dynamical models corresponding to the given density distribution.

REFERENCES

van Albada, T.S. and van Gorkum, J.H.: 1977, *Astron. Astrophys.*, 54, 121.
Binney, J.J.: 1978, *Comments on Astrophysics*, 8, 2.
Isaacman, R.B.: 1980, Ph. D. Thesis Leiden University.
Ostriker, J.P. and Peebles, P.J.E.: 1973, *Astrophys. J.*, 186, 467.
Richstone, D.O.: 1980, *Astrophys. J.*, 238, 103.
Schwarzschild, M.: 1979, *Astrophys. J.*, 232, 236.
Schwarzschild, M.: 1981, in *The Structure and Evolution of Normal
    Galaxies*, eds. S.M. Fall and D. Lynden-Bell (London, Cambridge
    University), p. 27.
Toomre, A.: 1964, *Astrophys. J.*, 139, 1217.
Vandervoort, P.O.: 1982, *Astrophys. J. Letters*, 256, 41.

THREE-DIMENSIONAL ORBITS IN TRIAXIAL GALAXIES

Louis Martinet
Observatoire de Genève

Tim de Zeeuw
Sterrewacht Leiden

Many elliptical galaxies may be slowly rotating, moderately triaxial systems. Their dynamics are probably characteristic of a 1:1:1-resonance between the frequencies of oscillation along the three principal axes.

For a study of the 1:1:1-resonance it suffices to consider a model potential of the form

$$V = \tfrac{1}{2}\kappa_1^2\, x^2 + \tfrac{1}{2}\kappa_2^2\, y^2 + \tfrac{1}{2}\kappa_3^2\, z^2 + \tfrac{1}{4}\{C_1 x^4 + C_2 x^2 y^2 + C_3 x^2 z^2 + C_4 y^4 + C_5 y^2 z^2 + C_6 z^4\}$$

where the three harmonic frequencies κ_1, κ_2 and κ_3 are assumed to be nearly equal. A general analytic investigation of motion in such a potential, nonrotating as well as rotating, has been made by de Zeeuw (1982). We are preparing a detailed comparison of these results with numerical computations (Martinet and de Zeeuw, in preparation). Here we limit ourselves to an inventory of the three-dimensional simple periodic orbits that may occur, and a discussion of their importance for real galaxies. (The orbits in the equatorial plane of triaxial models have been already considered by, e.g., de Zeeuw and Merritt (1982) and Martinet (1982)).

First we consider the case of no rotation. We start with κ_1, κ_2 and κ_3 as well as all coefficients C_i (i=1,...,6) equal to one. Three different types of three-dimensional simple periodic orbits exist for all energies:
  (i) orbits which are nearly straight lines through the center, inclined to all principal axes;
  (ii) elliptic orbits lying nearly in a plane containing one of the principal axes and tilted relative to two principal planes;
  (iii) plane elliptic orbits that intersect none of the principal axes.

Making the system triaxial by changing κ_2 and κ_3 we find that these three types of orbits still exist, but not for all energies. The larger the difference between κ_1, κ_2 and κ_3, the further away from the center these orbits first occur. The existence range also depends on the C_i, of course. Some of the periodic orbits are stable, others are unstable in one perpendicular direction or in two independent perpendicular

287

E. Athanassoula (ed.), Internal Kinematics and Dynamics of Galaxies, 287–288.

directions, depending on the values of the parameters in the potential V. In all cases considered, orbits of type (iii) are stable.

In the rotating case, direct and retrograde motion are distinct due to the Coriolis force. Orbits that in the absence of rotation are in the principal planes perpendicular to the equatorial plane, now tip out of these planes and become true three-dimensional orbits as Magnenat (1982) and Heisler et al (1982) have already found for other potentials. We find that all three-dimensional periodic orbits mentioned above also have rotating counterparts. Their stability is influenced by rotation. Some of the orbits now are complex unstable (see Magnenat 1982 for a definition). However, in all cases considered the retrograde generalization of the type (iii) orbit is stable. Direct stable tilted orbits may also exist in small ranges of energy.

The above results show that the interpretation of dust lanes observed in elliptical galaxies as gas and dust orbiting in certain stable periodic orbit families in a triaxial potential (van Albada, Kotanyi and Schwarzschild, 1982) may be more complicated than envisaged up till now. The existence of the stable orbits of type (iii) is particularly relevant here. Numerical work is in progress to investigate this point in more detail.

Not all periodic orbits described here need exist in a realistic triaxial potential, and certainly not at all energies. In nearly ellipsoidal potentials some of the tilted three-dimensional orbits do not occur at all. We observe that doubly, or complex, unstable orbits are in general accompanied by large stochastic regions in phase-space. Potentials in which these orbits do not exist are therefore likely to admit three isolating integrals of motion for most orbits. At the same time, the presence of type (iii) orbits, with their associated family of tube orbits, provides an additional building block for the construction of triaxial models (Schwarzschild, 1979).

REFERENCES

van Albada, T.S., Kotanyi, C.G. and Schwarzschild, M.: 1982, *M.N.R.A.S.*, 198, 303.
Heisler, J., Merritt, D.R. and Schwarzschild, M.: 1982, *Astrophys. J.*, 258, 490.
Magnenat, P.: 1982, *Astron. Astrophys.* 108, 89.
Martinet, L.: 1982, Proceedings of the CECAM Workshop on *Structure, Formation and Evolution of Galaxies*, held in Paris, August 1981 (eds. J. Audouze and C.A. Norman).
Martinet, L. and de Zeeuw, P.T.: 1983, in preparation.
Schwarzschild, M.: 1979, *Astrophys. J.* 232, 236.
de Zeeuw, P.T.: 1982, Proceedings of the CECAM Workshop on *Structure, Formation and Evolution of Galaxies*, held in Paris, August 1981 (eds. J. Audouze and C.A. Norman).
de Zeeuw, P.T. and Merritt, D.R.: 1982, submitted to *Astrophys. J.*

SCALE-FREE MODELS OF ELLIPTICAL GALAXIES

Douglas O. Richstone
Department of Astronomy, University of Michigan

In this note I will report the general character of all scale-free solutions of the collisionless Boltzmann and Poisson equations for a specific shape. In this case, the density falls as r^{-2} along any radial ray, the potential is logarithmic with oblate spheroidal level surfaces, and the flattening corresponds to an E5 or E6 density distribution (see Richstone 1980, hereafter Paper I). Schwarzschild's (1979) method was used. Toomre (1982) has recently discussed the properties of scale free models of this sort with distribution functions dependent on only the two classical isolating integrals. The method used here automatically incorporates the third integral.

The linear programming techniques used here in determining the orbit occupation numbers have certain noteworthy features. First, the models produced are "basic" (have the minimum possible number of orbits to produce a solution and lie on the very edge of the solution set. This may mean that they are astrophysically implausible because they are almost impossible. Second, the solution set is convex, hence, a number of basic solutions delineate the <u>complete</u> solution set, and any solution can be constructed via a weighted mean of basic solutions. Completeness of the solution set is achieved only if the orbits surveyed have complete or representative coverage of phase space (Richstone 1981).

With the above remarks in mind we display overleaf the projection of the complete solution set on to planes defined by the total z axis angular momentum ℓ_z and the non-trivial first and second moments of the velocity distribution. All orbits are assumed circulating in the same sense. All moments are computed on a thin spherical shell any distance from the center. The cross-hatched region on each plot corresponds to the ordinate label on the right hand side of the plot. The moment subscripts correspond to the usual cylindrical coordinates. The dispersion tensor is approximately aligned with the cylindrical coordinates.

These plots indicate that the choice of total ℓ_z <u>approximately</u>

E. Athanassoula (ed.), Internal Kinematics and Dynamics of Galaxies, 289–290.
Copyright © 1983 by the IAU.

determines the other global dynamical parameters, but that there is
additional freedom.   The velocity ellipsoid is always approximately
aligned with the principal coordinate axes.   At low ℓ_z the third
integral plays an important role in flattening the models.   The models
do locally satisfy a tensor virial theorem, but as pressure stresses are
very important Binney's (1981) estimate of an anisotropy parameter as a
function of v/σ doesn't work for them.

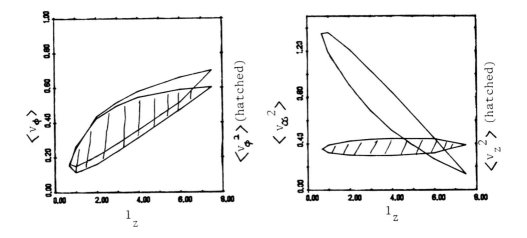

        Like their predecessor in Paper I, the low ℓ_z models described here
exhibit a tendency toward increasing <u>observed</u> first and second velocity
moments with distance from the equator.   This effect may not be present
in non basic low ℓ_z models, and is not a common feature of high ℓ_z
models.   Nonetheless, variation of $\langle v_{obs} \rangle$ and $\langle v_{obs}^2 \rangle$ with latitude may
be a valuable clue to the importance of the third integral in galaxian
structure.

REFERENCES

Binney, J. J.: 1981, in The Structure and Evolution of Normal Galaxies
        (ed. by Fall and Lynden-Bell) Cambridge University Press, pp. 55.
Richstone, D. O.: 1980 , Astrophys.   J. 238, pp.103.
                    : 1982, Astrophys. J. 252, pp.496.
Schwarzschild, M.: 1979, Astrophys. J. 232, pp.236.
Toomre, A.: 1982, Astrophys. J. 259, pp.535.

AXISYMMETRIC MODELS OF ELLIPTICAL GALAXIES WITH ANISOTROPY

Maria Petrou
Institute of Astronomy, Madingley Road,
Cambridge CB3 OHA, England

It is well known by now that models of elliptical galaxies based on Jeans theorem require the inclusion of a third integral in the distribution function.

When we assume axial symmetry, we have two exact integrals of the motion, the energy E and the angular momentum about the symmetry axis h_z. If we had spherical symmetry, the third integral would have been the modulus of the total angular momentum. In the case of axial symmetry, a plausible assumption is that the third integral is a generalization of the total angular momentum, say of the form:

$$I_3 = h^2/2 + g(r,\vartheta) \tag{1}$$

For I_3 to be an exact integral, $g(r,\vartheta)$ must satisfy the two equations

$$g = g(\vartheta) \tag{2}$$

$$g = -r^2 \Psi(r,\vartheta) + C(r) \tag{3}$$

where $\Psi(r,\vartheta)$ is the potential of the system and $C(r)$ a function of r. It can be shown that this happens only if $\Psi(r,\vartheta)$ can be written in the form

$$\Psi(r,\vartheta) = F_1(r) + F_2(\vartheta)/r^2 \tag{4}$$

where $F_1(r)$, $F_2(\vartheta)$ are any functions of r and ϑ respectively. Then

$$g(r,\vartheta) = r^2(\Psi(r,\vartheta_o) - \Psi(r,\vartheta)) = F_2(\vartheta_o) - F_2(r) \tag{5}$$

where ϑ_o is some constant value of ϑ, say $\pi/2$.

In general, however, an axially symmetric potential cannot be written in form (4). Instead it can be expanded into an infinite series of even order Legendre polynomials;

$$\Psi(r,\vartheta) = A_o(r) + A_2(r)P_2(\cos\vartheta) + \ldots \tag{6}$$

E. Athanassoula (ed.), Internal Kinematics and Dynamics of Galaxies, 291–292.
Copyright © 1983 by the IAU.

Suppose now that the following two assumptions hold for this expansion:
1) Only the first two terms are significant. 2) $A_2(r) \sim r^{-\alpha}$ where $\alpha \sim 2$. Then one can construct an approximate third integral by choosing $g(r,\vartheta)$ as:

$$g(r,\vartheta) = (\Psi(r,\pi/2)-\Psi(r,\vartheta))/A_2(r) \qquad (7)$$

Some models of elliptical galaxies were computed using the above technique and the validity of the above two assumptions was checked a posteriori to be true. Further, the models satisfied Jeans hydrodynamic equations to a good accuracy. The distribution function of the model presented here is:

$$f(E,h_z,I_3) = e^{-\Psi(0,0)}(2\pi)^{-3/2}(e^{-E}-1)(1+\Omega h_z/\sqrt{1+h_z^2})e^{-\Gamma I_3} \qquad (8)$$

where $\Omega=0.3$, $\Gamma=10^{-5}$, $I_3=(h^2-h_z^2)/2+g(r,\vartheta)$ and $g(r,\vartheta)$ is given by (7).
The modification in the form of I_3 (by the inclusion of h_z) was done to achieve the depopulation mainly of the orbits of high inclination so that the galaxy looks flat. The observational properties of the model when seen edge on are shown in figures 1,2 and 3. The lengths are in units of core radius r_c where $r_c=\sigma_o/\sqrt{G\rho_o}$ where σ_o, ρ_o are the central dispersion velocity and central density respectively.

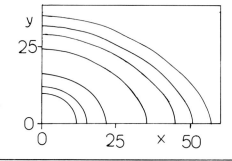

Fig.1:Equidensities.
The axial ratio is
0.82,0.8,0.75,0.68,
0.64,0.63,0.6(from
inside outwards).

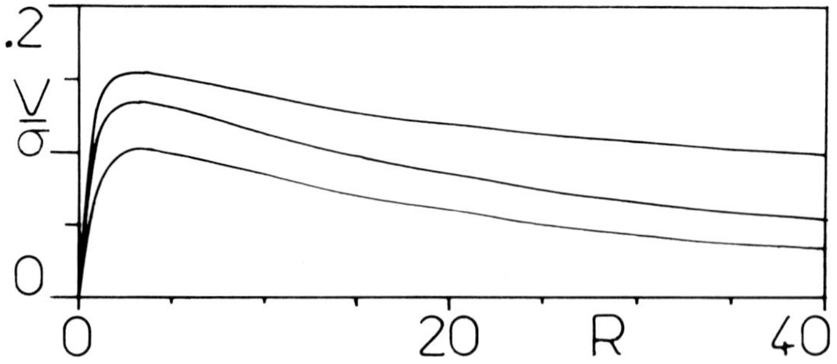

Fig.2:Observed
velocity curve for
position angles 0°,
42°,64° measured
from the major axis
(from top to bottom).

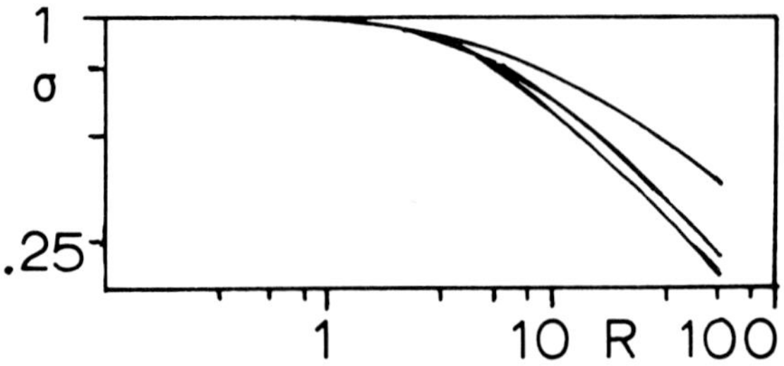

Fig.3:Observed
dispersion velocity
for position angle
0°,42°,90° from top
to bottom.

NEW PHENOMENA IN TRIAXIAL STELLAR SYSTEMS

P. Magnenat and L. Martinet
Geneva Observatory, CH-1290 Sauverny

For a number of years the Geneva Observatory stellar dynamics group has undertaken numerical investigations of stellar orbital behaviour in conservative dynamical systems with three degrees of freedom. Compared to results for 3-D axisymmetrical models (2 degrees of freedom), this work displayed the advent of several new phenomena which may appear in rotating or non-rotating stellar systems with three unequal axes (ellipticals, bulges, bars). Some of these phenomena deserve particular attention.

I. TILTED ORBITS

A. In 1:1:1: resonance regions of a nearly spherical potential, oblique plane periodic orbits (p.o.) can exist,

- which cross each main axis (generally unstable)
- which cross no main axis : the retrograde families are always stable, the direct are generally unstable (Martinet and de Zeeuw, 1982).

B. In a system symmetrical with respect to the fundamental planes, the elliptical p.o. whose orbital plane contains the Z-axis in the non-rotating case are tilted by the rotation around Z. They can be either retrograde or direct, depending on whether the rotation axis is the minor or the major axis (Magnenat, 1982a ; see also Heisler et al, 1982).

II. INSTABILITY TYPES (Magnenat, 1982b)

Three different instability types of p.o. may exist, to which correspond three kinds of behaviour of the neighbouring orbits in a four-dimensional space of section. These types are determined by the characteristics of the eigenvalues and eigenvectors of the linearized transformation matrix T connecting the consequents in the space of section.

293

E. Athanassoula (ed.), Internal Kinematics and Dynamics of Galaxies, 293–294.

A. Semi-instability (The orbit is unstable in the orbital plane or
   perpendicularly to it).

   T has two complex and two real eigenvalues : λ_1, λ_1^*, λ_2, λ_2^{-1} ;
$|\lambda_1| = 1$. The motion of points in the neighbourhood of the p.o. (in the
space of section) is dominated by a contracting and a dilating manifold
of dimension 1, as in the systems with 2 degrees of freedom.

B. Total instability (Instability in the orbital plane and perpendicularly to it).

   Four real eigenvalues corresponding to one contracting and one
dilating manifold of dimension 2 in the space of section.

C. Complex instability

   Four complex eigenvalues with $|\lambda| \neq 1$: λ, λ^*, λ^{-1}, λ^{*-1}. The
neighbouring motion in the space of section is dominated by two planes
with inward and outward spiralling, respectively.

   To conclude, let us recall that a stable p.o. is characterized by
4 complex eigenvalues on the unit circle. The motion of points in the
vicinity of the p.o. in the space of section consists then of two independent elliptical rotations around the p.o., leading to toroidal invariant surfaces.

III. THE NUMBER OF ISOLATING INTEGRALS

   - Totally or complex unstable oblique orbits are those generally
accompanied by large stochastic regions in phase space. Potentials
symmetrical with respect to the three main axes in which these orbits do
not exist, are therefore likely to admit 2 isolating integrals of motion
besides the Hamiltonian for most orbits.
   - In a system without the previously mentioned symmetries, two, one
or zero isolating integrals besides the Hamiltonian can exist. This fact,
already observed by inspecting invariant surfaces in the space of section
(Martinet and Magnenat, 1981), was recently confirmed by means of the
Lyapunov characteristic numbers (Magnenat, 1982c).

REFERENCES

Heisler, J., Merritt, D. and Schwarschild, M. : 1982, Astrophys. J. 258,
     p. 490.
Magnenat, P. : 1982a, Astron. Astrophys. 108, p. 89.
Magnenat, P. : 1982b, Celes. Mech. 28, p. 319.
Magnenat, P. : 1982c, Proceedings of the VII Sitges Conference "Dynamical
     systems and chaos".
Martinet, L. and Magnenat, P. : 1981, Astron. Astrophys. 96, p. 68.
Martinet, L. and de Zeeuw, P.T. : 1983, present symposium.

KINEMATIC MODELLING OF NGC 3379

Gary A. Mamon
Princeton University Observatory

Giant elliptical galaxies are now known to be supported by anisotropic pressure rather than by rotation (cf. Binney, 1981). This anisotropy can be derived from observable quantities for spherical systems as
was shown by Binney and Mamon (1982) in their study of M87. We investigate here the velocity anisotropy of the E1 galaxy NGC 3379, a giant elliptical whose surface brightness constitutes an excellent illustration
of the $r^{1/4}$ law.

For a spherical system, the kinematics are described by the equation
of stellar hydrodynamics:

$$df/dr + 2\beta f/r = -(M/L)F. \tag{1}$$

Here, $f=\ell\sigma_r^2$ is the stellar 'pressure', $\beta=1-\sigma_\theta^2/\sigma_r^2$ is the anisotropy
and $F=G\ell L(r)/r^2$; where σ_r and σ_θ are two principal components of the velocity ellipsoid, ℓ is the space-luminosity density, and $L(r)$ is the integrated luminosity. The radial velocity dispersion can then be projected
to obtain the line-of-sight velocity dispersion σ_v:

$$\int_R^{R_t} \frac{fr\,dr}{(r^2-R^2)^{1/2}} - R^2 \int_R^{R_t} \frac{\beta f\,dr}{r(r^2-R^2)^{1/2}} = \frac{1}{2}\sigma_v^2(R)\Sigma(R). \tag{2}$$

The knowledge of $\Sigma(R)$ provides us with ℓ and hence with F. In our first
scheme, we choose $\beta(r)$ and a constant M/L, compute σ_r from equation (1),
and project onto the line-of-sight with equation (2) to obtain $\sigma_v(R)$. In
the second scheme, we start directly with the observable quantities $\Sigma(R)$
and $\sigma_v(R)$ to derive $\beta(r)$ and a constant M/L following the solution to the
system of equations (1) and (2) for f and β from Binney and Mamon (1982).

We adopt for NGC 3379 the 2-component model of de Vaucouleurs and
Capaccioli (1979) for $\Sigma_B(R)$ (an $r^{1/4}$ law with a gaussian core superposed),
and truncate it at 1201". We begin with scheme #2 and fit three model
profiles through the velocity data of Sargent et al. (1978) and of Davies
(1981), and a fourth model (D) through the unpublished data of Davies,
Illingworth and McElroy (1982; Illingworth, private communication). Fig. 1
shows $\beta(r)$ as computed from the $\sigma_v(R)$ models plotted in Fig. 2. Models
A, B, and C all exhibit positive anisotropy, while model D is isotropic

E. Athanassoula (ed.), Internal Kinematics and Dynamics of Galaxies, 295–296.

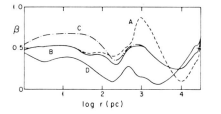

Figure 1. Anisotropy profiles for NGC 3379 (1"=38.8 pc).

in the main body of the galaxy.

These results are better understood when we compute the velocity dispersion profiles from constant β and M/L, using the first scheme. Since the innermost measurements are affected by atmospheric seeing and by averaging over the slit, we convolve our results in two dimensions with a gaussian PSF of σ^*=1" and the appropriate slits (Fig. 2). The anisotropic model (β=0.5,M/L=8.5) clearly fits better the older two sets of data, whereas the isotropic model (M/L = 11) presents an adequate fit to the more recent data.

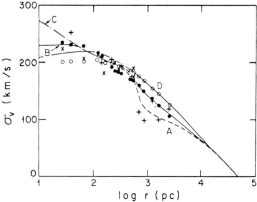

Figure 2. Line-of-sight velocity dispersion for NGC 3379. Crosses and plus signs are measurements by Sargent et al. (1978) and Davies (1981) respectively. The curves labeled A, B and C are three model fits to these data, while the D curve is a fit to the measurements of Davies et al. (1982; not shown in the figure). The open and filled circles are the seeing and slit convolved profiles from β=0, M/L=11 and β=0.5, M/L=8.5, respectively.

In summary, the strong scatter (±25%) in the velocity dispersion measurements beyond 15" in NGC 3379 prevents one from distinguishing between isotropic and anisotropic models. This puts the mass-to-light ratio in the interval (8.5,11). However, M/L may not be constant, and isotropic models with M/L increasing in the core can be made to fit the data. The answers to all of these questions must therefore await higher quality spectra in the inner arcsec and beyond 15" for NGC 3379.

REFERENCES

Binney, J.:1981, in Normal Galaxies, NATA ASI, p. 55, eds. S.M. Fall and D. Lynden-Bell, Cambridge University Press.
Binney, J. and Mamon, G.A. :1982, Mon. Not. R. astr. Soc. 200, p. 361
Davies, R.L., Illingworth, G. and McElroy, D. :1982, preprint.
de Vaucouleurs, G. and Capaccioli, M.:1979, Astrophys. J. Suppl.40, p.699
Sargent, W.L.W., Young, P.J., Boksenberg, A., Shortridge, K., Lynds, C.R. and Hartwick, F.D.A. :1978, Astrophys. J. 221, p. 731.

INTERSTELLAR MATTER IN ELLIPTICAL GALAXIES

G.R. Knapp
Princeton University Observatory, Princeton, NJ08544, U.S.A.

I. INTRODUCTION

This contribution concentrates on the properties of the large gas structures in elliptical galaxies, those seen in dust and HI. The distribution and kinematics of ionized gas, which is generally confined to a small region in the center of the galaxy, are discussed at this conference by Illingworth and by Ulrich.

The study of the interstellar matter in ellipticals has been approached from two directions over the past few years. The study of the global gas content of normal ellipticals using sensitive observations of the HI 21-cm line has been carried out by several groups. These searches have resulted in the detection of small amounts (several x 10^8 $M_\odot$) of gas in a few galaxies and in the setting of low upper limits for a great many more. At the same time, attention has been drawn (Bertola and Galletta 1978; Hawarden et al. 1981) to a class of galaxies containing dust lanes but no stellar disks (e.g. NGC5128); for these galaxies, the dust lanes are often not aligned with the major axis of the galaxy.

In this contribution, objects from both of these studies will be considered to represent examples of the same general phenomenon, and I am going to be somewhat casual with the morphological classification of the galaxies - galaxies with dust lanes are frequently classified as SO and IO, but will herein be considered along with the ellipticals. This paper is also going to concentrate on the observational aspects - the theory of orbits in these galaxies is discussed in this symposium by Binney and by de Zeeuw.

II. THE HI-RICH ELLIPTICALS

As is well known, the HI-rich ellipticals contain but small amounts of HI and the certain detection of the gas is correspondingly difficult. A great deal of observing time has been spent in an effort to provide convincing detections. A census of all the ellipticals with

E. Athanassoula (ed.), Internal Kinematics and Dynamics of Galaxies, 297–304.

well-established detections of HI in emission is given in Table 1; the list is roughly in inverse order of discovery date. Also given are indications of whether the object is a known radio continuum source, whether HI absorption is also seen, and whether dust is known to be present.

TABLE 1.   HI DETECTIONS FOR ELLIPTICAL GALAXIES

NGC4278 (c,na,d)	NGC3998 (c,na,nd)	NGC205 (nc,-,d)
NGC1052 (c,a,d)	NGC5173 (nc,-,?)	A1230+09
NGC5128 (c,a,d)	(?)NGC4105 (nc,-,nd)	NGC4370
UGC01503 (nc,-,?)	UGC09114 (c,na,?)	NGC3265
NGC5363 (c,a,d)	NGC2768 (c,na,nd)	NGC3608
NGC5506 (c,a,d)	NGC185 (nc,-,d)	

c = radio continuum   a = HI absorption   d = dust

Table 1 contains the respectable number of seventeen detections In addition to providing the last four detections in Table 1, Lake and Schommer (1982) find possible emission from NGC4318, 4551 and 5576. HI absorption at or near the intrinsic redshift is found for a further three galaxies, 3C293, NGC315 and UGC06671 (Shostak et al. 1982; Dressel et al. 1982a,b). There are, in addition, several HI-rich galaxies of quite uncertain type which may be related to the class of HI-rich ellipticals, including NGC5666, 1800, 520, 3773, 1315, 1275, IC5063 and possibly M82. Finally, there are several ellipticals suspected at one time of being HI-rich whose detections have not been confirmed by subsequent observatio These are NGC1587/88, 2974, 3156, 3226, 3904, 3962, 4636, and 5846.

Good upper limits have been set for many more galaxies; some interesting examples are NGC4472 (M(HI)/L_B <0.005), NGC3665 (<0.006), and NGC4374 (<0.005). As a class, then, the ellipticals have a great variation in HI content. Most are undetectable; those which are contain only small amounts of HI (typically between 10^8 and 10^9 M$_\odot$).

III. THE NATURE OF THE INTERSTELLAR MEDIUM

There is not yet much information on this topic, but what there is carries the interesting implication that the ISM in these galaxies is not too different from that seen in spirals. For several radio galaxies, absorption lines are seen against the continuum source, and are resolved into discrete components such as are produced by diffuse clouds in the Galaxy (Haynes and Giovanelli 1981; Thuan and Wadiak 1982). Synthesis observations of NGC4278 (Figure 1) also suggest that the HI in this galaxy is clumpy (Raimond et al. 1982). That the gas is not primordial in all cases is demonstrated by the presence of dust; the observation of NGC5128 by van Gorkom et al. (this conference) suggests that the HI and dust are well mixed. There is a qualitative suggestion in the survey data that the gas-to-dust ratio may vary among galaxies, in that HI is not detected from every dusty galaxy, and vice-versa.

The study of the molecular gas content of ellipticals has barely gotten underway, and few CO observations exist except for the work of Johnson and Gottesmann (1979). No CO emission has yet been detected. An upper limit of $M(H_2) < 5 \times 10^4\ M_\odot$ is inferred for NGC185 (Knapp, unpublished). OH, H_2CO and CH are seen in absorption against the nucleus of NGC5128 (Gardner and Whiteoak 1976; Whiteoak et al. 1980) and OH in NGC5363 (Rickard et al. 1982) with the molecular gas being in discrete clouds similar to those in the Galaxy.

IV. DISTRIBUTION AND DYNAMICS

To date, synthesis observations of the HI distribution exist for six of the galaxies listed in Table 1. NGC185 and 205 contain unresolved clouds near, but not coincident with, the centers of the galaxies (Gottesmann, this symposium). NGC5128 (van Gorkom et al., this symposium), and NGC3998 (Knapp et al. 1982) have their HI in roughly ring-like structures close to the minor axis, while NGC4278 (Raimond et al. 1981, 1982) and NGC5173 (Knapp and Raimond 1982) have the HI in disks.

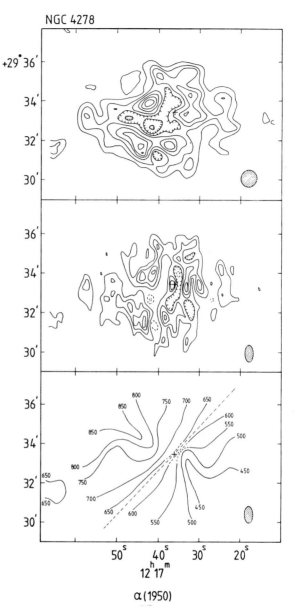

Figure 1. HI synthesis map of NGC4278 (Raimond et al. 1982), showing the smoothed (50"x50") total HI map, the unsmoothed (30"x50") map and the velocity field. The position of the radio continuum source is marked by a cross.

In NGC3998, the HI is in a ring inclined at ∿70° to the optical major axis. The ring

appears to be in prograde rotation with respect to the stellar component;
the relatively rapid rotation of the latter suggests that the HI is in a
polar distribution around an oblate galaxy (Blackman et al. 1982). The
position angle of the HII emission lies between that of the stellar
component and that of the HI (Blackman et al. 1982; Bertola, this
conference).

    The HI in NGC4278 is distributed in a disk inclined to the
stellar bulge by about 40⁰, again in prograde rotation. The center of
the disk is empty (Figure 1) and no absorption is seen against the central continuum source (Shostak et al. 1982). The velocity field in the

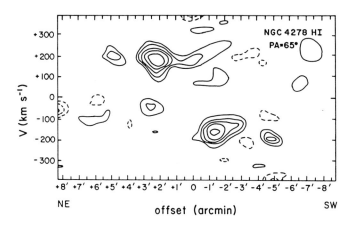

Figure 2. HI velocity-position map along the kinematic major axis,
centered on the position and systemic velocity of the galaxy.

HI disk (Figures 1 and 2) suggests that the circular velocity is constant
with radius, so that, at least out to a radius of 10 kpc (D = 10 Mpc),
the mass-to-light ratio of the galaxy is increasing with radius. The
position angle of the extended HII emission (Ulrich, this conference)
is intermediate between that of the stars and of the HI, suggesting
that the mass distribution in NGC4278 is triaxial.

    Hawarden et al. (1981) have shown that, for a much larger
sample of galaxies (those with dust lanes but no disks) misalignments
between the stellar and gaseous components are common. In Figure 3 is
plotted a histogram of the relation between the position angles of these
two components, using the sample of Hawarden et al. plus some twenty
additional galaxies from the literature. The notation and classification
is that of Hawarden et al., with the addition of the class M (mixed), i.e
galaxies such as NGC2685 with both polar and equatorial rings. The
situation is strikingly different from that in disk galaxies and suggests
that the gas in ellipticals takes up more or less random orbits. Orbits

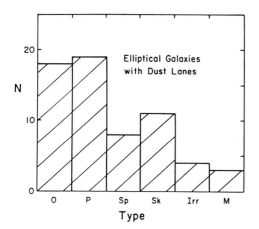

Figure 3. Distributions of relative orientations of dust/gas lanes and the galactic major axis in ellipticals. O = lane along major axis. P = lane along minor axis. Sp = dust lane circular in projection. Sk = lane inclined to major axis. Irr = irregular dust lane M = both perpendicular and parallel lanes present.

for gas in these systems are discussed by Richstone and Potter (1982) and Merrit and de Zeeuw (1982).

V. RADIO ACTIVITY AND HI ABSORPTION

It is by now reasonably well established that nuclear activity in early-type galaxies and the presence of extended structures of cold gas are closely related, at least in nearby galaxies (Hummell 1980; Dressell et al. 1982a). Kotanyi and Ekers (1979) have shown that the extended radio structures in a sample of active galaxies such as Cyg A are roughly perpendicular to the dust lanes in these galaxies, in good agreement with the predictions of models of the radio activity involving accretion onto a central massive engine (e.g. Blandford 1979). The difficulties of accreting hot gas onto the engine are much discussed; Gunn (1979) showed that cold gas from a surrounding disk could be accreted in adequate amounts. Recent observations of HI absorption lines towards the nuclei of some of these active galaxies have shown the presence of redshifted components, interpreted as due to individual cold clouds falling towards the nucleus (Thuan and Wadiak 1982; Dressell et al. 1982b; Shostak et al. 1982; van der Hulst et al. 1982); accretion rates of $\sim$0.2 $M_\odot$/yr are estimated from these observations, enough to fuel the radio source. Thus the causes of intense activity in radio galaxies (and quasars?) seem, at least in a first-order sense, to be understood.

VI. THE ORIGIN OF THE GAS

The lack of correspondence between the dynamics and distribution of the gas and stars argues against the material having its origin as mass loss from evolving stars, as does the bimodal distribution of the gas contents of ellipticals (Sanders 1980 - though see Sanders 1981).

These considerations also argue against ejection from the nucleus. The
most tenable explanation still seems to be the accretion of gas from
the outside (c.f. Silk and Norman 1979), although the presence of gas
in so many ellipticals strains this somewhat, and there appears to be
no correspondence between gas content and environment (Dressell et al.
1982a; Gallagher et al. 1982). However, if Larson, Tinsley and Caldwell
(1980) are correct in their inference that spirals must be continuing
to accrete gas, then early-type systems are presumably also accreting
material. The low gas content of the early-type systems may be due to
self-sweeping by the stellar component - note that the gas distributions
are generally rings around the light systems. A similar situation is
found for the S0 galaxies (van Woerden et al., this symposium). The
"disks" of ellipticals could then cover a wide range of ages, and, in a
sense, like the outer disks of spirals, be still in the process of
formation.

        I would like to thank Ernst Raimond, Hugo van Woerden and Wim
van Driel for permission to quote our work in progress, and George Lake,
Jaqueline van Gorkom, Bob Sanders, Bob Schommer, Renzo Sancisi, Seth
Shostak, Thijs van der Hulst, Linda Dressell and Steve Gottesmann for
providing results before publication. This work is supported by
NSF grant AST-8009252.

References

Bertola, F., and Galletta, G. 1978, Ap.J. 226, L115.
Blackman, C.P., Wilson, A.S., and Ward, M.J. 1982, MNRAS (in press).
Blandford, R.D. 1979, in 'Active Galactic Nuclei' ed. C. Hazard and
    S. Mitton, Cambridge University Press, p. 241.
Dressell, L.L., Bania, T.M., and O'Connell, R.W. 1982a, Ap.J. (in press).
Dressell, L.L., Bania, T.M., and Davis, M.M. 1982b, Ap.J. (in press).
Gallagher, J.S., Faber, S.M., and Knapp, G.R. 1982 (in preparation).
Gardner, F.F., and Whiteoak, J.B. 1976, Proc. Astron. Soc. Aust. 3, 63.
Gunn, J.E. 1979, in 'Active Galactic Nuclei' ed. C. Hazard and S. Mitton,
    Cambridge University Press, p. 213.
Hawarden, T.G., Elson, R.A.W., Longmore, A.J., Tritton, S.B., and
    Corwin, H.G. 1981, MNRAS 196, 747.
Haynes, M.P., and Giovanelli, R. 1981, Ap.J. 240, L87.
van der Hulst, J.M., Golisch, W.F., and Haschik, A.D. 1982, Ap.J. (in pre
Hummell, K. 1980, Ph.D. Thesis, University of Groningen.
Johnson, D.W., and Gottesmann, S.T. 1979, in 'Photometry, Kinematics and
    Dynamics of Galaxies' ed. D.S. Evans, University of Texas press.
Knapp, G.R., van Driel, W., and van Woerden, H. 1982 (in preparation).
Knapp, G.R., and Raimond, E. 1982, (in preparation).
Kotanyi, C.G., and Ekers, R.D. 1979, Astron. Astrophys. 73, L1.
Lake, G., and Schommer, R.A. 1982 (in preparation).
Larson. R.B., Tinsley, B.M., and Caldwell, C.N. 1980, Ap.J. 237, 692.
Merritt, D., and de Zeeuw, T. 1982, in preparation.
Raimond, E., Faber, S.M., Gallagher, J.S., and Knapp, G.R. 1981,
    Ap.J. 246, 708.
——————————————, 1982, in preparation.

Richstone, D.O., and Potter, M.D. 1982, Nature (in press).
Rickard, L.J., Bania, T.M., and Turner, B.E. 1982, Ap.J. 252, 147.
Sanders, R.H. 1980, Ap.J. 242, 931.
——————————— 1981, Ap.J. 244, 820.
Shostak, G.S., van Gorkom, J., and Sanders, R.H. 1982, (in preparation).
Silk, J., and Norman, C.A. 1979, Ap.J. 234, 86.
Thuan, T.X., and Wadiak, E.J. 1982, Ap.J. 252, 125.
Whiteoak, J.B., Gardner, F.F., and Höglund, B. 1980, MNRAS 190, 17p.

DISCUSSION

RICHSTONE : One remark about the polar ring type galaxies : it is very doubtful that the gas comes from the stars because the angular momentum vector of the gas is orthogonal to that of the stellar component for the three cases in which both are known. It seems much more likely that the gas comes from outside the system.

DRESSLER : A paper by you, Balick, Faber and Gallagher showed that SO galaxies with relatively large M_{HI}/L_B have a wide range of optical colors, and are often much bluer than gas-free SO's. Is there a similar effect for this sample of relatively gas-rich ellipticals ?

KNAPP : No. Such data as exist suggest no systematic color differences between gas-rich and gas-free ellipticals.

VAN WOERDEN : Thonnard's Arecibo map of HI in NGC4636, refutes 3 earlier single-dish detections. At Westerbork, Huchtmeier and Shostak also failed to find the HI. Is there a new Green Bank observation ?

KNAPP : Yes ; Gallagher, Faber and I have a new 91m observation with about 50 % better sensitivity than our previous observations and find no trace of emission.

KENNICUTT : Is there any correlation between the HI detection rate and the presence of nearby gas-rich companion galaxies ?

KNAPP : No. Statistical studies by Dressell et al. suggest no correlation with nearby galaxy density. Indeed, Haynes and Giovanelli (1980, Astrophys. J. 240, L87) found HI emission in isolated early-type systems. NGC4278 and 5173 lie in small spiral-containing groups, but the synthesis observations give no sign of tidal distortions and the like. The only possible counter-example is the HI-SO galaxy NGC2859 described by Shane at this symposium, which is surrounded by HI-containing dwarfs.

SANDERS : If the gas in these systems is rotating, then it is hard to argue that in any system the gas has come from the stars in the ellipticals, since the specific angular momentum of the gas is much larger than the specific angular momentum of the stars. Is it possible that the gas in well-observed galaxies like NGC4278 could have a substantial component of non-circular motion-inflow, for example ?

KNAPP : The best information on this question comes from the absorption-line observations. A strong, narrow component is always seen at the systematic velocity of the galaxy, which, with the large-scale velocity fields seen in NGC4278 and 5128, suggests rotation. But the absorption-line observations also show redshifted components –at the implied accretion rates of a few tenths of 1 $M_\odot$ per year the HI disks are not going to last long ! So the answer is yes, but I believe that rotation is dominant.

SENSITIVE SEARCH FOR HI IN E AND S0 GALAXIES

Norbert Thonnard
Department of Terrestrial Magnetism
Carnegie Institution of Washington

ABSTRACT. Do elliptical and S0 galaxies in which type I supernovae (SNI) were detected contain more gas than those without SNI detections? Thirteen E and S0 galaxies in the Virgo and Pegasus I clusters, seven with SNI detections and six without, were mapped well beyond the optical image at the 21-cm neutral hydrogen line. No HI was detected. In Virgo, the upper limit to M_{HI}/L_B is between 0.0005 and 0.0024.

There are now many sensitive searches for 21-cm neutral hydrogen emission from elliptical galaxies, but still, less than one-tenth of the galaxies observed have been detected. As the HI content of E galaxies is almost three orders of magnitude less than that of Sc's, while the type I supernovae (SNI) rate is only a factor of four less, it had been assumed that the progenitor stars of SNI's were long lived stars. Oemler and Tinsley (1979), instead, have suggested that the progenitor stars for SNI's are short lived ($\sim 10^8$ yr) stars, implying that there should be sufficient gas to form them. My colleague, C.K. Kumar, noticed a hint (from a small number of published detections), that a detection of HI was more likely in those elliptical galaxies having had a SNI.

To test the hypothesis that E and S0 galaxies which produced SNI's have more gas than those that didn't, Kumar and I observed 13 galaxies in the Virgo and Pegasus I clusters at Arecibo. Six of the galaxies had not had SNI detections, but were otherwise identical to the SNI producing galaxies, and were included as controls. As recent observers have suggested that HI in E and S0 galaxies is extended, the detectability of HI was increased by mapping the galaxies over an area larger than the optical image. One to seven positions on each galaxy were observed. After smoothing to 42 km s^{-1} resolution, the individual profiles were examined for evidence of HI. To increase sensitivity to extended gas, the profiles (for each galaxy) were combined, yieling an RMS between 0.4 and 0.8 mJy. No neutral hydrogen was detected. Adopting $H_o = 50$ kms^{-1} Mpc^{-1} and an HI profile width of 400 kms^{-1}, the 3σ limit to the neutral hydrogen content of the Virgo galaxies is $\leqslant 4 \times 10^7$ M$_\odot$. We summarize the observations in Table I below.

305

E. Athanassoula (ed.), Internal Kinematics and Dynamics of Galaxies, 305–306.

TABLE I

Cluster	Distance adopted Mpc	Galaxies observed NGC	Diameter range arc min	$M_H/M_\odot$ mean (min-max)	M_H/L_B mean (min-max)
Virgo	20	4365,4382,4472 4526,4564,4578 4621,4636,4762	3.1-8.9	$\leq 4.5 \times 10^7$ (2.6-6.4)	≤ 0.0015 (.0005-.0024)
Pegasus I	80	7619,7626,7634 7785	1.3-2.8	$\leq 8.3 \times 10^8$ (3.0-12)	≤ 0.0043 (.0011-.0085)

We note that the HI content of NGC 4278, a "normal" E galaxy, is $\sim 4 \times 10^8 M_\odot$ (Knapp, Kerr and Williams, 1978), while for five isolated E's and S0's, Haynes and Giovanelli (1980) detected $\sim 8 \times 10^9 M_\odot$.

Of particular interest (Fig.1.), is NGC 4472, brightest E in Virgo, showing no HI except for emission from UGC 7636, 6' to the SW, and NGC 4636, earlier reported as detected (Knapp, Faber and Gallagher, 1978; Bottinelli and Gouguenheim, 1977,1978), also showing no HI at 1/20 the level of previous observations.

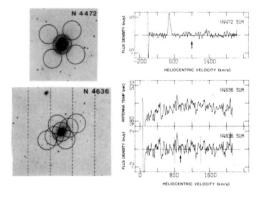

Fig.1. Left, reproduction of NGC 4472 and 4636 from the POSS (© 1957, National Geographic Society-Palomar Observatory Sky Survey). Solid circles indicate 3'.4 beam positions; dashed vertical lines, position of the Nancay observations. Right, the combined 21-cm profiles. For NGC 4636, we also show the unbaselined result, typical of the Virgo observations.

We have lowered the limit to the neutral hydrogen content of 13 E and S0 galaxies, and showed that HI is not "hiding" at the outskirts of these galaxies. No difference is seen (to our limit) in the HI content of those E and S0 galaxies having, and those not having had SNI's.

REFERENCES

Bottinelli, L. and Gouguenheim, L.: 1977, Astron.Astrophys. 60, p.L23.
----- : 1978, Astron.Astrophys. 64, p.L3.
Haynes, M.P. and Giovanelli, R., 1980, Ap.J. 240, p.L87.
Knapp, G.R., Faber, S.M., and Gallagher, J.S.: 1978, Astron.J. 83, p.11.
Knapp, G.R., Kerr, F.J. and Williams, B.A.: 1978, Ap.J. 222, p.800.
Oemler, A., Jr. and Tinsley, B.M.: 1979, Astron.J. 84, p.985.

NEUTRAL HYDROGEN OBSERVATIONS OF THE DWARF ELLIPTICAL GALAXIES NGC 185 AND NGC 205

S. T. Gottesman, Department of Astronomy, University of
Florida, Gainesville, Florida, U. S. A. and D. W. Johnson,
Battelle Observatory, Battelle Pacific NW Labs, P.O. Box 999,
Richland, Washington, U. S. A.

In this note we report the detection and mapping of neutral hydrogen
in two dwarf elliptical galaxies NGC 185 and NGC 205. Both are companions
to M 31 and both are classified as peculiar owing to the presence of
obscuring dust patches near their nuclei. Both galaxies also contain a
small population of blue, presumably young stars (Hodge 1963, 1973).

Eighteen antennae of the VLA facility of the National Radio
Astronomy Observatory were used with a 32 channel spectrometer to obtain
an angular resolution of ∿ 1.15 and a velocity resolution of 6.28 km/sec.

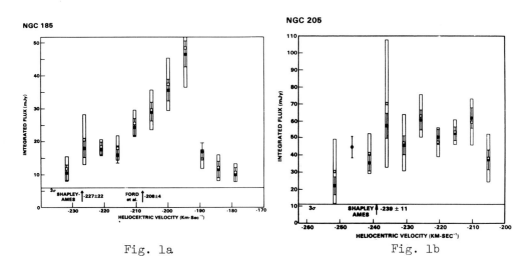

Fig. 1a   Fig. 1b

Figures 1a and 1b show the global HI emission profiles for each
system. We are unable to detect any significant rotation associated
with the hydrogen in NGC 185. However, the HI in NGC 205 exhibits many
of the properties related to a rotating disk. Thus, it is not surprising
that the profile shapes are markedly different.

E. Athanassoula (ed.), Internal Kinematics and Dynamics of Galaxies, 307–308.

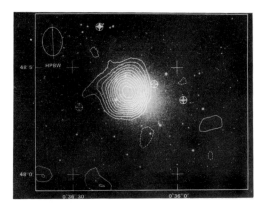

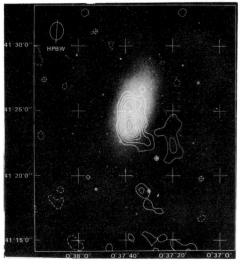

Fig. 2a   NGC 185   Fig. 2b   NGC 205

     Figures 2a and 2b show the distribution of atomic hydrogen in both
galaxies, assuming the gas to be optically thin. A single, not highly
resolved feature is seen in NGC 185. In the case of NGC 205, an extended feature with some structure is observed. For these systems,
$M_H \sim 2 \times 10^5 \, M_\odot$, and $M_H/L \sim .0013$.

     In each system the centroid of the gas is significantly offset from
the center of the galaxy. Furthermore, the gas in NGC 185 does not coincide with the well known dark clouds, for the centroid is displaced by
$\sim 30"$ to the northeast of the galaxy center. In the case of NGC 205 the
gas is offset by about 1' southwest of the center, but by comparison with
NGC 185 there is not such a large discrepancy in the association of the
HI with the dust patches. Also, in this system the material rotates $\sim$
with a maximum observed velocity of ~ 29 km/sec at radial distance of
$\sim 1\overset{.}{.}5$. The observations of NGC 205 confirm some of the findings reported
by Unwin (1980).

     It is very difficult to understand how $\sim .01\%$ of the total mass of
these objects has failed to coalesce at the bottom of the galaxien potential well. Perhaps the explanation is associated with the current
burst of star formation that is occurring in these peculiar elliptical
systems.

REFERENCES
Ford, H. C., Jenner, D. C., and Epps, H. W. 1973, Astrophys. J. (Letters),
      183, L73.
Hodge, P. W. 1963, Astron. J., 68, 691.
------------ 1973, Astrophys. J., 182, 671.
Sandage, A. and Tammann, G. A. 1981, in "A Revised Shapely-Ames Catalog
      of Bright Galaxies". (Carnegie Institution: Washington, D.C.).
Unwin, S. C. 1980, MNRAS, 190, 551.

EXTENDED GASEOUS EMISSION IN NORMAL ELLIPTICAL GALAXIES

M. H. Ulrich
European Southern Observatory, Karl-Schwarzschild-Str.2,
8046 Garching bei München, Federal Republic of Germany

It has been known for several decades that about 15% of elliptical galaxies contain ionized gas yet very little data is available in the literature on the properties of this gas such as spatial distribution, velocity field, and abundances. The properties of this gas are related to current problems about elliptical galaxies: (1) origin of the gas (stellar mass loss vs. accretion from a nearby gas-rich galaxy or an intergalactic cloud); (2) structure of the galaxy itself, because the locus of the stationary orbits of the gas depends on whether the galaxy is prolate or oblate and on whether the galaxy is rotating or not; (3) the relationship between the presence of gas in elliptical galaxies and their radio properties.

I give a brief report of a study done in collaboration with Alec Boksenberg (RGO) and Harvey Butcher (KPNO) of the gas in normal ellipticals. The spatial distribution of the gas has been observed using the Video Camera attached to the KPNO 4m telescope. Maps of the ionized gas in ten ellipticals have been obtained from frames taken through narrow-band filters centered at the redshifted Hα + NII lines and in the adjacent continuum. This is complemented by spectrography of the 6 galaxies with the brightest emission regions using the UCL IPCS attached to the Cassegrain Spectrograph of the ESO 3.6m telescope.

Extended emission is found in nearly all the galaxies observed. Typically the detected emission has a dimension of 2kpc. The isophotes are circular in the inner regions probably because the motion of the gas is dominated by turbulence near the centre. They become elongated further out, roughly approximating an ellipse or, in a few cases, they become irregular and very asymmetric. The spectra show that in contrast with the fairly irregular distribution of the gas, the velocity field is well ordered and is consistent with the gas being in a disk in rotation. There is, however, no clear relationship between the direction of the angular momentum of the gas and the major or minor axes of the stellar population. The emission line intensity ratios are compatible with the gas being excited by shocks but other processes such as photoionization by a weak central source, ultraviolet radiation from stars, kinetic energy of supernovae ejecta may also contribute to the excitation of the gas.

E. Athanassoula (ed.), Internal Kinematics and Dynamics of Galaxies, 309.
Copyright © *1983 by the IAU.*

GAS AND STAR KINEMATICS IN ELLIPTICAL GALAXIES

F.Bertola, D.Bettoni and M.Capaccioli
Istituto di Astronomia, Università di Padova
35100 PADOVA, ITALY

Since the time it became evident that ellipticals are not rotationally supported, there has been an increasing interest on their dynamical properties, and new models have been formulated. However, the spectroscopic, as well as the photometric and morphological observations, do not provide yet a clear understanding of the structure of these objects, just proving that they are much more complicated than thought before. An obvious way to attack the problem seems to look at the motions of the gaseous component in those few cases where it is present, because of the different response to the same potential field of the gas and star components.

In this paper an account is given of the spectroscopic observations of the gaseous and stellar components in four E galaxies, namely NGC 1052, NGC 2749, NGC 3998 and NGC 4125, to be published in full elsewhere. We present velocity dispersion profiles and radial velocity curves from both absorption and emission (λ 3727-29) lines, based on photographic spectra taken with the 4m Kitt Peak and 5m Palomar telescopes. The reduction procedure is based on the Fourier Quotient Method. The radial velocity curves for NGC 4125 (the only galaxy observed also along the minor axis) and NGC 3998 were also derived by measurements with a Grant machine. The velocity dispersion profiles of the inner regions (typically $0\overset{''}{.}1$ from the nucleus) from the absorption lines are constant or peaked at the center. The same profiles for the gas obtained from the [OII] λ 3727-29 accounting for the blend, have similar trends but are scaled down by a factor about 0.5 in the mean.

The ratio between the slopes of the emission and absorption velocity curves along the major axis is about 1 for NGC 1052, larger than 3 for NGC 4125 and 0.7 for NGC 3998. In the latter case, the low value is accounted by the fact that our measurements, together with those of Blackman et

311

E. Athanassoula (ed.), Internal Kinematics and Dynamics of Galaxies, 311–312.

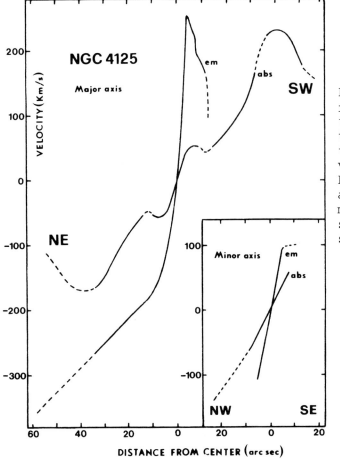

FIGURE 1. Smooth representation of the radial velocity curves (scaled to the systemic velocity) of the E4 galaxy NGC 4125 along the major and minor axes from absorption and emission lines.

al.(1972) indicate that the maximum slope of the gas velocity curve is found along a direction which lies at about 45° from the isophotal major axis.

A quite exceptional behaviour is found in NGC 4125. This fairly flat galaxy (E4) exhibits significant amount of rotation. The absorption velocity curve is fairly symmetric, while that derived for the gas is strongly asymmetric and extends much more on the NE side. In addition a strong velocity gradient is measured along the minor axis both for the gaseous and stellar components (Fig.1). Among the various possibilities which can be considered to explain this object we mention that of a merging process.

REFERENCE

Blackman,C.P., Wilson,H.S., Ward,M.J.:1982,M.N.R.A.S.in press.

OPTICAL KINEMATICS OF STRONG RADIO GALAXIES

S. M. Simkin
Department of Physics and Astronomy,
Michigan State University

Recent spectroscopic measurements of 5 strong radio galaxies showed that approximately 50% of these objects had internal gas velocities which were notably non-circular [1]. To investigate this, more detailed observations were done for the galaxies in question. Examination of these results suggests that the non-circular velocities tend to fall into a pattern. All of the objects have one kinematical component which fits the classical picture of "circular rotation". This occurs in BOTH the stellar and ionized gas and displays MUCH HIGHER peak velocities than are found in E galaxies [1]. In addition, all of the objects studied possess a second ionized gas system at a different velocity from that of the rotating gas. Finally, at least one of the objects has an envelope of stars which is either not rotating or has a very different kinematic major axis from that of the stellar population in its inner regions.

The observations were done with the Hale 5m telescope and SIT spectrograph during Jan 1980 and 1981. The measurement and reduction procedures (including effects of instrumental errors) are fully discussed in reference 1. The objects observed were 3C98(ED3), 3C184.1(D3-4), 3C218(cD2), and 3C390.3(cD?).

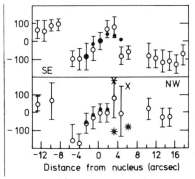

Fig. 1, Hydra A.
(a) ionized gas (b) stars

Fig. 2, 3C98, filled circles-stars; other symbols, gas

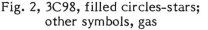

E. Athanassoula (ed.), Internal Kinematics and Dynamics of Galaxies, 313–314.

To date, the most complete analysis available is for the cD galaxy Hydra A (3C218). The velocity field of this object along the "major axis" is shown in Fig. 1. The two sets of data (stars and gas) have been normalized to coincide at the point 0,0. The ABSOLUTE velocities of these two systems are determined to no better than 75 km/sec. The RELATIVE accuracy of the velocities in one group (emission line or stellar) is shown by the error bars on the plots. The three kinematical features noted above are all present in Fig. 1. Both the stars and gas appear to be in rapid rotation near the nucleus with the second ionized gas system redshifted with respect to the nucleus. Since the absolute values of the absorption line and emission line velocities may differ by as much as 75 km/sec (at the one sigma level), it is possible that the "rotating" gas system (the lower velocity points in Fig. 1a) is associated with the group of low stellar velocities on the SE side of the galaxy (Fig. 1b). These points represent the systemic velocity of a second "knot" in the envelope of Hydra A which may be a "secondary nucleus"[2] or may be simply another cluster galaxy.

The other galaxy for which detailed information is available is 3C98. Velocities for the stars and gas in this object are shown in Fig. 2. Fig. 2a is for PA 115 deg and Fig. 2b for PA 180 deg. ("major axis" is 130 deg[1]). Again there is a region of rapid rotation near the nucleus and an outer region where the gas velocities seem to reverse sign. There are no stellar velocities available yet for this outer region. Finally, preliminary reductions show that two other objects (3C184.1 and 3C390.3) exhibit the same behavior, to wit: an inner region in rapid rotation and a second gaseous region with distinctly non-circular velocities. In addition, such anomalous velocities are also seen in NGC 1275.[3,4]

The complex velocity structure of these objects is so different from that usually found in rapidly rotating (Sa) systems without powerful radio activity that it seems likely that it is associated with the nuclear activity of these cD galaxies. Schweizer (this Symposium) has championed the view that such velocity fields are the signature of two galaxies in the process of merging. Indeed, Kent and Sargent[4] interpret the velocity field of NGC 1275 in this way. However, there are no obvious signs of recent mergers for any of the radio galaxies discussed here. In addition, some barred galaxies also have similar velocity fields. Regardless of the precise mechanism producing these anomalous velocity fields, it is clear that they imply an underlying galaxy structure which is similar to that needed to induce gas infall into the nucleus. The anomalous velocities, then, may well be evidence that the "Monster" is being fed.[5] Thus, more careful study of the phenomenon may well provide clues about the chronology and ecology of this activity.

REFERENCES

1.   Simkin, S. M. 1979, Astrophys. J., 234, 56.
2.   Ekers, R. D., and Simkin, S. M. 1982, Astrophys. J. (in press).
3.   Rubin, V. C., Ford, W. K., Peterson, C. J., and Lynds, C. R. 1978, Astrophys. J. Suppl., 37, 235.
4.   Kent, S. M. and Sargent, W. W. 1979, Astrophys. J. 230, 667.
5.   Gunn, J. E. 1979, "Active Galactic Nuclei", C. Hazard and S. Mitton (eds), Cambridge Univ. Press, Cambridge, pp. 213.

DUST AND GAS IN TRIAXIAL GALAXIES

George Lake
Bell Laboratories

Colin Norman
Huygen's Laboratorium

There are many reasons to consider the dynamics of triaxial galaxies. Elliptical galaxies are not rotationally flattened (Illingworth, this conference) and their isophotal profiles are often twisted. N-body simulations of collapse have produced a number of final state figures which are triaxial. Even those still skeptical will have to grant that barred spirals (a category which seems to include the Galaxy) are certainly not axisymmetric.

These arguments do not demonstrate convincingly that ellipticals are triaxial. Illingworth (this conference) has shown that low luminosity ellipticals appear to be consistent with oblate, rotationally flattened models and has argued that the continuity in all properties of faint and bright ellipticals suggests that all ellipticals are oblate. Clearly, one needs a clean, decisive probe. Gas and dust provide just that. Details of the work discussed here may be found in a paper submitted to the Astrophysical Journal.

We have analyzed two model Hamiltonians designed to approximate the inner and outer regions of a triaxial galaxy. These enable us to: (1) find isolating integrals of motion, (2) integrate the equations of motion in one subclass, (3) find the periodic orbits and assess their stability using reduced potentials or Liapunov functions, (4) associate families of orbits (box, tube, shell, etc.) with perturbations of the stable periodic orbits, and (5) apply Melnikov's method to deduce the fate of dissipating gas (or the effect of dynamical friction on the distribution of globular clusters) in these potentials (stable periodic orbits "attract," whilst unstable ones "repel").

One important feature is the structural instability of axisymmetric systems. In these potentials there is only one orbit family: all orbits circulate about the axis of symmetry tracing out a "tube." The only exceptions are the neutrally stable radial orbits lying in the plane and the unstable orbits which go over the pole. At the slightest introduction of nonaxisymmetry, new orbit families emerge. A class of radial orbits becomes stable to perturbation (parenting box orbits), as does one of

315

E. Athanassoula (ed.), Internal Kinematics and Dynamics of Galaxies, 315–316.
Copyright © 1983 by the IAU.

the polar orbits (yielding a family of tubes in a different orientation).

With these results, the unusual configurations of dust and gas seen in many ellipticals become compelling evidence for triaxiality. Particular examples are NGC 2768, NG3 5363, Cen A and especially NGC 2685, the Spindle galaxy which has two loops of gas. Recently, there have been proposals to explain polar rings as "quasi-stable" orbits in oblate con-figurations. "Quasi-stability" is a way of proposing to find gas in an unstable orbit which should actually behave as a "repellor." There are clear signs that such orbits are not responsible for these configurations. Figure 1 shows the difference in appearance of dust and gas started in

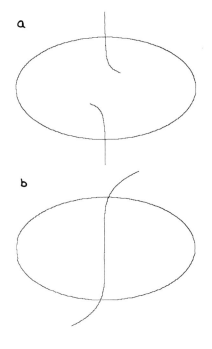

Figure 1: Appearance of a dust lane (open curve) relative to the stellar figure (ellipse) formed by a) capture in the unstable polar orbit of an oblate figure, b) capture in the stable periodic orbit of a triaxial figure.

an unstable polar configuration and allowed to evolve (1a) and that being attracted to a stable periodic orbit (1b). Cen A is a good example of the latter. We challenge advocates of the former to find something which looks like Figure 1a. A greater challenge is presented by NGC 4125 (Bertola, this conference). In this galaxy the stellar rotation is max-imized at an angle of 30° with respect to the major axis, a position coincident with the dust lane. This is impossible to reconcile with an axisymmetric model and provides compelling evidence that this is indeed a strongly triaxial galaxy.

The combined study of gas and stars in elliptical galaxies provides a powerful probe of their three dimensional shapes and dynamics. Observa-tions of elliptical galaxies with skewed dust lanes are of particular interest in this context.

VI

MERGERS

OBSERVATIONAL EVIDENCE FOR MERGERS

François Schweizer
Department of Terrestrial Magnetism
Carnegie Institution of Washington, Washington, DC 20015

Theory has long suggested that dynamical friction between colliding galaxies must lead to mergers. The problem for observers has been to find which galaxies are mergers. I shall first review the available evidence for mergers in various kinds of galaxies, then propose a tentative classification scheme for mergers, and finally discuss mergers in giant ellipticals and their relation to the evolution and perhaps even the formation of ellipticals.

I shall here define merger more restrictively than did Toomre (1977), who in his list of "eleven prospects for ongoing mergers" included pairs of interacting galaxies that can still be discerned separately, such as NGC 4038/39 (the "Antennae") and NGC 4676 (the "Mice"). Even NGC 3256, where it is already difficult to distinguish the two interacting disks, I shall not yet call a merger. Rather, with few exceptions, I shall consider systems where the two participants have merged into one single remnant, even though their individual debris, such as tidal tails, may still be visible. This restricted definition of merger would leave only three candidates in Toomre's list. Here, I shall describe or at least mention over two dozen such mergers.

1. MERGERS IN cD GALAXIES

The large extent of cD envelopes (>100 kpc) and the existence of dynamical friction suggest strongly that cluster galaxies which cross the envelopes will eventually fall in and merge (Ostriker & Tremaine 1975). The direct observational evidence for the occurrence of this process is, however, meager. Various observations of colors and color gradients in cD galaxies have been used to claim support for the theory; yet such claims seem inconclusive because we know little about the meaning of color differences and gradients even in nearby, normal ellipticals. Large core radii have also been quoted as evidence for mergers in cD's. However, they are ill determined observationally and it is not even clear that merging, which seems to occur nonhomologously, ought to lead to bloated cores (White 1978; Schweizer 1981).

319

E. Athanassoula (ed.), Internal Kinematics and Dynamics of Galaxies, 319–329.

It seems to me that two promising pieces of observational evidence
in favor of mergers in cD galaxies are (1) the presence of multiple
nuclei and (2) the existence of asymmetric envelopes. Hoessel (1980)
found that 28% of brightest cluster galaxies have multiple nuclei, and
derived merger rates from this number. Some of the "nuclei," however,
may be other, superposed cluster members: Minkowski (1961) found a
velocity spread of 1400 km/s among the three "nuclei" of NGC 6166.
Hence, a careful statistical study of multiple nuclei ought to include
velocity measurements and corrections for foreground and background
contamination. Asymmetries in cD envelopes have been noted by various
authors (e.g., van den Bergh 1977). Since they indicate deviations from
symmetric equilibrium configurations, they are probably transient and
may be the result of interactions with cluster galaxies or of mergers.

2. THREE WELL ESTABLISHED MERGERS

NGC 1316 (= Fornax A) is a classical Morgan D galaxy, which means
it has an elliptical-like body embedded in an extensive envelope. Figure
1b shows this envelope and two major protruding filaments discovered by
Arp. Schweizer (1980) interpreted them as the tails of small disk
galaxies that fell in ~1 Gyr ago. The case for a merger there rests also

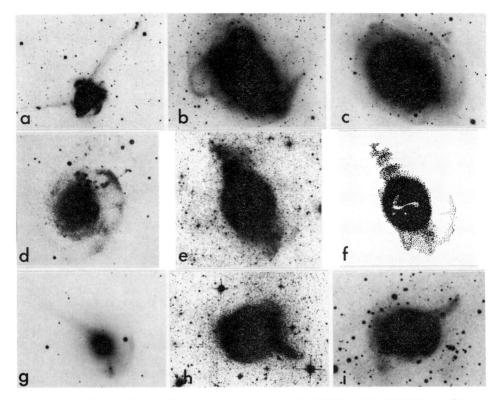

Figure 1. Tails and ripples in mergers: (a) N7252; (b) N1316 tails and
(c) ripples; (d) N3310; (e) + (f) N5128; (g) N7135; (h) M89; (i) N5018.

on the presence of (1) a highly inclined gas disk that rotates much faster than the stars and (2) "ripples," which are the arc-like features shown in Fig. 1c (enhanced photographically by unsharp masking). These ripples pervade the main body, consist of moderately old stars, and are a frequent signature of infallen disks, as we shall see. The small galaxy just to the north in Fig. 1b is NGC 1317, a normal barred spiral seen nearly face-on; it seems unlikely to have caused the tails and ripples because it looks quite unperturbed itself.

NGC 5128 (= Cen A) is probably the best known case of a recent merger, even though it was widely regarded to be an explosive galaxy until a few years ago. The general change of opinion has been caused by Graham's (1979) observation that the gaseous disk rotates fast, as found by the Burbidges in 1959, whereas the stellar body rotates barely at all. The fast rotation of the gas excludes nuclear ejection as its source, and the slow rotation of the spheroid excludes the possibility that the gas was shed by the stars. The most likely remaining source for the gas seems to be infall from outside of the galaxy, and models based on the assumption that a small gas-rich companion fell in have indeed been successful at modeling the observed warp of the gaseous disk (see comments after the paper by K. Taylor). Luckily for my contention that ripples are signatures of infallen disks, NGC 5128 features some ripples, too. On the deep photograph by Cannon (1981) shown in Fig. 1e, they appear as luminous steps in the faint NE and SW extensions, much as they were seen and sketched by Johnson (1963), here Fig. 1f. (After this Symposium, J.A. Graham showed me a beautiful new photograph, on which the ripples can be seen around the whole periphery of the galaxy.) Finally, the curved filament that emerges from the SW extension and turns northward may be either the remaining tail of the captured disk galaxy or, perhaps, just another ripple. Whatever they are, the various appendices of NGC 5128 seem to be part of the mess associated with the merger of two galaxies.

NGC 7252 is a particularly well established merger because of the lucky coincidence of five characteristics (Schweizer 1982): (1) It has two long tails (Fig. 1a) indicative of a strong interaction between two disk galaxies of about equal masses. (Such interactions produce only one tail per disk; see NGC 4038/39 and Toomre & Toomre 1972.) (2) The galaxy is so isolated that there is no external source for the observed tidal perturbations. (3) The tails move in opposite directions relative to the nucleus, as required by any tidal model. (4) Despite the two tails, there is one single body and nucleus. (5) Spectroscopy of the gas in the body reveals two surviving motion systems, which cause velocity reversals in the rotation curve. Apparently, two large disk galaxies began merging 1.0 Gyr ago, the time being calculated from the tail lengths divided by the velocities. The result is a single, rather symmetric, though messy looking remnant with an A-star-type spectrum and a respectable $M_V = -22.8$ ($H_0 = 50$). Ripples again pervade the body. The mean surface brightness follows an $r^{1/4}$ law quite closely, indicating that the remnant has a structure similar to an elliptical, presumably because it relaxed violently (Lynden-Bell 1967). Obviously, this merger of two disks has been very nonhomologous, and I believe it provides a first solid piece of evidence for the delayed formation of some ellipticals envisaged by the Toomres (1972).

3. A TENTATIVE CLASSIFICATION OF MERGERS

Before discussing individual candidate mergers, it seems useful to establish a rough classification for them. One possible scheme is to order them by the combined Hubble types of the merged components. To simplify matters, assume that there are only three structural types of galaxies: E = ellipticals, D = disks (incl. S0's), and G = gas clouds (incl. Irr's). If one distinguishes the more massive and less massive components by capital and small letters, respectively, and combines them in this order, the resulting classification has nine classes: Ee, Ed, Eg, De, Dd, ..., Gg. Figure 2 (to the right) represents these classes by boxes arranged in a square; their dimensions are roughly proportional to the fractions of galaxy types in the field. Written into the boxes are the names of merger candidates, e.g., NGC 1316 and 5128 as types Ed, and NGC 7252 as DD (where I use two capital letters to indicate that the disks were of roughly equal masses). We now describe the various classes in turn.

	e	d	g
E	N 750 I5250	N 596, 1316, 3923, 5018, 5128, M89	N42– 78
D	?	N3310, 4650A, 7135, Arp 230, Smith 60, AM 0044-2437 Equal disks (DD): N3921, 6052, 7252. N2685 ↔	?
G	?	?	II– Zw40

Dd mergers predominate in the field, where ~80% of all galaxies are disks. Among their most telling signatures are tidal tails, ripples (see §4), and motions in crossed planes. They display an amazing variety of forms, some resembling NGC 7252, and others not at all. NGC 3921, e.g., is a close kin in most respects (Schweizer 1978); its two long tails suggest that here, too, two similarly massive disks have just completed merging. NGC 6052 (= Mrk 297) may be a third example of a DD merger, although it has not yet run to completion; it features two nuclei and motion systems, and has been interpreted as a probable collision of two late-type spirals (Alloin and Duflot 1979). Certainly, its tremendous burst of star formation is compatible with this view, and I would guess that the stubs extending north- and southward are two beginning tails. NGC 3310 has a totally different appearance; it was first noted by Walker and Chincarini (1967) for its nucleus, which is offset from the rotation center, and the "bow and arrow" signature shown in Fig. 1d. Its beautiful inner spiral arms are associated with a strong density wave (van der Kruit 1976). Balick and Heckman (1981) found a sharp and unique (among 20 observed Sb galaxies) drop in metallicity within 1 kpc from the center and suggested a merger; a small, low-metallicity galaxy would have fallen in 100 Myr ago, depositing fresh gas, displacing the nucleus, and generating the density wave. It seems to me that the strongest signatures in favor of a recent Dd merger there are the bow and arrow, which I would reinterpret as the ripple and tail of the small disk intruder. Many more such mergers of unequal mass may have taken place in nearby spirals, but may remain unknown because of the rapid (<2 Gyr) disappearance of all tracers. Finally, let me mention a recently discovered class of probable Dd mergers: SO galaxies with polar rings or disks, such as NGC 2685 (Schechter & Gunn 1978), NGC 4650A (Laustsen & West 1980) which clearly is an edge-on old disk with a polar ring rather than a prolate elliptical (Schechter, private comm.), A0136-080 that Rubin, Whitmore, and I have been studying, and a half dozen others. They all have two crossed disks, of which one is massive and old, whereas the other seems to consist of the debris of a small victim (disk?) galaxy.

Ee mergers are hard to catch for two reasons. First, ellipticals being rare in the field, Ee interactions are (rare)^2. Second, because of their hot nature, interacting ellipticals produce faint, diffuse fans that dissipate rapidly, rather than bright, narrow tails. The only two candidate EE mergers that I know of are NGC 750/751 and IC 5250, both still distinctly double and hence not mergers in the restricted sense.

We postpone the discussion of Ed mergers until §4. Of the remaining six classes, Gg is perhaps the most interesting. The type example there is II Zw 40, a classical "isolated extragalactic H II region" discussed by Sargent and Searle. Recently, Baldwin et al. (1982) have shown that it has two tidal tails (the "fan jets S and SE" noticed by Zwicky), probably consists of two gas-rich dwarfs in collision, and seems to have many kins among other isolated H II regions. Given the brief duration and visibility of these events, large numbers of Gg remnants must exist. Among the Eg's, the elliptical NGC 4278 (see G. Knapp and M.H. Ulrich, this volume) is considered prototypical, although I suspect that an intruding small disk cannot be excluded as an alternative to a gas cloud. The Ge, Gd, and Dg classes have no known members yet, presumably because massive gas clouds are scarce and small gas clouds are difficult to detect when falling into large disks. Finally, small ellipticals punching through larger disks can temporarily create ring galaxies, but we do not yet know of any case where an eventual De merger seems likely.

4. MERGERS IN ELLIPTICALS

Ed mergers receive currently much attention, and for good reason since they tell us something about the evolution of ellipticals. The discovery of a "jet" and "shells" in M89 (Malin 1979; here Fig. 1h) and of "giant shells" in four more normal ellipticals (Malin and Carter 1980) has made it clear that such tails and ripples, as I prefer calling these structures, are common in giant ellipticals. We know already that tails are tidal structures, but what are the ripples? Based on their frequent association with tails in merging galaxies (Fig. 1), Schweizer (1980, 1982) suggested that they result from disk mergers. This empirical conclusion is now getting theoretical support from model calculations by P. Quinn (see this volume) and by A. Toomre.

The models suggest that ripples are the distorted remains of a disk galaxy accreted by an elliptical. The top two rows of Figure 3, kindly

provided by Toomre, illustrate the accretion process. In this simplified
calculation, a disk of 2000 noninteracting test particles (top row, time
zero) is "dropped" (in the direction of the arrow) into a fixed Plummer
potential of characteristic radius <u>a</u> (marked by the dashed circle) from
an initial distance 4<u>a</u>. The orbiting disk gets bent and stretched by
differential gravity, and wraps itself around the center of the
potential (t=20-80). As the wrapping progresses over several revolutions
(second row, scale slightly reduced and 6000 disk particles), more and
more "ripples" appear wherever the distorted disk surface is seen nearly
edge-on. They last for over ten initial orbital periods and resemble the
ripples observed in ellipticals to a surprising degree: The bottom row
of Fig. 3 shows NGC 3923 first <u>unmasked</u>, then masked unsharply by Malin
to emphasize the faint ripples, and Frame t=300 of the above model
sequence rotated by 180° and mirrored. Note that the model has not been
matched to the galaxy in any detail and that it shows "ripples" from
more than half of all possible viewing directions. Further model
calculations by Toomre suggest that even <u>part</u> of a disk, torn off during
a close E-D encounter, suffices to produce ripples in the elliptical.

To determine the frequency of ripples, I have photographed 28
bright field ellipticals (-19.5 > M_B > -22) with the CTIO 4-m telescope.
I find that 15 of them show ripples, though mostly not as spectacular
ones as NGC 3923 does. Among the ellipticals with newly discovered
ripples are <u>NGC 596</u>, well known for its strong isophotal twist (Williams
1981) and relatively fast rotation in the core (Schechter & Gunn 1979);
<u>IC 3370</u>, with crossed "streamers" forming an X and ripples filling two
opposite quadrants of this X; and <u>NGC 5018</u>, a "true" elliptical (i.e.,

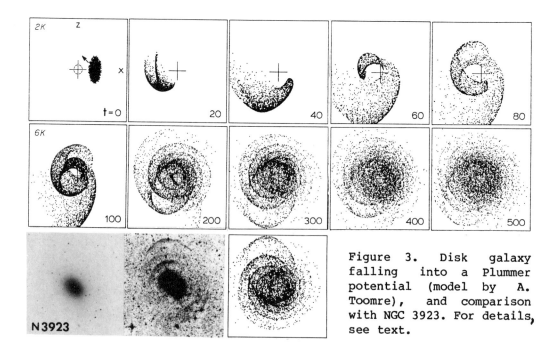

Figure 3. Disk galaxy
falling into a Plummer
potential (model by A.
Toomre), and comparison
with NGC 3923. For details,
see text.

classified as E by both de Vaucouleurs and Sandage), yet featuring dust, ripples, and two tail-like filaments (Fig. 1i).

The conclusions from all this work on Ed and other mergers are: (1) One quarter to one half of the bright field ellipticals show signs of having accreted disks or parts of disks recently. (2) If ripples last $\lesssim 2$ Gyr, as the model calculations suggest, then a typical bright field elliptical accretes at least 2-5 (parts of) disks over a Hubble time. After correction for higher interaction rates in the past (Toomre 1977) and aspect dependance, this number may increase to 4-10 (parts of) disks accreted over the age of the Universe. (3) This accretion process must have tended to reduce the number of disk galaxies and increase the luminosities of bright ellipticals. And (4), in at least one case, NGC 7252, we seem to witness the delayed formation of an elliptical from two merged disks. In a more speculative vein, this work also suggests that (a) random accretion of disks may have led to the slow apparent rotation of bright ellipticals; (b) varying admixtures of disk stars may explain the varying UV brightnesses of ellipticals; (c) disk wrapping may be an efficient mechanism for gas "removal" from ellipticals; and (d) ellipticals in clusters may be accreting disks also (M89 in Virgo did it).

To facilitate further work, Table 1 (next page) gives a list of currently known candidate mergers with their characteristics (isolation, ripples, tails, velocity anomalies, etc.) and types.

REFERENCES

Alloin, D., and Duflot, R.: 1979, Astron.Astrophys. 78, L5.
Baldwin, J.A., Spinrad, H., & Terlevich, R.: 1982, M.N.R.A.S. 198, 535.
Balick, B., and Heckman, T.: 1981, Astron.Astrophys. 96, 271.
Cannon, R.D.: 1981, in "Proc. Second ESO/ESA Workshop, Munich," p.45.
Graham, J.A.: 1979, Astrophys.J. 232, 60.
Hoessel, J.G.: 1980, Astrophys.J. 241, 493.
Johnson, H.M.: 1963, Publ. NRAO 1, 251.
Laustsen, S., and West, R.M.: 1980, J.Astron.Astrophys. 1, 177.
Lynden-Bell, D.: 1967, M.N.R.A.S. 136, 101.
Malin, D.F.: 1979, Nature 277, 279.
Malin, D.F., and Carter, D.: 1980, Nature 285, 643.
Minkowski, R.: 1961, Astron.J. 66, 558.
Ostriker, J.P., and Tremaine, S.D.: 1975, Astrophys.J. 202, L113.
Schechter, P.L., and Gunn, J.E.: 1978, Astron.J. 83, 1360.
Schechter, P.L., and Gunn, J.E.: 1979, Astrophys.J. 229, 472.
Schweizer, F.: 1978, in IAU Symposium No. 77, "Structure and Proper-
    ties of Nearby Galaxies" (Dordrecht: Reidel), p.279.
Schweizer, F.: 1980, Astrophys.J. 237, 303.
Schweizer, F.: 1981, Astrophys.J. 246, 722.
Schweizer, F.: 1982, Astrophys.J. 252, 455.
Toomre, A.: 1977, in "The Evolution of Galaxies and Stellar Popula-
    tions" (New Haven: Yale University Obs.), p. 401.
Toomre, A., and Toomre, J.: 1972, Astrophys.J. 178, 623.

Table 1: CANDIDATE MERGERS

Name	Other name	Decl.	Isol.	Ripples	Tails	Veloc.	Other features	Type	Note
N1316	For A	-37°		x	x	x		Ed	1
N5128	Cen A	-43	x	x	(x)	x		Ed	1
N7252	Arp 226	-25	x	x	x	x		DD	1
N 596		-7°		x	x	(x)	isophotal twists	Ed	2S
N1344		-31	x	x			dust	Ed	2M
N2685	Arp 336	+59	(x)			x	inner disk	Dd,Dg	2
N3310	Arp 217	+53	x	x	x		strong dens. wave	Dd	2
N3509	Arp 335	+5	x		x			Dd	2
N3921	Arp 224	+55			x		sloshing isoph.	DD	2
N3923		-29	(x)	x			dust	Ed	2M
N4552	M89	+13	x	x				Ed	2M
N4650A		-41				x	SO disk + ring	Dd	2
N5018		-19	x	x			dust	Ed	2S
N6052	Mrk 297	+21	x		(x)	x	two nuclei	DD	2*
N7135		-35	x	x	x			Ed,Dd	2
AM 0044-2437		-25				x	two rings	Dd	2
Arp 230		-14	x	x	x		N5128-type dust	Dd,Ed	2
Smith 60		-34	x		x	x		Dd,Ed	2
N750/51	Arp 166	+33°	(x)		x			EE,ED	3*
N1395		-23		x	(x)		isophotal twists	Ed	3M
N3585		-27	x		x			Ed	3S
N3656	Arp 155	+54	x	x	(x)		N5128-type dust	Dd,Ed	3
N4194	Arp 160	+55	x	x	x			Dd?	3
N7585	Arp 223	-5		x	(x)			Dd,Ed	3
N7727	Arp 222	-12	(x)	(x)	x			Dd	3
I1575	Arp 231	-4		x				Dd,Ed	3
I3370		-39	x	x			Crossed streamers	Ed	3S
I4797		-54	x					Ed	3M
I5250		-65			x			EE,ED	3*
ESO 293-IG37		-42	(x)	(x)	(x)			Dd	3
ESO 341-IG04		-38	x		x			Dd,Ed	3B
II Zw 40		+3	x		x			GG,DD	3*

Notes: 1 = established merger;   2 = probable m.;   3 = candidate m.;   B = discovered by Bergvall; M = Malin (1979, or with Carter 1980);   S = from Schweizer's survey; * = two component galaxies are still discernible.

van den Bergh, S.: 1977, Publ.A.S.P. 89, 746.
van der Kruit, P.C.: 1976, Astron.Astrophys. 49, 161.
Walker, M.F., and Chincarini, G.: 1967, Astrophys.J. 147, 416.
White, S.D.M.: 1978, M.N.R.A.S. 184, 185.
Williams, T.B.: 1981, Astrophys.J. 244, 458.

DISCUSSION

RICHSTONE : Kirk Borne and I had a first try at reconstructing NGC7252 as a wreck of two disks using a numerical restricted three-body technique. The spatial match isn't perfect and can probably be improved. The match to the rotation curve is better. I would like to make three comments :
a) It does become elliptical
b) It has a rather high V/σ for a luminous elliptical
c) It is an isolated system with considerable angular momentum. If galactic angular momentum comes from tidal torque, where is the object that supplied the torque in this case ?

SANDAGE : You leave me with the impression that all E galaxies are formed by merging disks. Your evidence in this case is the transient ripples in the outer envelopes. If E galaxies are secondary structures rather than primary, then the fundamental distinction between disk and spheroidal galaxies at the time of formation would be in doubt. Hence many of the ideas of formation via collapse with and without dissipation would be wrong. As your conclusion is based on the outer ripples, the following question arises : how many of your E galaxies with ripples are in the field, and how many are in groups or clusters ? If any are in the field, is there not a problem with the rate of mergers of field disks in the past 10^9 years, given the present low density of neighbors and the low mean random velocities ?

SCHWEIZER : I have made two specific claims : (1) In NGC7252, we seem to be witnessing the formation of one future giant elliptical from the merger of two disks ; and (2) between 1/4 and 1/2 of all giant ellipticals in the field have accreted disks in the past 1-2 Gyr. This accretion is an evolutionary process, and I have avoided making statements concerning how and when these field ellipticals were formed. The conclusion that they accreted 4-10 disks or parts of disks in the past is, however, difficult to avoid. In answer to your question, about half of my field ellipticals are as isolated as one can find, which means they have no neighbor within 500 kpc projected distance. Even in the "field", galaxies are often in pairs or very loose groups ; the crossing time for 500 kpc at v = 250 km/s is only 2 Gyr. It is well known that the observed number of interacting galaxies is significantly larger than one would expect from random encounters with the low mean random velocities that you advocate ; this implies that either these velocities are higher or, more likely, the majority of galaxies are members of multiple systems, just like stars are.

SANDAGE : I do not understand your answer in terms of binary galaxies. You state that the age of these ripples (or shells) is only 10^9 years. Hence they will not occur again in any given rippled galaxy because the companions no longer exist. Since less than 25 % of the field disk galaxies have companions now, I don't see how you can get a present fraction of 25 % of E's which have ripples and thus, according to you, recently merged, unless you believe that at the end of the next 10^9 years,

the % will be much less (i.e. will be at the present frequency of binary
field disk galaxies which is certainly much less than 10 %).

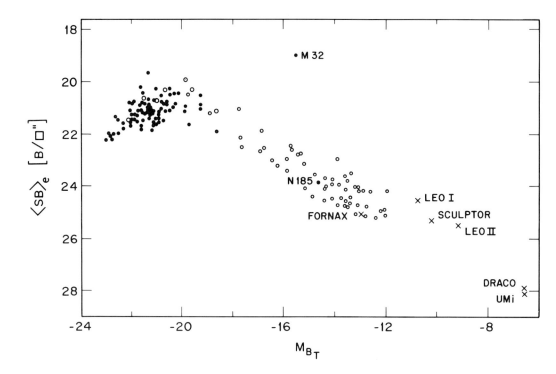

That E galaxies are generic to the Hubble sequence and are not due to
secondary processes as you suggest, follows from the continuity of
properties of effective surface brightness, effective radius, cut-off
radius, and central density over more than 10^6 in intensity from
$M_B = -23$ to $M_B = -8$. The evidence for this continuity in effective
surface brightness is shown in the diagram which is from a study of
Virgo cluster E galaxies (open circles) supplemented with field galaxies
(closed circles) and with 7 local group spheroidals. The Virgo cluster
data is from work done with Bruno Binggeli of Basel and Massimo Tarenghi
of ESO.

SCHWEIZER : It seems mysterious to me how you can draw conclusions
concerning the formation of ellipticals from your graph, unless you claim
to understand the maximum that occurs at $M_B = -20$ and the differing
slopes on either side. Furthermore I note that all field galaxies lie
on one side of the maximum and most cluster galaxies on the other. This
diagram, which is not understood at all, cannot compete with signatures
such as the rails and ripples, which we understand and know how to model.

SIMKIN : The interesting rotation curve you showed for NGC7252 has the
same structure as that found in the gas of 3 of the 5 cD radio galaxies
I have measured. If it is a sign of recent mergers then it must persist

much longer than any morphological signs of such a merger, since none of
these cD galaxies show any "shells", "tails", or double nuclei.

OORT :   Is there a satisfactory interpretation of the ripples ?

SCHWEIZER :   Peter Quinn will give us more details about the models in a
moment and will show a movie. Let me add that if ripples are distorted
and wrapped-around disks, then spectrograms taken with a long slit across
the ripples should show velocity "jumps" at the ripples. These local
deviations from the velocity of the underlying elliptical should have
opposite signs on opposite sides of the nucleus. Bosma (priv. comm.) has
some evidence for such jumps at the innermost ripple in NGC 1316 in one
slit position, but does not know yet for the ones further away.

DRESSLER :   Fornax A certainly cannot be a prototype of how bright ellipticals are formed. It fails to satisfy the Faber-Jackson relation -its
central velocity dispersion is much too low for its luminosity (220 km/sec
instead of $\sim$ 350)- and its V/σ (as determined by Bosma) places it above
the oblate line in the V/σ - ellipticity diagram. These results are
extremely atypical of luminous ellipticals.

SCHWEIZER :   I have claimed only that Fornax A swallowed one or two small
disks recently. One ought to consider how such infalls might affect the
measured velocity dispersion : the line profiles must be composite, i.e.,
formed by two different stellar population types with different velocity
dispersions. It is not clear to me how Fourier programs derive velocity
dispersions from such composite, presumably non-Gaussian line profiles.

RUBIN :   In two pairs of interacting spirals which I have studied, the
rotation curves fall to zero (and below) in all galaxies. On the basis of
these observations as well as your own, I would suggest that steeply
falling rotation curves are a diagnostic for interacting galaxies.

SCHWEIZER :   I agree.

TAURUS OBSERVATIONS OF THE EMISSION-LINE VELOCITY FIELD OF
CENTAURUS A (NGC 5128)

Keith Taylor
    Anglo-Australian Observatory & Royal Greenwich Observatory
Paul D. Atherton
    Imperial College, London

ABSTRACT

    Using TAURUS - an Imaging Fabry Perot system in conjunction with
the IPCS on the AAT, we have studied the velocity field of the Hα
emission line at a spatial resolution of 1.7" over the dark lane
structure of Centaurus A. The derived velocity field is quite
symmetrical and strongly suggests that the emission line material is
orbiting the elliptical component, as a warped disc.

1. OBSERVATIONS

    Recently much interest and speculation has centered on the origin
and dynamics of the gas and dust which constitute the dark lanes across
the giant elliptical radio galaxy Centaurus A. In an attempt to clarify
the nature of this structure we have used TAURUS - an imaging Fabry
Perot system (Taylor 1978, Taylor and Atherton 1980, Atherton et al
1982) to study the velocity field of the Hα emission over the central
9 x 5 arc mins.

    The Fabry Perot etalon was used at a finesse of 25 in the 332nd
order of interference to study the Hα line, thus giving us a velocity
resolution (FWHM) of 36 km sec^{-1} and a Free Spectral Range of 903 km sec^{-1}
An integration time of 2 hours yielded approximately 3.10^3 independent
velocity determinations over the surface of the dust lane, with an
internal consistency of ∿ 6 km s^{-1} r.m.s. Indeed our data is in
excellent agreement with Graham's spectroscopic velocity determinations.
However we see not a simple ring of HII regions, as claimed by Graham
(1979), but instead a very symmetrical S-shaped envelope of emitting
regions, similar in shape to the picture developed by Tubbs (1980) from
theoretical considerations. Our spatial coverage agrees well with the
map of HII regions derived by Dufour et al (1979) from UBV photometry,
giving the velocity of almost all the region presented by them, and also
showing the presence of an underlying diffuse component.

331

E. Athanassoula (ed.), Internal Kinematics and Dynamics of Galaxies, 331–334.
Copyright © 1983 by the IAU.

Perhaps the most striking characteristic of the velocity field of the excited gas is its symmetry about the centre of the underlying elliptical component. The velocities of the warped regions in the NW and SE are farily constant, consistent with material rotating differentially about the centre of the elliptical component. Across the centre of the object the velocity field shows a steep gradient, probably induced by the deep potential well of the elliptical, consistent with a highly inclined rotating gaseous disc. The central hole in the observed Hα emission prevents measurement of this gradient in the innermost regions and the steepest measured gradient is 170 km sec^{-1} kpc^{-1} at a radius of 0.5 kpc (D = 5 Mpc), which leads to a Keplerian estimate of 10^{10} M$_{\odot}$ inside 1 kpc.

2. KINEMATIC MODELS

We have approached the problem of interpreting this wealth of kinematic data by attempting to fit simple geometrical models suggested by the appearance of the isovelocity contours, shown on an unsharp-masked print of Cen A (courtesy of David Malin, AAO), in Figure 1.

By deprojecting an expanding/rotating circular anullus of varying radius onto the data we are able to identify the centre of rotation of the system to an accuracy of < 2 arcsec which coincides to within measurements error with the position of the radio point source, IR and X-ray nucleus. Furthermore the derived heliocentric systemic velocity of 544 ± 4 km s^{-1} is in excellent agreement with Graham's value. Most striking is the azimuthal continuity of the velocity data for each annulus indicating that despite the ∿ 70° orientation (β) of the rotation axis with the line of site, we are seeing emission material at the far side of the disc. Using however the orientation parameters used by

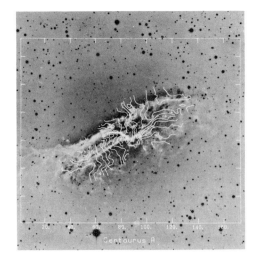

Figure 1.   25km/s isovelocity contours beginning at 330km/s (SE)

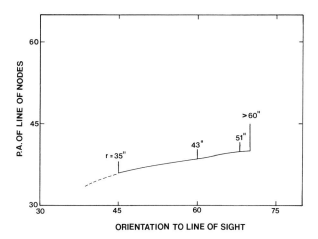

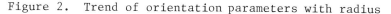

Figure 2.   Trend of orientation parameters with radius

Graham (1979) we consistently obtain a significantly non-zero expansion
component to the kinematic model, indicating either strong non-circular
motions or an incorrect value for the PA of the line of nodes (α).

Prompted by a change in the apparent expansion term with radius we
attempted to optimise the two orientation parameters α, & β at each radius
for a model which allowed for no expansion term. The results are
uncertain for radii < 30 arcsec but outside this we were able to see a
clear change of orientation of the disc with radius. This result is
depicted in Fig. 2 from which it can be seen that despite the visible
appearance of the dust lane, the warp of the disc is most prominent in
the orientation parameter, β.

3. THE KINEMATIC WARP

The slight change in P.A. of the line of nodes is in the wrong
sense to account for the warped appearance of the outer dust lane in
Cen A, and we take this as evidence that the appearance of a warp is a
result of viewing a spiral disc-like geometry nearly edge-on, so that
the trailing arms seen to wrap around the elliptical component of Cen A.
However the main change is seen in β which indicates that the tumbling
of the elliptical component seen in the absorption lines along its major
axis is duplicated in the inner dust lane.

Independent of these considerations however is our result that the
predominant value of α, the kinematic P.A. of the line of node is
130° ± 2° contrary to the visible dust lane which lies at 120°. This
difference is reinforced by the fact that the inner radio lobes, the
inner optical filaments and the diffuse X-ray structure all are orthogonal
to the kinematic α rather than the angle of the visible dust lane. Of
course the X-ray jet is even further from the visible dust lane orien-

tation (i.e. orthogonal to 143°). This suggests that all but the most
recent activity represents ejection along the rotation axis of the inner
disc, a view which may include the ejection of the X-ray jet if we assume
a gradual precession of the disc.

The reason for the descrepancy between the kinematic α and that
implied from the visible appearance of the dust lane may have to do with
the possibility that the disc is not uniformly populated with gas but
posesses a spiral-type geometry suggested also by the apparent warping
of its outer contours.

Ref.   Atherton, P.D.,Taylor,K.,Pike,C.D., Porker,N.M.,Harmer,C.F.W.,and
       Hook,R.N. 1982. Mon.Not.R.astr.Soc. In Press.
       Dufour,R.J.,Van den Bergh,S.,Hanvel,C.A.,Martin,D.H.,Schiffer,F.H.,
       Talbot,R.J.,Talent,D.L.,Wells,D.C. 1979. A.J.,84,p284.
       Graham,J.A. 1979. Astrophys.J.,232,p60.
       Taylor,K., 1978. 4th. Int.Coll.on Astrophys(Triests),p469(eds.M.Hack
       Taylor,K.,Atherton,P.D. 1980. Mon.Not.R.astr.Soc.,191,p675.
       Tubbs,A.D. 1980. Astrophys.J.,241,p969.

DISCUSSION

SCHWEIZER : I would like to point out that three models have recently
been proposed for NGC 5128, each involving a slightly different mechanism.
The first and best known model is by Tubbs (Astrophys. J. 241, 969, 1980),
who modeled the aftereffects of the infall of a small gas-rich galaxy into
NGC 5128 by studying the behaviour of an inclined disk of non-interacting
test particles in a fixed, prolate potential. Differential precession at
various radii leads to a strongly distorted disk that can be made to
resemble the observed gas-dust disk of NGC 5128. However, in a detailed
comparison with all existing observations, G. Simonson (Ph.D. thesis,
Yale Univ. 1982) has shown that the Tubbs model errs on which is the
"upper" side of the disk that we see, and that it cannot be made to agree
with the observations. (We see the south pole of the disk in NGC 5128).
He improved on the Tubbs model by introducing cloud-cloud collisions in
the model disk, which lead to "viscous" damping and a flat central disk
that grows outward as time elapses. He places the transition region
between this damped central part and the still precessing outer parts at
about r = 4 kpc, where Graham observed a ring of HII regions. Simonson's
model reproduces many observations in considerable detail. Finally, a
third and very different model has been proposed by van Albada, Kotanyi,
and Schwarzschild (Monthly Notices R. Astron. Soc. 198, 303, 1982), who
studied the equilibrium deformation of a gaseous disk in the potential
of a slowly tumbling triaxial galaxy. However, unpublished measurements
of the rotation of the NGC 5128 spheroid by Danziger and collaborators
suggest rotation in the sense opposite from that required by the model.
If confirmed, these measurements would seem to rule out any steady-state
model as an explanation for the observed disk warp. The beautiful new
observations presented here by M. Marcelin and K. Taylor ought to be
compared with the Simonson model, which for the moment seems to be the
most detailed and promising for NGC 5128.

KINEMATICS AND EVOLUTION OF NGC 5128

M. Marcelin
Observatoire de Marseille
2, place Le Verrier
13248 Marseille Cedex 4, France.

Recently Marcelin et al (1982) published some results of Fabry-Perrot interferometry of NGC 5128. We would like here to draw attention to several more aspects of this enigmatic galaxy.

The following photograph is a H_α plate of NGC 5128 taken by Comte and Georgelin with a focal reducer and an image-tube attached at the Cassegrain focus of the 3.6 m ESO telescope. Arrows point at bubble-like HII regions.

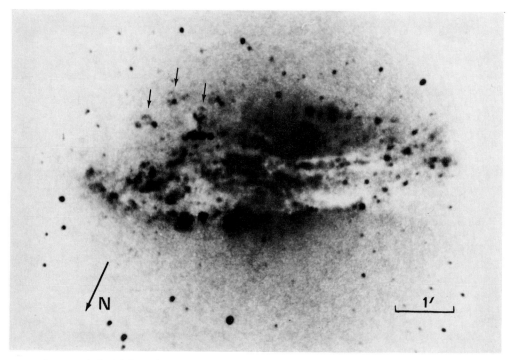

Several bubble-like HII regions have been thoroughly studied in M31 and M33 (Pellet et al. 1978). Others have been detected recently in NGC 925 (Marcelin et al. 1982b) and NGC 1313 (Marcelin and Gondoin, 1983).

E. Athanassoula (ed.), Internal Kinematics and Dynamics of Galaxies, 335–336.

The best defined ring has a diameter of 11", i.e. 250 pc at 5 Mpc, which
is consistent with other observations of HII rings. However, according to
de Vaucouleurs (1978), the diameter of the largest HII ring would suggest
here a distance closer to 4 Mpc, which is the distance to the Centaurus
group.

    The figure below shows a comparison between the observed rotation
curve (including the outermost points, see Marcelin et al. 1982a) and a
theoretical model with two components (Monnet and Simien, 1977). The
photometric parameters adopted respectively for the bulge and disk are
those given by Dufour et al. (1979) and Freeman (1970).

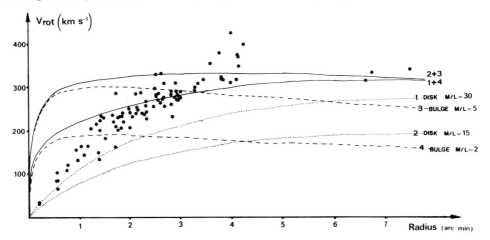

    It is clear that the observed rotation curve surprisingly looks
like a pure disk rotation curve (see curves 1 and 2). Even if one takes
into account some projection effect due to matter moving out of the plane,
(Marcelin et al. 1982a) it is not possible to explain the discrepancy
between observations and model when the bulge is supposed to have a
normal M/L around 5 (curve 2+3, quadratic sum). A value of M/L = 2 for
the bulge is far more acceptable (curve 1+4) although it implies a large
M/L = 30 for the disk.

    NGC 5128 thus appears to have a young massive dark disk surrounding
an old light bright bulge.

REFERENCES

de Vaucouleurs, G. : 1978, Ap. J. 224, 14.
Dufour, R.J., van den Bergh, S., Harvel, C.A., Martin, D.H., Schiffer,
    F.H., Talbot, R.J., Talent, D.L., Wells, D.C. : 1979, Astron. J. 84,285.
Freeman, K.C. : 1970, Astrophys. J. 160, 811.
Marcelin, M., Boulesteix, J., Courtès, G., Milliard, B. : 1982a, Nature,
    297, 38.
Marcelin,M., Boulesteix,J., Courtès,G. : 1982b, Astron.Astrophys.108, 134.
Marcelin,M., Gondoin,P. : 1983, Astron.Astrophys.Suppl. to be published.
Monnet, G., Simien, F. : 1977, Astron. Astrophys. 56, 173.
Pellet, A., Astier, N., Viale, A., Courtès, G., Maucherat, A., Monnet, G.,
    Simien, F. : 1978, Astron. Astrophys. Suppl. 31, 439.

SIMULATIONS OF GALAXY MERGERS

Simon D. M. White
Department of Astronomy and Space Sciences Laboratory,
University of California, Berkeley, CA   94720, U.S.A.

ABSTRACT   The last few years have seen a considerable amount of effort devoted to the problem of simulating the coalescence of galaxies. After a discussion of the merits and limitations of the N-body techniques that have been used, I summarise the insight this research gives into the mechanisms driving strong interactions in galaxy collisions and into the structure of the remnants such collisions produce.

1 INTRODUCTION

Strongly interacting pairs of galaxies have been known for many years, but Toomre & Toomre (1972; hereafter TT) were the first people to emphasise that such systems are likely to degenerate rather rapidly into single ellipsoidal star piles, suggesting that at least some elliptical galaxies formed by the merger of pre-existing spirals. Galaxy mergers also came into prominence in at least two other contexts during the 1970's. Lecar (1975), Ostriker & Tremaine (1975) and White (1976) independently noted that dynamical friction would cause bright galaxies to collect at the centre of rich galaxy clusters, and that these galaxies might coalesce to form the supergiant cD's often oberved there. Ostriker & Tremaine (1975) and Tremaine (1976) considered how an analogous process would cause satellite galaxies to merge with their parents. Aspects of all these processes are discussed in the reviews by Toomre (1977), Tremaine (1981) and White (1982). Little progress can be made on the problem of galaxy coalescence by analytic methods. However, the recent rapid increase in available computing power and the development of efficient N-body programs have made it possible to simulate the interaction and merging of galaxies directly; as a result considerable effort has been put into merger simulations over the last five years. While the problem is still far from fully explored (for example there are few published simulations of mergers of unequal systems) a number of general results are beginning to emerge.

2 NUMERICAL METHODS

The collision and merging of galaxies is difficult to simulate

E. Athanassoula (ed.), Internal Kinematics and Dynamics of Galaxies, 337–345.

realistically with present techniques. During galaxy merging large
global changes in structure occur and can only be followed correctly by
a program which solves self-consistently for the gravitational field at
all times. In all but a few atypical situations the mass distribution
is fully three-dimensional and possesses no usable symmetry properties.
In addition the program must have a large dynamic range. A resolution
of a few hundred parsecs is needed to model the central regions of a
galaxy, while the orbit of a pair prior to collision should be several
hundred times bigger. The orbital period of stars in a galaxy core is
shorter than the time required for complete merging by a similar factor.
The ideal simulation method should thus have temporal and spatial
resolution which can vary strongly with particle position.

The simplest kind of N-body scheme is a direct integrator which
calculates the force on each particle by summing the forces exerted on it
by all other particles. The most efficient such schemes use two
timesteps for each particle; one for the rapidly changing force due to a
few nearby particles and one for the more slowly changing force due to
the rest of the system (Ahmad & Cohen 1973). Such codes are clearly
Lagrangian. Their resolution limit is a fraction $1/N$ of the mass of the
system. Their spatial and temporal resolution vary with position so that
the orbits of a random sample of stars are followed with uniform
accuracy. These properties are ideal for the merger problem, but the
time-consuming force calculation limits the number of particles to
$N \lesssim 10^3$. As a result, the fluctuating force field due to particle
discreteness is much more important in such a simulation than in a real
galaxy. Discreteness effects can be partially overcome by softening the
interaction potential on small scales. This amounts to replacing the
particles by fuzzy blobs of finite size; the price paid is the
introduction of a resolution limit given by the softening length. When
using such a code to simulate hot ellipsoidal systems (as in the merger
simulations of White 1978, 1979, and Roos & Norman 1979) the effects of
discreteness are limited to an orbital diffusion which drives a gradual
change in structure. Galaxies with a cold disk component are much harder
to simulate and severe disk instabilities can only be suppressed by using
a large softening length, a careful setup technique and a stabilizing hot
spheroidal component. Attempts to make "spiral" galaxies for use in
merger experiments have met with considerable difficulty (Gerhard 1981,
Farouki & Shapiro 1982, Quinn 1982, Negroponte 1982).

Transform methods can speed up the force calculation considerably.
The simplest technique is to tabulate a smoothed particle density on a
cubic grid and then use Fast Fourier Transforms to obtain the potential.
Miller & Smith (1980, this conference) have used such a method to follow
10^5 particles on a 64x64x64 grid. Such a large calculation completely
suppresses all discreteness problems and allows a very detailed sampling
of phase-space. Unfortunately, the method has linear resolution
independent of position and requires an equal timestep for all particles.
When applied to merger simulations it is actually less able to model
centrally concentrated galaxies and has more difficulty in following a
merger to completion than the direct codes discussed above. More

efficient transform codes can be constructed using spherical harmonic
expansions centred on each galaxy; the radial gridding can then be
designed to provide greatly improved resolution in regions where it is
needed. Such codes are difficult to implement because two grids are
needed whose centres are accelerated with respect to each other and to
an inertial frame. Impressive results from a two-dimensional code of
this type have been published by van Albada & van Gorkom (1977). A
technique which is fully Lagrangian and avoids all gridding has been
introduced by Villumsen (1982). The force field within a galaxy is
calculated using the first few terms of a tesseral harmonic expansion of
the potential of each particle about the galactic centre. The
acceleration of the particles can be computed by ordering them in radius
and looping through them twice. In principle the resolution of this
technique is limited only by the number of particles, although Villumsen
found it useful to introduce a softening length into the force
calculation. Once again, the merger problem is complicated by the two
accelerated centres that are needed. Villumsen (1982) studied mergers
between equal and unequal spherical systems using 2000 particles.

All the above methods simulate the evolution of purely stellar
systems. A first attempt to include a dissipative interstellar medium
in merger calculations has been made by Negroponte (1982) using a
modified direct N-body code. He includes a population of "gas clouds"
which have finite radii and can undergo dissipative collisions. His
calculations show that a relatively small gas fraction can significantly
enhance the interaction in galaxy collisions, and that much of the
interstellar gas will fall to the centre of the remnant of a merger
between spirals and will presumably form stars.

3 THE INTERACTION OF COLLIDING GALAXIES

When two galaxies collide, tidal forces couple their orbital energy
to their internal degrees of freedom. The result of this coupling is
most easily estimated in the impulsive approximation which assumes the
orbital velocity to be rapid compared to internal velocities; the
individual galaxies are heated at the expense of the orbit (Alladin
1965; see also the reviews cited above). In the strong collisions of
greatest astronomical interest the impulsive condition is not satisfied
however, and the dynamical situation is much more complex; the tidal
field couples quasi-resonantly to the orbital motion of stars within
each galaxy and can induce a strong collective response. One such
response is the coherent bounce phenomenon seen in near head-on
collisions between spherical systems (van Albada & van Gorkom 1977,
White 1978, Miller & Smith 1980). When two galaxies overlap, the stars
of each feel a strong inward force from the extra mass interior to their
orbits; one dynamical time later when they bounce back the galaxies have
separated and the extra mass has gone. The result is a forward spray of
particles producing a characteristic "bowtie" morphology. Another
example of this quasi-resonant coupling is the bridge and tail making
process investigated by TT. In a slow encounter stars couple strongly
to the tidal field if their galactocentric angular momentum is

approximately aligned with the orbital angular momentum of the galaxy
pair. Nearside stars tend to be captured onto the intruder while
farside stars are thrown into extended orbits which lag behind the
galaxy. The restricted 3-body experiments of TT showed that when a cold
disk is subjected to such perturbation, its kinematic coherence can
produce dramatic morphological structures. Except in a brief footnote
these authors did not try to estimate the backreaction of the
tail-making process on the binary orbit. Palmer & Papaloizou (1982)
found from a linear calculation that in a direct parabolic encounter
energy is transferred to the orbit from the disk. This result is not
confirmed by fully self-consistent simulations of encounters between
disk systems (Quinn 1982). White (1979) showed that strong spin-orbit
coupling occurs in encounters between rotating spherical galaxies;
corotating objects interact much more strongly than counterrotating
ones. This dependence has also been found in all attempts to simulate
the collision of disk or disk/halo systems (Gerhard 1981, Farouki &
Shapiro 1982, Negroponte 1982, Quinn 1982). Farouki & Shapiro (1982)
find an orientation dependence associated with the overlap of stellar
distributions; the interaction between disk/halo systems correlates
strongly with the degree to which the disks pass through each other.

A simulation of interacting spiral galaxies similar to that used by
TT to model NGC 4676 (the "Mice") is shown in Fig. 1. Each diagram is a
projection of a subset of the particles near the apocentre of the
post-collision orbit. The top set shows particles from the spheroidal
component of the initial galaxies while the lower set shows particles
from their disks. The left four diagrams include particles from both
galaxies, the centre four show particles from the corotating system and
the right four show particles from the other system. In each set of six
diagrams the top three show projections onto the orbital (x-y) plane and
the lower three show projections onto the x-z plane. The responses of
the two spheroids are quite similar; a few particles transfer allegiance
and a broad tail forms behind each. The response of the disks is,
however, extremely asymmetric. The corotating disk forms a massive tail
and transfers many nearside particles to its opponent; the other disk
makes a smaller tail and transfers only one particle. Tail formation
and particle transfer are similar in self-consistent calculations and in
the restricted 3-body experiments of TT (see Gerhard 1981, Negroponte
1982, and Quinn 1982); they account for most of the tidal coupling and
strongly affect the later evolution of colliding systems.

4 MERGER REMNANTS

Interpenetrating collisions from near parabolic orbits lead to
rapid coalescence; the central regions of the remnant then settle to a
more or less steady state after a few dynamical times. Since the
merging of two stellar systems is a dissipationless process in which
little mass or energy is lost in escaping stars (e.g. White 1979,
Negroponte 1982), several properties of merger remnants can be estimated
very simply. The remnant mass, luminosity (before correction for aging
of the stellar population) and binding energy are equal to those of the

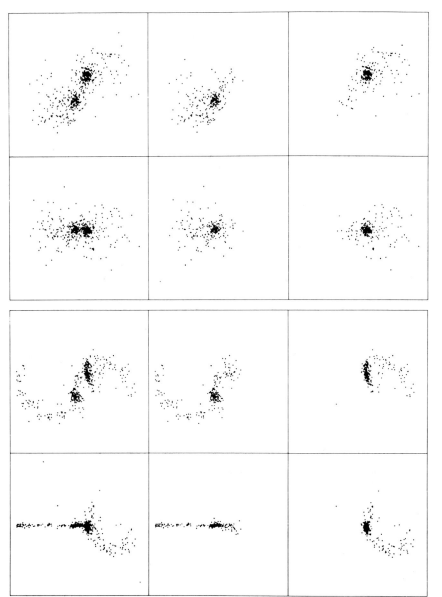

Fig. 1.   Snapshots of a 1000-body simulation of a collision between two
galaxies.   Each system initially contained a near-exponential disk of 80
"gas clouds" and 170 "stars" and a non-rotating centrally concentrated
spheroid made up of 250 "stars".   Identical systems were placed on a
parabolic orbit with pericentric distance 1.4 times their individual
half-mass radius and oriented so that one disk corotated with the orbit
and lay almost in its plane while the other was inclined at 80 degrees
with its line of nodes along the pericentre vector.   (Data from
Negroponte 1982).

pair of progenitors. Thus if two similar systems merge from a parabolic orbit, the mass, the uncorrected luminosity and the gravitational radius of the remnant are twice those of each progenitor; the overall rms velocity dispersion is the same, and the mean surface brightness is half as big. Note, however, that the remnant will not usually be homologous to its progenitors so that these scalings need not apply to quantities such as the central surface brightness and central velocity dispersion. Large-scale simulations have demonstrated several general properties of merger remnants. Provided the relaxation process during merging is sufficiently violent, the density profile of the remnant follows a power law of index near -3 over up to two decades in radius (White 1978, 1979, Gerhard 1981, Villumsen 1982, Farouki & Shapiro 1982). In mergers of disk/halo systems both components end up with a density profile of this form. Its ubiquity may be due to its logarithmic divergence in mass on both large and small scales which allows non-homologous behaviour even when both mass and energy are conserved. Profiles of different form can occur in mergers of diffuse objects from bound initial orbits, in mergers of unequal systems, and in other less violent situations.

In general merger remnants tend to be ellipsoidal, although remnants of disk-disk mergers may have box-shaped isophotes (Quinn 1982). Near head-on collisions of spherical galaxies produce prolate (E3-4) remnants with weak streaming motions; more oblique collisions lead to oblate remnants with significant rotation (White 1978, 1979, Villumsen 1982). This result carries over to the more complex case of mergers between disk/halo systems according to Gerhard (1981), Farouki & Shapiro (1982) and Negroponte (1982), but Quinn's (1982) simulations of high angular momentum encounters produce rapidly rotating bar-like structures. The most likely source of this discrepancy would seem to be the very large potential softening used by Quinn, but the situation is still unclear. All workers agree that small pericentre encounters of disk/halo systems lead to triaxial bar-like remnants in which figure rotation and internal streaming depend on disk orientation and on the pericentre of the initial orbit. Negroponte also finds a dependence of remnant flattening on the inclination of the initial disks in the sense that high inclination disks produce rounder remnants.

Fig. 2 shows how the apparent shape and apparent rotation of merger remnants depend on viewing direction and on orbital angular momentum. These remnants all resulted from parabolic encounters of similar disk/halo systems. The upper panel plots points derived for 5 small pericentre mergers with roughly random initial disk orientations; the lower panel plots points for 5 large pericentre mergers. Each remnant was projected along 20 random directions and its apparent axial ratio, maximum rotation velocity and central velocity dispersion were measured. These data were then plotted up in the $V/\sigma - \varepsilon$ diagram so beloved of observers. Most of the remnants in the upper panel of Fig. 2 are prolate; they have large V/σ and large ε, large V/σ and small ε or small V/σ and large ε depending on the direction of viewing. Typical values of V/σ are quite small and are comparable with those seen in giant elliptical galaxies. The remnants in the lower panel are all oblate and

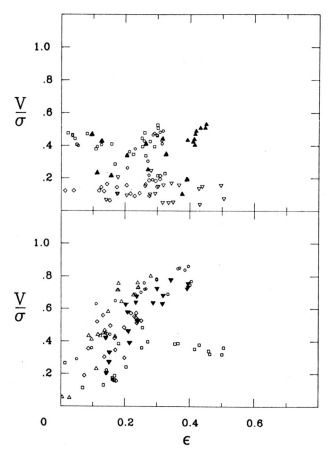

Fig. 2. V/σ – ε diagrams for the remnants of mergers between 250-body disk/halo systems. The initial galaxies had equal amounts of mass in a disk and in a non-rotating halo. All initial orbits were parabolic and all particles were counted when calculating the quantities shown. In each diagram all points plotted with the same symbol refer to the same remnant. (Data from Negroponte 1982.)

scatter along and somewhat below the "oblate line" discussed by Binney (1978); they have considerably higher values of V/σ. If any observed slowly rotating ellipticals formed by mergers of similar spirals, the angular momenta of the initial orbits must have been quite small. Aarseth and Fall (1980) argue that such an impact parameter distribution occurs naturally in a hierarchically clustering universe. Such slowly rotating merger remnants will be prolate objects.

REFERENCES

Aarseth, S.J. & Fall, S.M.: 1980, Astrophys.J. 236, 43.
Ahmad, A. & Cohen, L.: 1973, J.Comp.Phys. 12, 389.

Alladin, S.M.: 1965, Astrophys.J. 141, 768.
Binney, J.: 1978, Mon.Not.R.astr.Soc. 183, 501.
Farouki, R.T. & Shapiro, S.L.: 1982, preprint.
Gerhard, O.E.: 1981, Mon.Not.R.astr.Soc. 197, 179.
Lecar, M.: 1975, Dyn.Stell.Syst. (ed. Hayli, A.) 161.
Miller, R.H. & Smith, B.F.: 1980, Astrophys.J. 235, 421.
Negroponte, J.: 1982, Ph.D. thesis, Univ. of Calif., Berkeley.
Ostriker, J.P. & Tremaine, S.D.: 1975, Astrophys.J. 202, L113.
Palmer, P.L. & Papaloizou, J.: 1982, Mon.Not.R.astr.Soc. 199, 869.
Quinn, P.J.: 1982, Ph.D. thesis, Austral. Nat. Univ.
Roos, N. & Norman, C.A.: 1979, Astron.Astrophys. 76, 75.
Toomre, A.: 1977, Evol. of Gal.Stell.Pop. (eds. Tinsley, B.M. &
     Larson, R.B.) 401.
Toomre, A. & Toomre, J.: 1972, Astrophys.J. 178, 623.
Tremaine, S.D.: 1976, Astrophys.J. 203, 72.
Tremaine, S.D.: 1981, Struct.Evol.Norm.Gal. (eds. Fall, S.M. &
     Lynden-Bell, D.) 67.
van Albada, T.S. & van Gorkom J.H.: 1977, Astron.Astrophys. 54, 121.
Villumsen, J.V.: 1982, Mon.Not.R.astr.Soc. 199, 493.
White, S.D.M.: 1978, Mon.Not.R.astr.Soc. 184, 185.
White, S.D.M.: 1979, Mon.Not.R.astr.Soc. 189, 831.
White, S.D.M.: 1982, Morph.Dynam.Gal. (eds. Martinet, L. & Mayor, M.)
     Geneva Observatory.

DISCUSSION

PALMER : What is the rotation curve in your disks, and how stable was
it over the timescale of the interaction ?

WHITE : The rotation curve of the disk/halo galaxies used by John
Negroponte for his thesis work rises rapidly over the inner 30 % by
mass of the disk and thereafter remains essentially flat. When one of
these "galaxies" is allowed to evolve in isolation, small clumps tend to
appear and disappear in the disk despite the fact that its velocity
dispersion corresponds to a value $Q = 1.4$ for Toomre's local stability
parameter. These clumps drive a slow evolution of the disk towards a
state of higher velocity dispersion and greater central concentration,
but changes are not large over the timescale on which collision takes
place in a merger simulation. Nevertheless, direct N-body codes are too
crude to be able to simulate the evolution of a disk/halo galaxy properly ;
we have to hope that the details of disk structure are of no great importance for the later evolution of a merging system.

SANDAGE : Is the effective surface brightness of the final daughter
higher or lower than that of two identical spheroidal parents ? Observations show that $<SB>_e$ of giant E galaxies (M_B brighter than -22) are
lower by ~ 1.2 mag than those of fainter E galaxies that are between
$M_B = -22$ and -19.

WHITE : As mentioned in my text, the mean surface brightness of the remnant (defined using the total luminosity and the half-light radius) is expected to be about a factor 2 smaller than that of the progenitors, in rough agreement with the numbers you quote. It is important to remember that this scaling may well not apply to the central surface brightness. Predicting the surface brightness of the remnant of a spiral-spiral merger is more tricky, but when allowance is made for the large change in morphology and the fading of the stellar population, the remnant surface brightness appears too low compared with that of normal ellipticals of the same luminosity ; a related and more serious problem is that the remnant velocity dispersion is predicted to be too low.

ON THE FORMATION AND DYNAMICS OF SHELLS AROUND ELLIPTICAL GALAXIES

P.J. Quinn
Mt. Stromlo and Siding Springs Observatories
The Australian National University

Sharp shell-like structures around elliptical galaxies can be modelled by the phase wrapping of a cold disk of stars in the fixed potential of a massive elliptical galaxy. Such structures can then be used as probes of the potential field of ellipticals and a diagnostic of the merger event.

Schweizer (1980) has suggested that the sharp, ripple-like structures around Fornax A may be the result of a merger involving a disk galaxy. Sharp ripple or shell-like structures can also be seen around several ellipticals in the Arp Atlas (1966) and many examples have been recently found by Malin and Carter (1980, 1981) around otherwise normal ellipticals. Photometry of the shell structures (Carter, Allen, Malin, 1982), indicates their extremely low surface brightness and colours which appear to be bluer than the central galaxy. Contrast enhanced prints of prime focus plates of shell galaxies taken on the Anglo-Australia Telescope have been used to characterise the shell structures. It has been found that the shells are incomplete, that is they do not completely encircle the central galaxy. They occur over a large range in radius (outer shell radius = 30 x inner shell radius, NGC3923) and the shells are interleaved in radius, that is the next outermost shell occurs on the opposite side of the nucleus. The "shells" can be considered as caps of stars with no two caps being at the same radius.

As an attempt to model such structures, experiments involving the radial encounter of self-gravitating, N-body disks and fixed spherical potentials have been conducted. Two dimensional encounters have produced sharp, incomplete "shell-like" structures which appear on either side of the post-collision system and spread slowly in radius. N-body models failed to produce large numbers of shell-like structures due to a lack of particles to populate an increasing number of shells, particularly in three dimensions.

The formation of shells can be viewed as a simple phase wrapping of a cold system in a fixed potential, (pig-trough dynamics), Lynden-Bell (1967). Figure 1 shows the evolution of a one dimensional system of test particles falling into a fixed Plummer potential with a scale length of one unit.

347

E. Athanassoula (ed.), Internal Kinematics and Dynamics of Galaxies, 347–348.

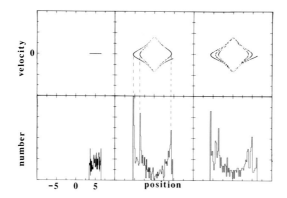

The difference in periods of the leading and trailing edge of the infalling
system causes the system to phase wrap. The maximum spatial excersion of
each phase wrap corresponds to a spatial density clump. These clumps
correspond to the shells. As the system evolves, the number of shells
increases at a rate determined by the range of periods available in the
initial system. Shells therefore occur between two well defined radii
set by the spread in energy of the initial system; the shells are
interleaved in radius and occur at different radii.

    The distribution of shells around several shell galaxies is not a
smooth distribution as would be expected from a simple wrapping of a
single system in a fixed smooth potential. It is not possible to rule out
the suggestion that the observed shell structures are the result of several
mergers with disk systems. However, the fact that the shells appear to be
predominantly interleaved over a large range in radius, argues against the
suggestion that they arise from the random infall of several disks. If
the shells are the product of the merger of a massive elliptical and a
single disk, the distribution of shells can be used to probe the potential
of the elliptical over many effective radii of the central luminous galaxy.
The shell distribution of two bright ellipticals with a large number of
shells ($\sim$20) suggests that the shells are moving in a potential that is
not as centrally condensed as the central luminous galaxy. A "halo-like"
component would be a natural component of an elliptical if it were to be
formed by or made move massive by a merger with a disk and its halo
component.

Shells represent the tell-tale evidence that the history of some elliptical
is linked to mergers involving disk galaxies. It is clearly very important
to investigate the statistics of shell ellipticals and identify the
importance of mergers to the current epoch distribution of ellipticals and
to use the shells as a probe of the mass distribution of ellipticals.

Arp, H., 1966, Ap.J., Supp. Ser., 14, 1.
Carter, D., Allen, D.A., Malin, D.F., 1982, Nature 295, 126.
Lynden-Bell, D., 1967, M.N.R.A.S., 136, 101
Malin, D.F., Carter, D., 1980, Nature, 285, 643.
Malin, D.F., Carter, D., 1981, private communication.
Schweizer, F., 1980, Ap.J., 237, 303.

MERGER CROSS-SECTIONS OF COLLIDING GALAXIES

K. Shankara Sastry and Saleh Mohamed Alladin
Centre of Advanced Study in Astronomy, Osmania University,
Hyderabad, India

ABSTRACT : The velocities of escape of colliding galaxies obtained under
the impulsive approximation are compared to those obtained from N-body
simulations.

A knowledge of merger cross-sections of colliding galaxies is very useful
in the studies of interacting galaxies. Because of the inelastic nature
of the galactic collisions, we have to distinguish between the parabolic
velocity of escape, $V_e(1)$ and the more accurate velocity of escape $V_e(2)$
obtained by taking the tidal effects into account.
We designate the value of $V_e(2)$ at closest approach, p, by V_{crit}, the criti-
cal velocity, obtained from the requirement that $|\Delta E|/E = 1$ where E is the
total orbital (translational) energy of the pair initially at infinite
separation, and ΔE is the change in this energy during the encounter due
to tidal accelerations of the stars in the two galaxies and the consequent
increase in their binding energies U_1 and U_2. Hence, $\Delta E = -(\Delta U_1 + \Delta U_2)$.
The initial velocity for which $|\Delta E|/E = 1$ is designated as V_{cap}, the
capture velocity. The two galaxies merge if $V_p < V_{crit}$ or $V_\infty < V_{cap}$,
where V_p and V_∞ are the relative velocities of the galaxies at closest
approach and at initial separation respectively.
It is the aim of the present paper to summarize the results of impulsive
approximation for the aforementioned escape velocities and compare them
with the results of N-body simulations.

The parabolic velocity of escape of a pair of polytropic model spherical
galaxies of masses M_1 and M_2 (not necessarily of the same mass distribution) superposed on each other so that their centers coincide,
$V_e^{(1)}(0)$, is obtained from

$$^1/_2 M_1 M_2 (V_e^{(1)}(0))^2 / (M_1 + M_2) = - W(0) = G M_1 M_2 / \bar{R}_{12}(0) \quad (1)$$

where $W(0)$ is the interaction potential energy of the galaxies at zero
separation and $1/\bar{R}_{12} = <1/r_{12}>$ where r_{12} is the distance between a star in
the galaxy of mass M_1 and a star in the galaxy of mass M_2. $\bar{R}_{12}(0)$ can
be obtained from the function $(\psi/s)_{s=0}$ tabulated in Alladin (1965).
If the separation r of the centers of the two galaxies is expressed in

349

E. Athanassoula (ed.), Internal Kinematics and Dynamics of Galaxies, 349–350.

units of $\bar{R}_{12}(0)$ and if the parabolic velocity, $V_e^{(1)}$ at this separation is expressed in units of $V_e^{(1)}(0)$, the relationship between $V_e^{(1)}$ and r is practically independent of the choice of models for the two galaxies. Let $x = r/\bar{R}_{12}(0)$ and $v_e = V_e^{(1)}/V_e^{(1)}(0)$. The following formulae describe the relationship quite well :

$$v_e = 1.00 - 0.00015\ x - 0.12\ x^2 + 0.025\ x^3 \ ; \ x \leqslant 3 \qquad (2a)$$
$$= x^{-\frac{1}{2}} \ ; \quad x \geqslant 3 \qquad (2b)$$

Assuming for simplicity that the energy change ΔE is symmetric with respect to the position of closest approach, we obtain V_{crit} from

$$\frac{1}{2} M_1 M_2 V_{crit}^2 /(M_1 + M_2) - \frac{1}{2} M_1 M_2 (V_e^{(1)}(0))^2/(M_1 + M_2) = \frac{1}{2}\Delta E \qquad (3)$$

Following Tremaine (1981), the assumption of impulsive approximation gives for galaxies of root mean square radius R_{rms} :

$$\Delta E = 4\ G^2 M_1 M_2 (M_1 (R_{rms})_1^2 + M_2 (R_{rms})_2^2) / 3\ V^2 p^4 \ ; \ p \gg \bar{R}_{12}(0) \qquad (4)$$

putting $V = V_{crit}$ in this equation and using it with Equation (3), we obtain V_{crit} for distant encounters. Sastry (1972) derived V_{crit} for penetrating collisions of spherical galaxies under the impulsive approximation. Let $v_{crit} = V_{crit}/V_e^{(1)}(0)$ and $y = p/(R_{h_1} + R_{h_2})$

where R_h is the radius containing half the mass of the galaxy. His results for identical polytropic $n = 4$ models of galaxies give :

$$v_{crit} = 1.15 - 0.42\ y + 0.06\ y^2 \ ; \ y \leqslant 3.5 \qquad (5a)$$
$$= 0.79\ y^{-\frac{1}{2}} \qquad\qquad ; \ y \geqslant 3.5 \qquad (5b)$$

Toomre (1977) obtained $V_{crit} = 1.16$ in a head-on collision of Plummer model galaxies under the impulsive approximation. Aarseth and Fall (1980) have summarized the results for V_{crit} obtained from N-body simulations. A comparison of our results with theirs shows that there is good agreement up to a distance of $y = 1.6$. Aarseth and Fall neglect merging beyond $y = 2$.

Capture velocities were derived by Alladin et al (1975) under the impulsive approximation for head-on as well as off-centre collisions. Let $v_{cap} = V_{cap}/V_e^{(1)}(0)$ and $z = p_\infty/(R_{h_1} + R_{h_2})$ where p_∞ is the impact parameter. We find :

$$v_{cap} = 0.54 - 0.079\ z + 0.0054\ z^2 - 0.00013\ z^3 \ ; \ z \leqslant 20 \qquad (6a)$$
$$= 0.69\ z^{-\frac{3}{4}} \qquad\qquad\qquad ; \ z \geqslant 20 \qquad (6b)$$

A comparison of these results with those of Roos and Norman (1979) shows that our values are somewhat higher.

REFERENCES

Aarseth, S.J. and Fall, S.M., 1980, Ap.J. 236, 43.
Alladin, S.M., 1965, Ap.J. 141, 768.
Alladin, S.M., Potdar, A. and Sastry, K.S., 1975, IAU Symp. 69, 167.
Ross, N. and Norman, C.A., 1979, Astr. Ap. 76, 75.
Sastry, K.S., 1972, Ap. Sp. Sci. 16, 284.
Toomre, A., 1977, "The Evolution of Galaxies and Stellar Populations"
     (eds. : B.M. Tinsley and R.B. Larson), Yale Univ. Obs., p. 401.
Tremaine, S.D. 1981, "The Structure and Evolution of Normal Galaxies"
     (eds. : S.M. Fall and D. Lynden-Bell), Univ. of Cambridge, p. 67.

ENVIRONMENTAL EFFECTS ON GALAXIES IN CLUSTERS

R. H. Miller*
European Southern Observatory

B. F. Smith
NASA Ames Research Center

This study is part of a program to investigate consequences to a galaxy of its orbiting within a cluster of galaxies. Numerical experiments have been conducted to study the tidal influences on the internal dynamics of a galaxy (Miller and Smith 1982). The present first cut at the general problem treats effects of the cluster's tidal potential on the pattern motion and observable properties of a galaxy (shape, density contours, velocity fields, velocity dispersions) for a fixed external potential. The first-order cluster force field is balanced by the galaxy's acceleration in the real cluster, leaving tidal terms as the leading terms in a Taylor series expansion of the cluster force field about the center of the galaxy. The cluster tidal force field is usually stretching along a line toward the cluster center and compressive at right angles.

The external potential used for the numerical experiments has the form $V_{ext} = (k_L x^2 + k_T y^2 + k_T z^2)$. This potential represents the dominant part of the potential seen by a galaxy orbiting in a cluster, with the cluster center located at some distance along the x-axis. The experiments were run with a barlike galaxy placed in this external field, and the initial bar rotates about the z-axis. The non-symmetrical tidal forces affect the galaxy's rotation if the galaxy's rotation axis is not toward or away from the cluster center. The difference $k_{ext} = /k_T - k_L/$ represents the anisotropy of the tidal force. Experiments were run with $/k_T - k_L/$ calibrated in terms of the galaxy's self-consistent potential: $k_L = \partial^2 V_{ext}/\partial x^2$; $k_{sc} = \partial^2 V_{sc}/\partial x^2$; etc. for the initial galaxy. Four different experiments were run with $k_{ext}/k_{sc} = 0.02, 0.05, 0.1,$ and 0.2.

Although weaker than the galaxy's internal force field, the tidal field continues to act over long periods of time so significant cumulative effects can build up. The time scale on which cluster influences act is the cluster crossing time, typically a few by 10^9

* Permanent address - University of Chicago

351

E. Athanassoula (ed.), Internal Kinematics and Dynamics of Galaxies, 351–352.
Copyright © 1983 by the IAU.

years (Gott and Turner 1977). The ratios, k_{ext}/k_{sc}, are roughly the squares of the ratios of crossing times: $(T_{sc}/T_{cluster})^2$. The ratios tested are somewhat on the high side so the experiments will run in a reasonable time, but they are in the observed range $(T_{sc}/T_{cluster}) \sim 1/10$ to $1/20$. More importantly, the experiments confirm that internal responses scale as expected (linearly in k_{ext}/k_{sc}). Results can be confidently extrapolated to weaker external fields with this scaling rule.

Results from these experiments were analyzed by means of motion pictures and by means of numerical summaries. The motion picture was used to analyze the way in which pattern motion of the galaxy responds to the external tidal field. Two possible consequences have been checked by these numerical experiments. (1) Observable rotation is reduced so initially rapidly rotating galaxies can be slowed to low values of (V/σ) compatible with observation for giant elliptical galaxies. The shapes of the galaxies were not significantly changed through tidal braking. (2) Pattern rotation is affected such that galaxies tend to align pointing toward the cluster center with a weak secondary maximum perpendicular to the line between the galaxy and the cluster center. This pattern of alignment has been reported by Adams, Strom, and Strom 1980. The important conclusion to be drawn from these results is that galaxies in a cluster respond to, and are affected by, their surroundings in significant, observable ways. The rotation of a galaxy may have been strongly braked in a cluster environment.

REFERENCES

Adams, M.T., Strom, K.M., and Strom, S.E. 1980, Astrophys. J. 238, 445.
Gott, J.R. and Turner, E.L. 1977, Astrophys. J. 213, 309.
Miller, R.H., and Smith, B.F. 1982, Astrophys. J. 253, 58.

COLLISIONS AND MERGING OF DISK GALAXIES

B. F. Smith
NASA Ames Research Center

R. H. Miller*
European Southern Observatory

Collisions between disk galaxies embedded in massive halos have been studied by means of numerical experiments. The large three-dimensional n-body programs used for numerical experiments in the dynamics of galaxies and galaxy collisions (Miller 1978; Miller and Smith 1980) have been used in this investigation of the collision and merging of galaxies. The use of the large n-body code is important in a systematic study of the physical processes involved. There are enough particles (10^5) and the detail in the force field is sufficient to give the necessary resolution. Dynamical and relaxation time scales are well separated. This is particularly important in investigations of the merging process and of the nature of the merger product.

The stable disks that are required as initial states for an experiment were produced by embedding the (visible) disk in a(n invisible) halo. The disk is a luminous tracer that represents but 1% of the total mass of the system. Stability of the initial state of the model was experimentally confirmed. We have investigated collisions with various combinations of initial orbital energy and angular momentum and of disk orientations. Two initially parabolic cases with different impact parameters were followed long enough in time until the interactions led to a merger. The interpenetration of the two galaxies causes a contraction and subsequent disruption of the entire system. A variety of responses has been found for the visible disks ranging from stretched-out nearly linear features to rapidly propagating ringlike patterns. These forms depend critically on initial disk orientations, but do not depend strongly on the sense of disk rotation. The ringlike patterns were found even for collisions with significant impact parameters.

The cases that were followed to merger were parabolic collisions. One was head-on and the other had an impact parameter of the half mass radius of the initial galaxy. During the merger process material initially in one of the two galaxies continues to oscillate past that

* Permanent address - University of Chicago

E. Athanassoula (ed.), Internal Kinematics and Dynamics of Galaxies, 353–354.

initially in the other for several complete oscillation cycles (5-10), long after gross features such as the external form seem to have quieted down. The time to a "complete" merger is long. For these cases it is roughly 20 crossing times of the original systems or about 2×10^9 years. The final galaxy in both cases is a prolate object. The projected density profiles fit well to an $R^{1/4}$ luminosity profile. For the off-center case the final merged system has a maximum V/σ of about 0.2 which is consistent with V/σ values observed in many large ellipticals.

REFERENCES

Miller, R. H. 1978, Astrophys. J. 223, 1922.
Miller, R. H., and Smith, B. F. 1980, Astrophys. J. 235, 421.

ENERGY TRANSFER DURING THE TIDAL ENCOUNTER OF DISC GALAXIES

Philip L. Palmer
Dept. of Astronomy, University of Athens,
Panepistimioupolis, Athens 621, Greece.

Numerical simulations of merging galaxies do not include a disc component due to bar instability modes. Analytic work is based upon the impulsive approximation which leads to energy loss by the perturber. However, for the perturber to become bound we need consider parabolic encounters. Here we present an analytic technique suitable for all types of encounters.

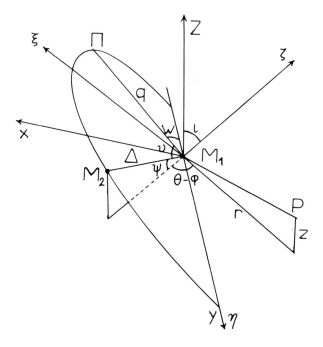

Fig.1:The geometry of the flyby.
M_1:Perturbed galaxy; M_2:perturbing galaxy in Keplerian orbit about M_1. P:massless star in circular orbit about M_1. (M_1,ξ,η):plane of orbit of M_2; (M_1,x,y): plane of unperturbed orbit of P; M_1y: line of intersection of orbital planes; υ: true anomaly of M_2 measured from $M_1\Pi$; (r,ϑ,z): cylindrical co-ordinates of P.

E. Athanassoula (ed.), Internal Kinematics and Dynamics of Galaxies, 355–356.

The reduced perturbing potential as seen from M_1 is:

$$U = \mu m(z\sin\psi + r\cos\psi\cos(\vartheta-\varphi))/\Delta^2 - \mu m(r^2+z^2+\Delta^2-2z\Delta\sin\psi-2r\Delta\cos\psi\cos(\vartheta-\varphi))^{-1/2}$$

where $\mu=GM_1$, $m=M_2/M_1$. This is then Fourier analysed taking $z=0$:

$$U = \sum_{n=-\infty}^{\infty}\int_{-\infty}^{\infty} a_n(r,\omega)e^{-i(n\vartheta-\omega t)}d\omega$$

The transforms a_n are calculated as integrals over the variable $Y=\tan\upsilon/2$.

We linearise the position of a star; $r=r_o+\delta$, $\vartheta=\vartheta_o+\Omega t+\lambda$ and $z=\zeta$ and define Fourier transforms for the perturbed quantities analogous to a_n. The linearised equations of motion are then solved for these transforms as functions of a_n and $\hat{a}_n \equiv \partial\hat{a}_n/\partial z_{z=0}$ ($\hat{a}_n$ is calculated as for a_n). The energy change of the disc is given by:

$$\Delta E = \int_{-\infty}^{\infty}\int_0^R\int_0^{2\pi} \partial U/\partial t \sigma(r)r\,dr\,d\vartheta\,dt$$

The 1st order terms average to zero and so we expand to 2nd order. The integral is then expressed in terms of the Fourier transforms. The ω contour is taken along the real axis with small indentations above the poles. Only the poles (resonant interactions) give a net energy transfer:

Corotation:

$$\Delta E = -\sum_{n=1}^{\infty}4n^2\pi^3\int_0^R a_n a_n^* \; d(\sigma/B)/dr \, dr \quad (\omega=n\Omega)$$

Lindblad:

$$\Delta E = \pm\sum_{n=1}^{\infty}\pi^3\int_0^R (n\Omega\mp\varkappa)(4n^2\Omega^2 a_n a_n^* - 2n\Omega r\varkappa d(a_n a_n^*)/dr + r^2\varkappa^2 a_n' a_n'^*)\sigma/(r\Omega\varkappa B)dr \quad (\omega=n\Omega\mp\varkappa)$$

$$\Delta E_z = +\sum_{n=1}^{\infty}4\pi^3\int_0^R \hat{a}_n \hat{a}_n^* \sigma(r)r\,dr \quad (\omega=(n\mp\eta)\Omega)$$

Radial:

$$\Delta E = +4\pi^3\int_0^R a_o' a_o'^* \sigma(r)r\,dr \quad (\omega=\varkappa)$$

$$\Delta E_z = +8\pi^3\int_0^R \hat{a}_o \hat{a}_o^* \sigma(r)r\,dr \quad (\omega=\eta\Omega)$$

($\varkappa^2=4\Omega B$; B=Oort's constant; $\eta\leqslant 1$). All energy exchanges are positive except the inner Lindblad term as $d(\sigma/B)/dr<0$.

In the impulsive limit the standard result can be rederived from the above formulae. The above integrals were evaluated for a disc with constant rotation velocity. The impulsive aproximation was found to be good for encounters with $V\geqslant 3V_p$, where V_p is the parabolic velocity. For parabolic encounters the inner Lindblad resonance was found to dominate over almost all the parameter space. The disc component therefore tends to accelerate the perturber and inhibits merging. The effect is weaker for large inclinations.

VII

GALAXY FORMATION

DYNAMICS OF GLOBULAR CLUSTER SYSTEMS

K.C. Freeman
Mt Stromlo and Siding Spring Observatories
Research School of Physical Sciences
The Australian National University

In the Milky Way, the globular clusters are all very old, and we are
accustomed to think of them as the oldest objects in the Galaxy. The
clusters cover a wide range of chemical abundance, from near solar
down to about [Fe/H] $\approx$ -2.3. However there are field stars with
abundances significantly lower than -2.3 (eg Bond, 1980); this implies
that the clusters formed during the active phase of chemical enrichment, with cluster formation beginning at a time when the enrichment
processes were already well under way.

We do not yet know what happened during this phase of chemical
enrichment, nor do we know how or why the clusters formed. Knowledge
of how the clusters move is essential for understanding this phase,
because the cluster kinematics reflect the conditions at the time of
their formation. Also, clusters are useful as probes of the underlying
galactic potential, so any knowledge about their orbital properties is
worth having.

I will now give a comparative review of the kinematical properties
of the globular cluster systems in the Milky Way, M31, and the LMC.
First, a warning: the dynamical properties of the cluster systems may
be different from the properties of the underlying diffuse stellar
component with which the clusters are usually identified. There are
two direct indications of such a difference: (i) the different range
in [Fe/H] for the Milky Way clusters and the halo field stars, mentioned above; (ii) in several elliptical galaxies, the (U-R) color of the
diffuse light at a given radius is significantly redder than the mean
color of the globular clusters at that radius (Strom et al, 1981,
Forte et al, 1981).

I. THE GALAXY

Two procedures have been used to study the orbital properties of the
globular cluster system in our Galaxy. The first is to use the cluster
radial velocities. These velocities are not usually a strong constraint

E. Athanassoula (ed.), Internal Kinematics and Dynamics of Galaxies, 359–364.

on the orbital properties of an individual cluster, but they are very
useful statistically. From the radial velocities of 66 clusters, Frenk
and White (1980) showed that the cluster system has a slow rotational
velocity of 60±26 km/s, and is kinematically isotropic, with a line of
sight velocity dispersion of 117 km/s. Recently Norris has made an
unpublished solar motion solution for the nearby halo field stars; his
values for the mean rotational velocity and velocity dispersion for
these field stars (in the appropriate abundance interval) are in
excellent agreement with those derived by Frenk and White for the
clusters. So, at this level at least, the clusters and the stars of
the diffuse stellar halo appear to be fairly similar kinematically.

The second procedure is to use the tidal radii of the globular
clusters. These tidal radii are set by the galactic tidal field at the
cluster's perigalactic distance R(min). Seitzer and I (unpublished)
have used the tidal radii to estimate R(min) for a sample of 48 clusters;
we assumed that the Galaxy has a spherical flat rotation curve potential
with V = 220 km/s. The ratio R(min)/R, where R is the present distance
of the cluster from the galactic center, is a statistical estimator of
the cluster's orbital eccentricity. It turns out that the <u>tidally</u>
estimated distribution of orbital eccentricities is consistent with
the <u>kinematical</u> isotropy derived by Frenk and White.

We looked also at the dependence of orbital eccentricity on
[Fe/H]. For clusters with R > 8.5 kpc, there is a wide range of [Fe/H]
at a given R, but there is no apparent abundance gradient in the
[Fe/H]-R plane (Zinn 1980). However for these outer clusters there is
a striking dependence of R(min) on [Fe/H] : all our clusters of intermediate [Fe/H] plunge in to small values of R(min) (2 to 3 kpc), while
the metal weak clusters have a wide range of R(min) values (2 to 15
kpc). This means that the intermediate abundance clusters with
R > 8.5 kpc are all in highly eccentric orbits, while the outer metal
weak clusters have a wide range of orbital eccentricities. This
dependence of orbital eccentricity on [Fe/H] for the outer clusters
has also been detected in a study that we have made of their kinematics.

Why should the outer intermediate abundance clusters be in highly
eccentric orbits with small values of R(min) ? It may be that these
clusters formed in the inner enriched regions of the early galaxy,
and were then flung out in the (roughly spherical) violent relaxation
of the spheroidal component. The metal weak clusters, on the other
hand, may have formed throughout the early galaxy (cf Zinn's 1980
discussion of the second parameter effect), and the orbits of those
metal weak clusters which formed in the outer parts of the system would
be relatively unaffected by the violent relaxation. Alternatively, we
recall that the <u>net</u> rotation of the metal weak halo field stars in
the solar neighborhood is low. If the intermediate abundance clusters
formed from accumulated enriched material lost by evolving stars of
the early halo, then we could expect their angular momenta to be low,
and their orbital eccentricities to be correspondingly high.

II. M31

From van den Bergh (1969) and Huchra et al (1982), we now have a sample
of about 60 globular clusters with abundance determinations and accurate
radial velocities. This sample covers a wide range in abundance, from
[Fe/H] > 0 to [Fe/H] = -2.2, and shows only a weak radial gradient in
abundance. The radial velocities can be used to estimate the change
of mean orbital eccentricity with abundance for these clusters.

The more metal rich clusters ([Fe/H] > -0.6) form a rapidly
rotating subsystem, with maximum rotational velocity of about 200 km/s
(the rotational velocity of the flat part of the HI rotation curve is
about 235 km/s), and velocity dispersion of about 90 km/s. There seems
little doubt that these metal rich clusters lie in a fairly disklike
(and therefore dissipated) system. It is not yet clear whether the
metal rich clusters in the Galaxy lie in a similar disklike distribution.
The metal weaker clusters ([Fe/H] < -0.6) show little net rotation, and
their line of sight velocity dispersion is again about 90 km/s. This is
surprisingly low, because the velocity dispersion measured for the
diffuse stellar bulge is about 180 km/s (Capaccioli 1979). This low
value may result from selection effects in the sample. For example,
if the cluster orbits are fairly eccentric, then the highest velocities
will be observed for clusters near perigalacticon, and these innermost
clusters are the ones most likely to be omitted from the sample (cf
de Vaucouleurs and Buta 1978).

From a simple histogram of radial velocities, we can in principle
estimate the eccentricity distribution of the cluster orbits (assuming
again a spherical flat rotation curve potential). However the probable
selection effects make this difficult to do reliably. It is much less
difficult however to estimate whether subsamples of the clusters have
different eccentricity distributions. Again it turns out that the
intermediate abundance clusters (-0.7 > [Fe/H] > -1.2) are in orbits
of significantly higher eccentricity than the metal weak clusters
([Fe/H] < -1.3), at the 92 percent confidence level.

To summarise this section: (i) the metal rich globular clusters
in M31 form a rapidly rotating disklike system; (ii) the intermediate
abundance clusters have orbits of significantly higher eccentricity
than the metal weak clusters. This latter effect was found earlier
for the galactic globular clusters, from their tidal radii and then
verified from their kinematics. Its appearance again, in M31, from the
cluster kinematics, increases our confidence in its reality and therefore in its importance for galaxy formation theory.

III. The LMC

In our Galaxy and probably also in M31, the globular clusters are all
old. Cluster formation took place during the phase of significant
chemical evolution and apparently ceased about 15 billion years ago.

The situation in the LMC is quite different. The LMC contains globular clusters of all ages. The oldest are similar in age to the halo clusters of the Galaxy. The youngest are only about ten million years old but, in structure and mass, they are similar to the old globular clusters. See Freeman (1980) for a brief review of their properties. We should ask why globular cluster formation is going on now in the LMC but not in the Galaxy: the answer may help us to understand the physical conditions in the Galaxy at the time of globular cluster formation.

Illingworth Oemler and I (in press) have made a study of the kinematics of the LMC globular cluster system. Combining our radial velocity data with similar data from Hartwick and Cowley, and Searle and Smith (which they kindly allowed us to use before publication), gives a sample of 60 clusters with ages from 10^7 to 10^{10} years.

The younger clusters (ages < 10^9 years) lie in a rotating disk similar to the HI/HII disk. The apparent rotation amplitude for this young cluster disk is V(rot) = 37±5 km/s, its galactocentric systemic velocity V(sys) = 40±3 km/s, its kinematic major axis is in position angle 1°±5, and the line of sight velocity dispersion about the rotation curve is 13 km/s. The older clusters (ages > 10^9 years and including the halo-type clusters) also lie in a disk system, with velocity dispersion of 14 km/s and V(rot) = 41±4 km/s. However this old cluster disk has V(sys) = 26±2 km/s and the kinematic major axis is in position angle 41°±5, very significantly different from the values for the disk defined by the young clusters and the gas. We do not yet fully understand this situation. We suggest that the kinematics of the old clusters reflects the gravitational potential of the LMC, while the kinematics of the gas (and the young objects that have recently formed from it) may have been affected by the interaction of the LMC, SMC and the Galaxy. Whatever the explanation, it seems clear that even the oldest clusters of the LMC are confined to a disk with a typical z-scaleheight of only about 600 pc, and there is no evidence for a kinematic halo population among the globular clusters of the LMC.

V. CONCLUSION

This comparison of the orbital properties of the globular clusters in the Galaxy, M31 and the LMC gives some useful insight into the formation histories of the cluster systems. In our Galaxy and M31, cluster formation began early, at a time when the chemical abundance was low. Cluster formation continued through the epoch of active chemical enrichment, and the mean orbital eccentricity of the cluster orbits increased as the chemical abundance increased. Cluster formation persisted, at least in M31, to a time when a fairly metal rich disk had already formed. In the LMC, the history was quite different. Although the oldest clusters appear to be metal weak, it seems that cluster formation did not begin until the LMC had already settled to a disk, and clusters have continued to form up to the present time.

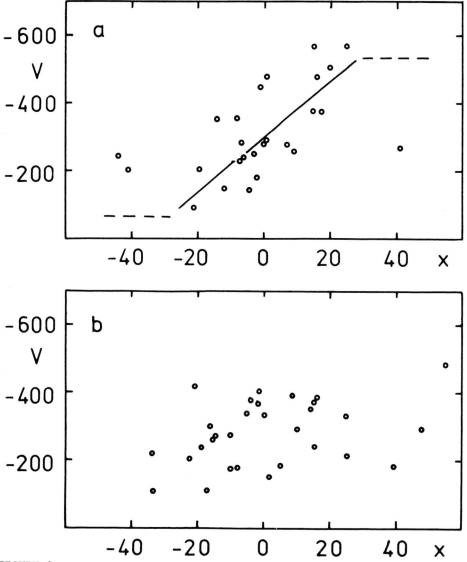

FIGURE 1

a) Rotation for the metal rich globular clusters ([Fe/H] > -0.6) in M31. V is the observed radial velocity (km/s) from Huchra et al (1982). X is the distance (arcmin) of the cluster from the minor axis of M31. The broken lines show the rotational velocity of the flat part of the HI rotation curve for M31. The unbroken line represents the mean rotation for these clusters. The rapid rotation of this subsystem is evident.

b) Similar diagram for the metal weak globular clusters ([Fe/H] < -0.6) in M31. These clusters show little net rotation, and their line of sight velocity dispersion is only about 90 km/s.

REFERENCES

Bond, H.E.   1980. Ap.J. Suppl. 44, pp 517-533.
Capaccioli, M. 1979. Photometry, Kinematics and Dynamics of
     Galaxies , ed D.S. Evans (Department of Astronomy, University
     of Texas at Austin), pp 165-176.
de Vaucouleurs, G. and Buta, R. 1978. A.J. 83, pp 1383-1389.
Forte, J.C., Strom, S.E. and Strom, K.M. 1981. Ap.J. 245, L9-13.
Freeman, K.C. 1980. Star Clusters (IAU Symposium 85), ed J.E. Hesser,
     (Reidel, Dordrecht), pp 317-320.
Frenk, C.S. and White, S.D.M. 1980. M.N.R.A.S. 193, pp 295-311.
Huchra, J., Stauffer, J. and van Speybroeck, L. 1982. Preprint.
van den Bergh, S. 1969. Ap.J. Suppl. 19, pp 145-174.
Zinn, R. 1980. Ap.J. 241, pp 602-617.

KINEMATICS OF CLUSTERS IN M33

C.A. Christian   R.A. Schommer
CFHT Corporation   Rutgers University
Kamuela, Hawaii   Piscataway, New Jersey

We have obtained 9.5 A resolution spectrophotometry of 10 clusters in M33 (Cf. Christian and Schommer 1982) with the KPNO 4m telescope. Velocities were derived by comparing the cluster spectra to data for template stars, using the "Fourier quotient" technique described by Sargent et al (1977). The positions and velocities of the clusters are shown schematically in Figure 1, where the HI velocity map (Rogstad, et al 1976) is also indicated. We find that the velocities of younger, bluer clusters are more closely matched to the disk-like motion of the gas, while the older redder clusters have more discrepant values (Table I). As shown in Figure 2, Velocity difference, Vel (HI − cluster), varies smoothly with the integrated (B-V), suggesting a slow smooth collapse of M33 to a disk galaxy. This fall, we intend to increase our sample of measured cluster velocities to probe more completely the history of the cluster and disk formation in M33.

REFERENCES

Christian, C.A. and Schommer, R.A. 1982, Ap. J. Supp., 49, pp. 405-424.
Rogstad, D., Wright, M., and Lockart, A. 1976, Ap. J., 204, pp. 703-716.
Sargent, W., Schechter, P., Boksenberg, A. and Shortridge, K. 1977, Ap. J.
    212, pp. 326-334.

E. Athanassoula (ed.), Internal Kinematics and Dynamics of Galaxies, 365–366.
Copyright © 1983 by the IAU.

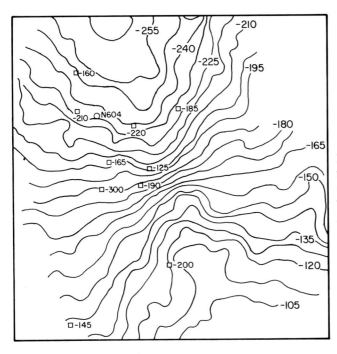

Figure 1: Schematic
representation of
cluster positions and
velocities relative
to HI gas.

Figure 2: ΔVel (HI –
cluster) vs. integrated
(B–V).

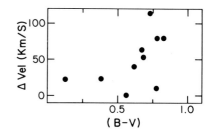

Table I

Cluster	Spt	B–V	Velocity	ΔVel
CL20	G1	0.77	−160	85
CL27	A5	0.37	−210	24
CL39	F5	0.56	−145	0
HII21	F5	0.61	−165	40
HII38	G2	0.83	−200	80
U49	G0	0.68	−185	55
U62	A	0.12	−220	20
M9	F8	0.72	−300	115
R12	G3	0.77	−190	12
R14	G3	0.68	−125	62

S0 AND SMOOTH-ARM Sa's WITHIN THE HUBBLE SEQUENCE

Allan Sandage
Mount Wilson and Las Campanas Observatories of the
Carnegie Institution of Washington

ABSTRACT

    S0 and smooth-armed Sa galaxies are generic to the Hubble sequence,
not formed by environmental stripping.

I. INTRODUCTION

    Because S0 galaxies are flattened disks without spiral arms,
speculation began already in the 1950's (Spitzer and Baade 1951) that
they are formed by the secondary processes of gas stripping from parent
spirals, leaving the daughters inert. The clearest statement of envi-
tonmental formation is given by van den Bergh (1976) where he extends
Baade's (1963)[1] suggestion that S0's and other "anemic" spirals form a
parallel sequence with normal gas-rich spirals, differing from their
"parent" spirals only in their gas and dust content.

    This view has received wide attention, and a large school of galaxy
strippers has arisen. If true, a major revision in the Hubble sequence
would be required.

    The contrary view (Sandage, Freeman, and Stokes 1970, SFS) is that
S0 and smooth-armed Sa galaxies naturally form the early terminus of the
disk systems. S0's are earlier than Sa types and are a transition to
the flattened E galaxies (Hubble 1936; Sandage 1961), initially placed
at that point in the sequence by the extreme values of some distributed
parameter (eg., density or specific angular momentum). Evidence in
support of this generic viewpoint is discussed in the next sections.

II. SURFACE BRIGHTNESS DISTRIBUTIONS

    That S0 and smooth-armed Sa's cannot be stripped Sb and Sc galaxies
is seen by looking through a telescope, an archaic activity evidently
not now practiced by strippers.

367

E. Athanassoula (ed.), Internal Kinematics and Dynamics of Galaxies, 367–372.

In work related to the Revised Shapley Ames Catalog (Sandage and Tammann 1981) photographs of all listed galaxies were taken with the Mount Wilson, Palomar, and Las Campanas reflectors. To check the telescope setting, most fields were visually inspected. The fact, known generally to old visual observers, was rediscovered that Sd and Sc galaxies were extremely difficult to see with large scale reflectors, Sb's easier, Sa's still easier, whereas all S0 and E galaxies stand like beacons in the focal plane.

This progressive change of surface brightness with Hubble type is shown quantitatively in Figures 1a and 1b where the average V_{mag} surface brightness within a photoelectric blocking diaphragm A/D(0) is plotted for various Hubble types. The photoelectric data are taken from standard literature sources (eg., de Vaucouleurs and de Vaucouleurs 1961; Sandage 1975; Sandage and Visvanathan 1978; Griersmith 1980). The shape of the line is from integrating the standard growth curve (Sandage 1975, Table B1), but forced through the E plot.

The E and S0 galaxies have nearly the same surface brightness distribution. However, as soon as evidence of arms appears (i.e., in S0/a types), the average surface brightness at any A/D(0) > 0.1 decreases. That dust is not a problem is shown in panels 4 and 5 of Fig. 1a by dividing the Sa's into dust-free and dust-present systems.

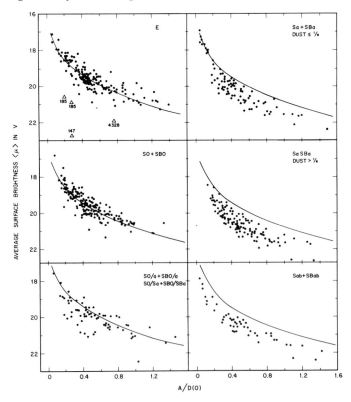

Fig. 1a. The surface brightness in V mags for E through Sab galaxies, averaged over the area inside a radius of A/D(0), where D(0) is ∿0.3 of the Holmberg radius. The mean line for E galaxies is repeated in every panel.

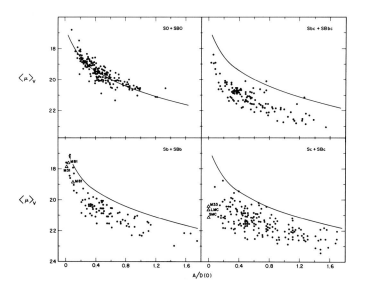

Fig. 1b. Same as Fig. 1a for Sb, Sbc, and Sc galaxies, with the data for S0 galaxies repeated. The absence of the bulge in Sbc and Sc galaxies is evident by inspection of the curves for A/D(0) < 0.2. Hence, if such galaxies are stripped, they will not form the S0 distribution.

To overcome objections about using isophotal D(0) diameters rather than a metric scale length, effective surface brightness, defined as that averaged over the circular radius that contains half the light, has been calculated, as shown in Figure 2. The line through the S0 types is repeated in all panels. S0 galaxies have the highest $\langle SB \rangle_e$ of all types. The progressive decrease with Hubble type is listed in the Table. As before, it can be shown that internal dust absorption is not the cause of the progressive decrease in surface brightness along the Hubble sequence.

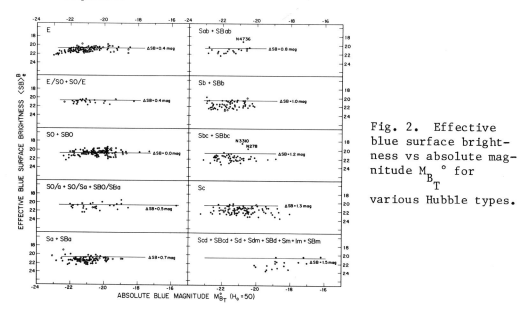

Fig. 2. Effective blue surface brightness vs absolute magnitude $M_{B_T}^{\circ}$ for various Hubble types.

Type	ΔSB Relative to S0's	Type	ΔSB Relative to S0's
E	+0.4	Sab + SBab	+0.8
E/S0 + S0/a	+0.4	Sb + SBb	+1.0
S0 + SB0	0.0	Sbc + SBbc	+1.2
S0/a + S0/Sa	+0.5	Sc + SBc	+1.3
Sa + SBa	+0.7	Scd and later	+1.5

Clearly, S0 galaxies cannot be formed by the stripping of late type galaxies of their star producing material because these prospective parent galaxies are already fainter in effective surface brightness than their resulting daughters--an impossibility.

Other evidence against environmental stripping to form S0's include (1) normal S0's occur in the isolated field where no stripping can be expected, and (2) S0's and spirals coexist in pairings that include simple binaries, loose groups, and spiral-rich clusters such as Hercules and the Virgo Cluster core. If S0's were once normal spirals and then stripped, why are their present-day companion spirals not also stripped?

Other interesting features of Fig. 2, useful for other problems are:

(1)   Over the absolute magnitude interval $M_B \sim -23$ to ~ -19 the $\langle SB \rangle_e$ is nearly independent of M_B which is Hubble's old relation $m \sim 5 \log \theta$. Hence there is no distance information in angular diameters over this intensity interval.

(2)   The spread in M_B within any type is large. Without doubt, it is even larger than shown in Figure 2 because there is an artificial cut-off at faint M_B values due to the bright apparent magnitude limit of the Shapley Ames at $B \simeq 13$. Clearly, the Hubble type earlier than Scd is hardly a function of absolute luminosity, and therefore cannot be used as a distance indicator.

(3)   The $\langle SB \rangle_e$ of E galaxies decreases for M_B brighter than ~ -21, possibly due to mergers.

III.   SMOOTH-ARMED Sa's

Because smooth-armed Sa's occur in clusters, it has sometimes been claimed (cf. Wilkerson 1980) that they too are a result of stripping. However, similar to S0's, they also occur frequently in the general field.

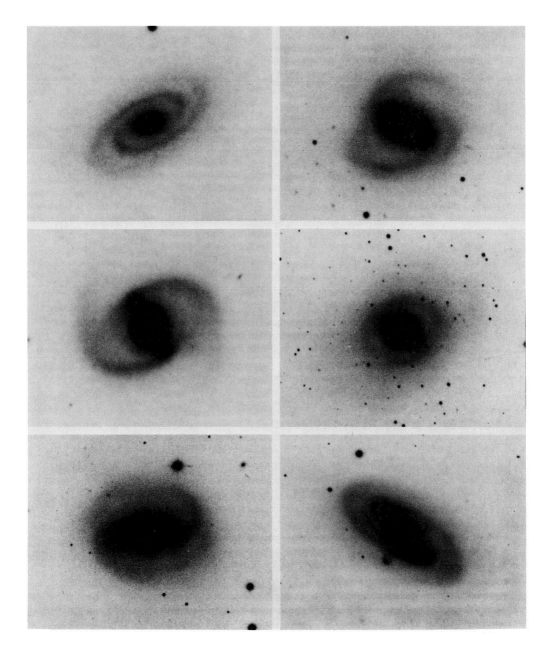

Fig. 3. Six smooth-armed Sa galaxies. Three are in the Virgo Cluster, three are in the general field. The field SBa galaxy, NGC 7743, in the lower left, is also in the Hubble Atlas (p. 43), and discussed there as the earliest part of the continuous Sa sequence. That smooth-armed Sa's occur in the general field also shows that they are not formed by stripping in clusters.

Figure 3 is a panel of six plates of smooth-armed Sa's. Three are in the Virgo Cluster core and three are in the general field.

A morphological study of the full Sa section of the Hubble sequence shows that galaxies in Figure 3 are only the earliest examples of a continuum, within which the arms become progressively less smooth as one moves through the Sa section to Sab and then to Sb types.

The conclusion is, then, that smooth-armed Sa galaxies are generic to the Hubble sequence. The evidence consists of (1) field examples, and (2) morphological continuity between the S0/a and the Sab types.

IV. CONCLUSION

The hypothesis (SFS) still seems tenable that a galaxy finds itself along the Hubble sequence according to the amount of gas left over in the disk after a collapse from a larger volume, with gaseous dissipation near the end of the collapse. Only little evolution subsequently occurs along the sequence. To be sure, copious star formation must have occurred initially in S0 and Sa disks to explain their high surface brightness now, due to the remaining old disk stars. But this gas must have mostly been converted into stars early to account for the low rate of star formation now in these systems.

---

[1]On p. 15 of the quoted reference, Baade incorrectly states the situation as it was known in the 1950's. Never did Hubble propose the division of S0's into S0a, S0b, and S0c types, nor was such a division made in the Hubble Atlas.

REFERENCES

Baade, W.: 1963, Evolution of Stars and Galaxies (Harvard University Press: Cambridge) p. 15.
de Vaucouleurs, G. and de Vaucouleurs, A.: 1961, Mem. R.A.S., 68, 69.
Griersmith, D.: 1980, Astron. J., 85, 789.
Hubble, E.: 1936, Realm of the Nebulae (Yale University Press).
Sandage, A.: 1961, The Hubble Atlas (Carnegie Institution Publication No. 618).
Sandage, A.: 1975, Astrophys. J. 202, 563.
Sandage, A., Freeman, K. C., and Stokes, N.: 1970, Astrophys. J., 160, 831 (SFS)
Sandage, A. and Tammann, G. A.: 1981, A Revised Shapley Ames Catalog (Carnegie Institution Publication No. 635).
Sandage, A. and Visvanathan, N.: 1978, Astrophys. J., 223, 707.
Spitzer, L. and Baade, W.: 1951, Astrophys. J., 113, 413.
van den Bergh, S.: 1976, Astrophys. J., 206, 883.
Wilkerson, M. S.: 1980, Astrophys. J. (Letters), 240, L115.

IS THE HUBBLE TYPE OF A DISK SYSTEM ESSENTIALLY DETERMINED BY ONE
PARAMETER: TOTAL MASS?

R. Brent Tully
Institute for Astronomy, University of Hawaii

The contentious title of this discourse was chosen to draw attention to some ideas expressed in a recent publication by Tully, Mould,
and Aaronson (1982, hereafter TMA). Two important correlations were
discussed in that article. One of those was a correlation between the
colors and the intrinsic luminosities or magnitudes of spiral galaxies.
The other was a relationship between the color and total mass of disk
systems. It is now reasonably accepted that there is also a tight connection between luminosity and total mass.

The color that I refer to is a construct of observations in blue
(B_T) and infrared ($H_{-0.5}$) passbands: hence it is sensitive to differences between young and old populations. When I speak of total mass,
it is really the deprojected global HI line profile (W_{20}) which is the
observed parameter. I take some licence with the implication of a
close association between total mass and profile width, but it is the
tightness of the luminosity-profile width correlation which encourages
me to do so.

The claim is that, for gas-rich disk galaxies, we have good correlations between each pair of the triad: mass, luminosity, and color.
One can ask if there is any additional information in the morphological
classification of individual galaxies, in the sense that part of the
scatter in these relations might be due to systematic variations with
type. It can be seen in the review by Rubin at this conference that
she and others are impressed by the differences in certain of the
above-mentioned relations that depend on the morphological type. By
contrast, I am impressed by the lack of differences.

Certainly, this is not to say that we believe there are no differences at all. The galaxies of type Sa-Sb do seem to deviate from the
later types in the color-mass relationship demonstrated by TMA, in the
sense that the early systems of a given mass are redder. From the
mass-luminosity relations, it would seem that most of the type variations are in the blue passband. However, in the samples that we

373

E. Athanassoula (ed.), Internal Kinematics and Dynamics of Galaxies, 373–374.
Copyright © 1983 by the IAU.

accumulated the variations are <u>small</u>; substantially smaller than those found in the samples employed by Rubin and her collaborators, for example. The color-mass (profile width) diagram is particularly useful in the investigation of type-variations because the observables are distance independent so a large and coherent sample can be assembled. In this diagram, as in the plots of luminosity-mass and luminosity-color, the evidence for type variations are marginal.

There is a qualification which might be important from the standpoint of the debate over type dependences. As one reaches the very earliest disk systems, distinctions in the color-luminosity-mass plots become apparent. There is a striking separation between lenticulars and spirals on the color-magnitude diagram in TMA. Huchtmeier has demonstrated the clear separation of Sa systems from later types in the (blue) luminosity-mass plane. It was proposed in TMA that disk systems are disposed to lie in either of two distinct branches in the color-magnitude diagram: either in a branch reserved for gas-rich spirals and irregulars or in one delineated by gas-poor lenticulars. Star formation is continuing in one branch but has essentially stopped in the other.

It remains to be resolved whether the Sa systems lie in the gas-rich or the gas-poor branch, or somewhere between. We have not been able to address that problem properly because early systems are difficult to detect in HI and because the Sa class is poorly represented in volume-limited samples. In summary, it seems probable that gas-poor systems have very distinct color-luminosity-mass properties from gas-rich systems, though the demarcation point between these two regimes is not well explored. At the same time, there are surely other galaxies with anomalous color-mass-luminosity combinations, such as the extreme low surface brightness spirals.

The proposition that is offered is that a gas-rich disk system with a given amount of mass knows how luminous it should be and what dimension it should have (the original correlations discussed by myself and Fisher), what global color it should have (TMA), and how centrally condensed its mass should be (Rubin and collaborators have demonstrated the trend of central velocity gradients with maximum rotation). In TMA, we presented a simple model as an explanation of the color-magnitude relationship. In the model, the total mass governs the galactic star formation rate and must also affect either the galactic metallicity or stellar initial mass function. These properties might be regulated by the amplitude of streaming motions and thus be tied to the dynamics of the galaxy. The amplitude of rotational motions, the scale of the system, the vigor of current star formation in contrast with its past history; these properties must produce consequences which are interpreted in the morphology of a galaxy.

Tully, R.B., Mould, J., and Aaronson, M.: 1982, Ap.J. 257, 527.

SYSTEMATICS OF BULGE-TO-DISK RATIOS

F. Simien   and G. de Vaucouleurs
Observatoire de Lyon   McDonald Observatory
Saint Genis Laval, France   University of Texas

The main results of a new analysis of the spheroidal (I) and disk
(II) components of 98 lenticular and spiral galaxies are :

(i) on the average, the magnitude difference between spheroid and
total luminosity, $\Delta m_I = B_T(I) - B_T$, varies smoothly along the Hubble
sequence from early lenticulars to late-type spirals (Fig. 1);

(ii) the trend of Δm_I confirms the concept of the lenticular class
as intermediate between E and S classes, not as a parallel sequence;

(iii) the large scatter at any given type, $\sigma(\Delta m_I) \simeq 0.7$ mag, is
still dominated by measuring and decomposition errors;

(iv) The effective surface brightness of the spheroid, $\mu_e^c(I)$,
corrected for galactic extinction A_B, decreases by ~ 2 mag from early to
late types, but with a large range (~ 3 mag) at T = const. (Fig. 2a).

(v) The effective surface brightness of the disks, corrected for
galactic extinction and inclination, $\mu_e^c(II) = \mu_e(II) - A_B + 3 \log R_{25}$,
is almost independent of type, with $\langle\mu_e^c(II)\rangle \simeq 23.5$ for spirals. This
implies a corrected central surface brightness $\mu^c(0) = \mu_e^c -1.82 \simeq 21.7$,
in good agreement with the Freeman rule, but with a large scatter.
However, the disks of lenticulars (T < 0) tend to be ~ 0.5 mag fainter
than the disks of spirals (Fig. 2b).

(vi) The linear effective radii of the spheroidal components are
largest, $\langle r_e(I)\rangle \gtrsim 1$ kpc, among the early type spirals, in agreement
with the Hubble classification criterion. The spheroid of lenticulars
and late-type spirals tend to be smaller, $\langle r_e(I)\rangle \simeq 0.5$ kpc, but with a
large scatter (Fig. 3a). There is no indication of systematic difference
between ordinary (SA) and barred (SB) spirals.

(vii) The linear effective radii of the disk components are largest
$\langle r_e(II)\rangle \gtrsim 5$ kpc, among intermediate type spirals. The disks of lenticular and late type spirals tend to be smaller (Fig. 3b).

(viii) The mean absolute magnitudes of the disk and spheroidal components depend on type (Fig. 4). On the average the disks are brighter
($M_{II} \simeq -19.5$) among types Sb-Sbc, spheroids ($M_I \simeq -19$) among types L^+ to
Sa, but, again, with a large scatter. Disks and spheroids are about
equally bright ($M_I \simeq M_{II} \simeq -19$) at stage S0/a (T = 0).

Applications to the systematics of rotation curves are in progress.

E. Athanassoula (ed.), Internal Kinematics and Dynamics of Galaxies, 375–376.

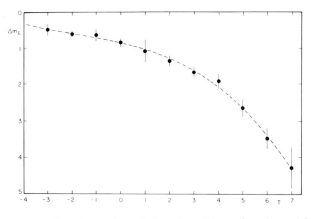

Figure 1. Fractional luminosity of spheroid

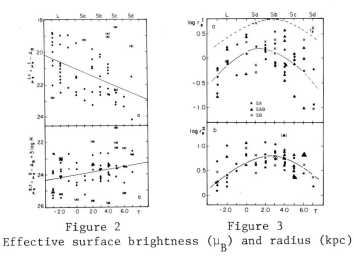

Figure 2   Figure 3
Effective surface brightness (μ_B) and radius (kpc)

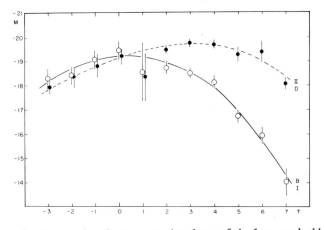

Figure 4. Mean absolute magnitudes of bulge and disk

THE MANIFOLD OF ELLIPTICAL GALAXIES

P. Brosche and F.-Th. Lentes
Observatorium Hoher List der Universitäts-
sternwarte Bonn, D-5568 Daun, F.R.Germany

ABSTRACT
The significant number of dimensions of the manifold is two.
Notwithstanding the contamination of the second dimension
by projection effects, there are arguments for an intrinsic
variation.

The manifold of spiral galaxies was studied by Brosche
(1973), Bujarrabal et al. (1981) and Mebold and Reif (1981)
with the main result of two significant dimensions and a
quite stable configuration of the contributing variables.
The kinematic quantities from the observed rotational motions
were essential for this result. The increasing number of
observed inner motions in elliptical galaxies enabled us to
undertake an analogous study for this class of galaxies. The
data of Tonry and Davis (1981) contain n=101 galaxies for
which the type parameter is known to be in the range -6 to
-4. Amongst them, for n=29 galaxies rotational motions are
known from other sources. We studied both samples; in both
cases with and without the only distance-dependent variable,
a photometric radius R. Since the incorporation of R did not
influence severely the other parameters, and since the results of all parameters are similar for both samples, we
present here only the smaller sample for which rotation velocities V are available. Other parameters involved are: the
velocity dispersion S, surface brightness L/R^2, ratio of
major to minor axis D/d (V and S in km/s, R in kpc, L in
solar units).
   In the figure we present the gradients of the contributing
observables (solid lines) and some secondary variables
(broken lines) which can be derived from the observables,
namely the mass M, the absolute luminosity L and the ratio
M/L. The observed rotational velocities and the apparent
axis ratios are both influenced by projection effects. But
the variation of logV is larger and the direction of its gradient is more different from that of log(D/d) than one would

377

E. Athanassoula (ed.), Internal Kinematics and Dynamics of Galaxies, 377–378.
Copyright © 1983 by the IAU.

expect if the second dimension were the inclination only.
There is a certain qualitative correspondence with the spirals' picture. From the many possible representations of L
we may quote $L(B) = 0.27 \cdot \log S - 1.70 \cdot \log(L/R^2) + 25.18$ as a
'Faber-Jackson-relation' generalized for the presence of a
second dimension. With S as only parameter, the correlation
becomes $L(B) = 3.81 \cdot \log S + 6.43$.

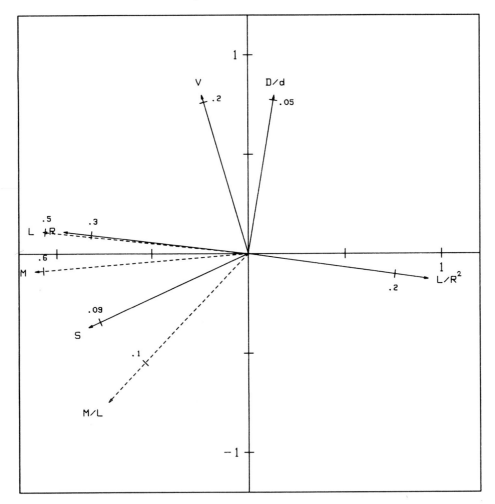

The support of the Deutsche Forschungsgemeinschaft is
gratefully acknowledged.

REFERENCES
Brosche,P.: 1973, Astron. Astrophys. 23, 259
Bujarrabal,Y., Guibert,J., Balkowski,C.: 1981, Astron.
       Astrophys. 104, 1
Mebold,U., Reif,K.: 1981, Mitt. Astr. Ges. 51, 143
Tonry,J.L., Davis,M.: 1981, Astrophys. J. 246, 666

THE FORMATION OF GALAXIES

James E. Gunn
Princeton University Observatory, Princeton, NJ08544, USA.

Abstract: The presently fashionable ideas for galaxy formation are
reviewed briefly, and it is concluded that the standard isothermal
heirarchy fits the available data best. A simple infall picture is
presented which explains many of the observed properties of disk
galaxies.

Galaxy formation is an extremely active field of research at
the present time, but if anything has grown more confused in the past few
years. This confusion is largely due to the impetus lent by new data on
the distribution of galaxies in space and the growth of grand unification
theories (GUTs) in particle physics which, at least in their most straightforward applications, put severe constraints on the allowed forms of perturbations which might form galaxies. There are still basically two
schools of thought on the subject, though the data look more and more as
if some synthesis will ultimately describe Nature better than either.

It is generally agreed that galaxies form from density (and
perhaps velocity) perturbations generated in a still-unknown way in the
very early universe, and in particular those perturbations which survive
until the epoch of the initial combination of the primeval plasma (de-
coupling) at a temperature of about 4000 K or a redshift of about 1500.
The perturbation amplitudes required to form galaxies are of the order of
a few percent at that epoch, and are large for low-density universes and
for scenarios in which galaxies form very early. The disagreement centers
around the nature of the perturbations in the earlier universe...are they
adiabatic (i.e. accompanied by no change in the specific entropy, in which
case $\Delta\rho = 3\Delta T$) or are they isothermal ($\Delta T=0$)? The crucial difference for
galaxy formation is that, as was first shown by Silk (13), adiabatic perturbations are damped by radiative viscosity for masses below 10^{13} to
10^{15} solar masses. If the universe is dominated by neutrinos of mass
about 30 eV, as now seems quite possible, an analagous phenomenon occurs
-- here the neutrinos simply run away from the density maxima because
they have large random velocities, and the lower mass limit on surviving
perturbations is set by the comoving distance they can travel once they
uncouple, which in proportional to the inverse of their mass. For neutrino

E. Athanassoula (ed.), Internal Kinematics and Dynamics of Galaxies, 379–386.

masses of interest, the cutoff masses are of the same order as the Silk mass. Since the neutrinos are generally believed once to have been in thermal equilibrium in the Universe (at times $\lesssim 1$ sec, temperatures $\gtrsim 10^{10}$ Kelvin), any perturbations in the neutrino density will of necessity be adiabatic. Since GUTs are able to account for the synthesis of baryons by thermal processes as well, it would seem that the most likely sort of perturbations would be the adiabatic ones (see the Silk and Weinberg papers in (1)).

Both adiabatic pictures (the baryons-only one and the baryons-plus-neutrinos one) have negligible perturbation amplitudes on the scale of galaxies at decoupling, and so in these pictures clusters (or super-clusters) form first, and galaxies form later by fragmentation processes. Gravitation plays an important role in these processes, to be sure, but the formation is NOT from collapse of perturbations present at de-coupling. An excellent review is contained in ref (18).

The isothermal perturbations trade the mystery of their genesis for the simplicity of their subsequent behaviour - they remain essentially frozen in during the early evolution of the Universe and emerge from de-coupling unscathed, and with negligible attendant velocity disturbances. It is possible, as shown by the work of Peebles and coworkers (see, for example, (11)) to account for the development of structure on all scales in the Universe today from a simple heirarchy of isothermal perturbations. The recent observations of Davis et al. (2) of the distribution of nearby galaxies and those of Kirschner et al. (8) pointing to the existence of very large voids in the galaxy distribution seem to indicate the existence of large amplitude disturbances on very large scales, as one would expect from the adiabatic picture, though on smaller scales ($\lesssim 3$ Mpc, say) the heirarchical (isothermal) picture works quite well.

What tests are there? It would seem that the crucial question is that of the relative formation times of galaxies and clusters, and in my opinion essentially all the data on this question suggests that galaxies form very much earlier than clusters. The data take several forms, but among the most persuasive are the remarkable uniformity of the colors and spectra of ellipticals, both in the "field" and in rich clusters, synthesis models for which indicate great (and uniform) ages (see, for example, (7)); in the same vein is the data on very distant ellipticals seen at redshifts near unity, the spectra of which are consistent with ratios of ages then to now the same as the ratio of the ages of the Universe then to now. (This is a rather more sensitive test than the absolute age alone, since many uncertainties disappear in the ratio comparison (Gunn, in (1)). To me the most telling piece of evidence comes from the Galaxy and the Local Group. The Galactic globular cluster system is very old, and must correspond to a formation time for the bulge of the Galaxy at a redshift of at least about two, when the mean density of the Universe was thirty times greater than now. The density in the collapsing perturbation must have been at least three to five times (depending on the geometry) denser than the mean at turn-around, and that grows by at least a like factor during collapse. The density contrast then grows at least as fast as $(1+z)^{-2/3}$. Thus the cluster/supercluster which collapsed to fragment and form the Galaxy in the adiabatic picture must be a structure of some 10^{13} to 10^{15} solar masses with a density some thirty times the mean, and there exists no such candidate structure. The local supercluster is the only aggregate in which the Galaxy is embedded which is massive enough, but its mean density interior to the Galaxy is only about two and a half times that of the surrounding, and it can at most be barely turning around around NOW. This situation is further discussed by Dekel in this volume.

Thus it would appear that individual galaxies and small groups are very unlikely to have had their origins in adiabatic pancakes. There are other scenarios (see, for example, Ostriker, in (1), which still require isothermal seeds) but the most straightforward is to fall back on the isothermal heirarchy. (It must in fairness be remarked that if one wishes the large-scale structure to be adiabatic and the small-scale to be isothermal, a remarkable coincidence in initial conditions is required in order that the amplitudes are comparable at decoupling - but there is no reason why the structure cannot ALL be isothermal with a suitable power spectrum.) We must only await a plausible mechanism for generating the isothermal perturbations.

Let us review briefly the growth of isothermal perturbations after z=1500. A perturbation (which we shall assume roughly spherical for simplicity) is characterized by a dimensionless amplitude $\delta(r)$, which is the ratio of the mean density interior to the critical density at that epoch. For reasonable ($\geqslant 0.1$) values of Ω, the mean density differs from the critical density by an amount which is small compared to the perturbations of interest for the formation of galaxies. The collapse time t_c for the material interior to r is simply related to $\delta(r)$, as is the ratio of the maximum expansion radius to the initial radius:

$$t_c = \frac{\pi}{H_i} \delta^{-3/2} \qquad\qquad R_m = R_i \delta^{-1}$$

If the RMS density perturbation is a power law in the contained mass, as is often assumed, one can make some simple predictions for the resulting structures. Let us assume that

$$\delta_{RMS} \; \alpha \; M^{-\frac{1}{2} - \frac{n}{6}}$$

(Here n=0 is white noise, and negative n corresponds to more power at large scales). If the further assumption is made that $\delta(r)$ is a stationary (in space) stochastic process with random phase and with the above spectrum to scales much smaller than those of interest, it follows that the distribution of densities is approximately normal. On thing to notice immediately is that galaxies, clusters, etc. form around peaks in the density, and it a tautological property of peaks that the surrounding mean density is lower than the peak value. Since the collapse time varies as the inverse three-halves power of the amplitude, it follows that late infall of material is a general property of this picture.

If there is no dissipation in the subsequent evolution through maximum expansion and collapse, the potential energy at maximum expansion is converted to potential and kinetic in the virial ratio in the final stationary configuration, (corresponding to a factor of two decrease of gravitational radius) and the collapse time is related by a constant ratio to the final dynamical time. Thus the maximum radius and collapse time are simple functions of the mass and velocity dispersion in the final structure, and since the RMS perturbation amplitude is related to the mass by the heirarchy, one can predict the mean radii and velocity dispersions as functions of mass (see, for example, (11)).

Now it is certainly NOT true that galaxies form without dissipation; that is an obvious enough statement for spirals, but equally good arguments can be made for ellipticals based on density contrast with their surroundings (see, e.g. (17)). It IS reasonable to assume that the dark halos of galaxies form dissipationlessly. It can be argued (Gunn, in (1)) that one can relate simply the rotational velocity V_c of spirals and the velocity dispersion σ_* of the stars in ellipticals to the velocity dispersion σ_H in their halos; $\sigma_H = \sqrt{3/2}\sigma_* = \sqrt{1/2}V_c$. For galaxies with both disks and dynamically important bulges, the last half of this relation can be checked empirically, and the fact that those same galaxies have flat rotation curves demands the satisfaction of the whole relation. A long series of arguments ((6), (15); Faber, Gunn in (1)) suggests that the ratio of dark "halo" stuff to ordinary matter is very nearly constant on (galaxy+halo) scales and larger, and that this ratio is about fifteen to on

Assuming this to be true, and using the semi-empirical mass-to-light ratios adopted by Faber and Gallagher (3), one can now plot a MASS-(either halo mass or "visible" mass) HALO VELOCITY DISPERSION diagram, in which one can draw, for example, lines of constant collapse time and the mean line for any heirarchy. The results of doing this for about fifty galaxies with well-determined dynamical properties is shown schematically in Figure 1, where also shown is the Coma Cluster (c), the mean line for an n=-1 heirarchy which has been placed correctly with respect to Coma at that mass scale, and two regions related to the cooling of material in the perturbation - the one vertical line to the right of which atomic cooling processes cannot cool the object on the collapse timescale (12), (17), and the $t_c = 3 \times 10^8$ yr line, above which any object can cool by Compton cooling against the microwave (at those epochs infrared) background in a time shorter than the collapse time. Galaxies should be found only above and to the left of the region in which structures cannot cool, and that is indeed where they obligingly lie. The ellipticals (black dots; spirals are open symbols) are systematically high, and indeed lie almost exclusively in the Compton-cooling dominated regime, many of them in the region where they cannot have cooled by atomic processes (note that MERGERS move points to the right and horizontally in the diagram, so these massive ellipticals cannot have been formed by merging lower-mass systems of any sort we see today).

This diagram suggests a simple picture for the formation of galaxies, which is discussed somewhat more fully in my paper in (1).

First, ellipticals and the bulges of spirals are formed in the Compton-
cooling regime by the densest parts of perturbations, which for the most
tightly-bound ellipticals involves essentially all the mass. The lower-
density outer parts both cool and collapse more slowly. The perturbation
is given angular momentum by the action of tidal torques throughout its
history, and acqires in the mean a dimensionless angular momentum λ =
$J|E|^{\frac{1}{2}}M^{\frac{5}{2}}G^{-1}$ of about 0.07. Fall and Efstathiou (4) have shown that that
value is roughly correct for disk galaxies when account is taken of dis-
sipation. If the perturbation is given roughly solid-body rotation by the
torques (as one might expect from simple but possibly naive dimensional
arguments), then a remarkable result follows. The angular momentum distribution of such a system, when translated into a disk density distribution
with a flat rotation curve, results in a distribution of surface density
which closely approximates an exponential over some four scale lengths.
Furthermore, as successive shells fall in, the disk grows self-similarly
with almost constant central surface density and always has the nearly
exponential form. It is a significant fact that the bulge population of
the Galaxy has a lower specific angular momentum than the disk (Gunn, (1))
which argues strongly against the hypothesis that the bulge and disk
formed from the same stuff, separated only by star formation rate - it
would appear that the bulge is a separate dynamical subsystem. In this
picture the bulge formed early and the disk later, with the latter still
forming.

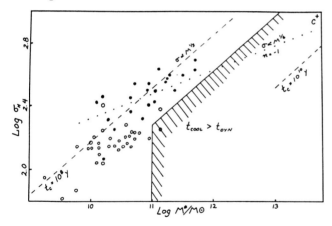

Figure 1. Halo velocity
dispersion versus
visible mass.

        The picture predicts a definite evolutionary history for the
disks of galaxies. If one looks at a fixed value of specific angular
momentum (which does not correspond to fixed radius because the mass
interior to the point in question grows slowly with time), the disk is at
first of very low surface density and is vertically non-self-gravitating.
The temperature of the gas will be about 15000 K and it will be mostly
neutral. The scale height will be H=σ_Dr/V$_c$, where σ_D is the equivalent
one-dimensional velocity dispersion, about 10 km/s. As more material
falls in, the disk becomes self-gravitating, and almost immediately becomes Jeans unstable (5,14). The only way to restore stability as more
mass is added is for the gas to heat, but it cannot because of the cooling
barrier. The mass motions driven become supersonic and rapid star form-

ation occurs, after which the instability can heat the stars. Thus as the disk grows it always remains marginally unstable, at a fixed appropriate value of Q.

One can predict from this picture the velocity dispersion and scale height as a function of radius, and the latter is shown for the present Galaxy, with the local surface density taken to be 70 $M_\odot/pc^2$ and a rotational velocity of 220 km/s, in Figure 2. Note the remarkable constancy of the scale height, observed, of course, in external systems (9). The rapid ballooning of the disk occurs where the disk becomes non-self gravitating, and the neutral hydrogen in the disk in the Galaxy exhibits this behavior, as it may in other systems (Sancisi, this volume). This phenomenon is probably the best evidence for the dark halo material being roughly spherically distributed (6).

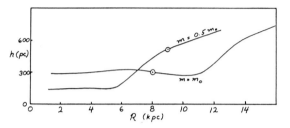

Figure 2. Predicted scale height for the present Galaxy and for an epoch when the disk was half as massive as now.

An added bonus of infall is that it explains the existence of spiral structure in a straightforward way. One would expect the fact that the disk is kept at the threshold of Jeans instability to keep it unstable to spiral modes as well, and I suggested that that would be the case in (1). Recently Carlberg and Sellwood have performed appropriate numerical experiments (described elsewhere in this volume) and the results are spectacularly positive. In addition, the spiral instability pumps the velocity dispersions of old stars, and the age-kinematics relations can probably be explained also, as they describe. Infall has been invoked for a long time, of course, to explain the age-metallicity relation (see, for example, (16)) and it appears to be as useful for dynamics.

Thus a large number of the properties of galaxies emerge in a natural way from the picture; on a rather more speculative level, the Hubble sequence finds a ready explanation. Galaxies which today have vigorous star formation are those for which the infall rates are high now, and those systems, on average, are ones which have long collapse times (systems with short collapse times have high infall rates at early times, but run out of material later). Thus the strong tendency for late-type systems to fall at the long-collapse-time end of the distribution in Figure 1 is explained. The correlation of bulge-to-disk ratio with Hubble type (defined in this instance by color or star formation rate) follows immediately – the bulge is a short-collapse-time subsystem, and perturbations which have long collapse times should, on average, have smaller such subsystems (if any at all) than denser ones. This picture is very schematic, but seems quite promising.

        This research was supported by the National Science
Foundation of the U.S.

REFERENCES

1. Bruck, H.A., Coyne, G.V., Longair, M.S. 1981, Astrophysical Cosmology,
    Pontifica Academia Scientiarum, Citta del Vaticano.
2. Davis, M., Tonry, J., Huchra, J., and Latham, D. 1982, Ap. J. 253, 423.
3. Faber, S.M., and Gallagher, J.S. 1979, A.R.A.A. 17, 135.
4. Fall, S.M., and Efstathiou, G. 1980, M.N.R.A.S. 193, 189.
5. Goldreich, P., and Lynden-Bell, D. 1965, M.N.R.A.S. 130, 97.
6. Gunn, J.E. 1980, Phil. Trans. Roy. Soc. London A., 246, 313.
7. Gunn, J.E., Stryker, L.L., and Tinsley, B.M. 1981, Ap. J. 249, 48.
8. Kirschner, R.P., Oemler, A., Schechter, P.A., and Schectman, S.
    1981, Ap. J. 243, L127.
9. van der Kruit, P., and Searle, L. 1981, Astron. Astrophys. 95, 105.
10. Peebles, P.J.E. 1969, ap. J. 155, 393.
11. Peebles, P.J.E., The Large-Scale Structure of the Universe, Princeton
    University Press.
12. Rees, M.J., and Ostriker, J.P. 1977, M.N.R.A.S. 179, 451.
13. Silk, J. 1967, Nature 215, 1155.
14. Toomre, A. 1964, Ap. J. 139, 1217.
15. Tremaine, S.D., and Gunn, J.E. 1979, Phys. Rev. Letters 42, 467.
16. Twarog, B.A. 1980, Ap. J. 242, 242.
17. White, S.D.M., and Rees, M.J. 1978, M.N.R.A.S. 183, 341.
18. Zel'dovich, Y., Einasto, J., and Shandarin, S. 1982, Nature, in press.

DISCUSSION

BLITZ : Wouldn't the star formation which takes place after the formation of the disk affect the evolutionary picture of the infall ? In our
own galaxy, the massive star formation is a strong function of radius,
and thus the effect on the interstellar medium via HII regions, stellar
winds and supernovae is also likely to be a strong function of radius.
Wouldn't you expect the dynamics of the ISM to seriously affect the
simple infall picture ?

GUNN : All that the "simple infall picture" demands is that star formation more-or-less keep up with the infall, and the Jeans instability
almost guarantees that this happen . Details will doubtless affect the
quantitative chemical evolution, for example, and may or may not through
transport effects, move the density distribution around in the disk.
Qualitatively, however, it is difficult to see how the picture could be
substantially changed unless the star formation activity were energetic
enough to unbind the infalling matter.

WHITE : I would like to make a comment on what "pancakes" actually look
like. Carlos Frenk, Marc Davis and I have run N-body simulations of the

growth of clustering from random. Those initial conditions are with a
short wavelength cut-off (the sort of initial condition advocated by
Zel'dovich and his colleagues). We find that at times when the correlation properties of such models are similar to those of the observed
galaxy distribution, they contain large coherent structures as much as
ten times larger than the scale at which the two-point correlation is
equal to unity. These "pancakes" are best displayed by equidensity
contours of the smoothed particle distribution. They are primarily prolate
objects and seem to link together at low density contrast in what may be
a space-filling network. Such large-scale structure is not seen in model
universes in which clustering has grown from "white noise" initial
conditions even though they have a two-point correlation function quite
similar to that of the pancake simulations.

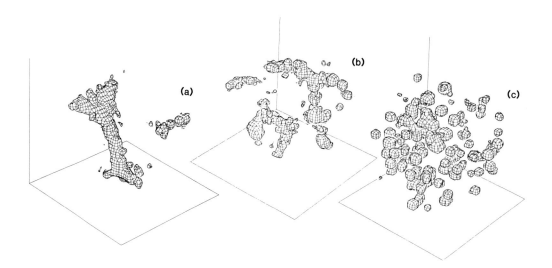

Equidensity contours at 3 times the mean density in N-body simulations of clustering in an Einstein-de Sitter universe. (a) and
(b) show the generic structure formed from random phase "adiabatic" initial conditions. In (a) the coherence length is such that
only one "pancake" forms in the region simulated, while in (b)
it is smaller, and several linked structures are apparent. (c) is
the analogous structure in a simulation started from white noise
"isothermal" initial conditions. The correlation function is
unity on similar scales in (b) and (c).   (From Frenk, White and
Davis, in preparation).

GUNN : This is very interesting, but the crucial question is whether
the universe looks like your pictures or not. It may be (or may not be)
that it looks more complicated than n = −1 clustering models, but it is
not clear either whether your models are closer than those to reality
or not.

WHAT MAY SUPERCLUSTERS TELL US ABOUT GALAXY FORMATION?

Avishai Dekel
California Institute of Technology and Yale University

ABSTRACT: The structure and dynamical properties of superclusters
(SC's) may help us distinguish between various scenarios for galaxy
formation. N-body simulations and analytic estimates show that (contrary
to a common belief) a nondissipative pancake scenario can explain the
observed flattening of SC's. Then, the disk-halo structure of the Local SC
(LSC) and its velocity field suggest that galaxies form before SC's and not
necessarily as a result of a dissipative gas collapse to SC pancakes. A
confrontation of the simulations with the data indicates that the LSC has
collapsed along its short axis only after $z \sim 1$.

Observations of SC's, and of the LSC in particular, are approaching a
stage where quantitative comparisons with formation theories may be
attempted, and let me mention here such an attempt. Two extreme scenarios
are commonly discussed: a) the clustering picture where structure evolves
hierarchically from small to large, starting from isothermal perturbations,
and where the clustering of galaxies is gradual and dissipationless, and
b) the dissipative pancake picture where structure evolves from large to
small via fragmentation of SC's. In the latter, large scale adiabatic
perturbations (of baryons and/or of massive neutrinos) that survive damping
are the progenitors of all structure, and formation of highly asymmetric
SC's is a natural outcome because of the large-scale asphericity of the
velocity field. If the collapsing material is gaseous, dissipation
produces a thin, dense pancake (or string) that fragments to form galaxies,
but it is hard to explain in this picture the presence of many galaxies
very far from the SC plane in which they are assumed to be born. Yet, a
non-dissipative pancake scenario, which is a combination of the two
scenarios, is also possible: galaxies may already exist during the
collapse of SC's as an outcome of small-scale perturbations, and they cross
the SC plane (or line) with little dissipation to form a thicker pancake
while some of them still populate the SC halo. By studying nondissipative
pancakes, and comparing with the LSC, we estimate the degree of dissipation
required to explain the observed flattening (Dekel, 1983a).

An N-body simulation of Sc formation is illustrated by a sequence of
snapshots in Fig. 1. An initially Hubble expanding, spherical system of a
comoving radius ℓ ($\ell_{18} \equiv \ell/18\text{Mpc}$), which corresponds to the critical

387

E. Athanassoula (ed.), Internal Kinematics and Dynamics of Galaxies, 387–388.

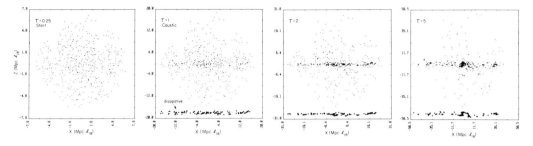

Figure 1.   Edge-on, comoving snapshots.

damping length, is perturbed adiabatically in one direction (a sin wave in Z) and is assumed to evolve in the linear regime according to the approximation of Zeldovich (1970). The slight flattening in the initial state (time τ=0.25) is associated with a systematic peculiar flow toward the XY plane which leads to focusing of trajectories of 20% of the particles in a thin pancake at τ=1. The relative flattening is not transient! It becomes even more pronounced after τ=1, even when the absolute thickness grows again (the thickness is eventually affected by subclustering).

The nondissipative flattening and cooling is primarily due to the expansion in the orthogonal directions. When the period of oscillation of a particle about the plane becomes small in comparison with the expansion time scale, there is an <u>adiabatic invariant</u> approximated by $h(\tau)v(\tau)\cong$const., where h and v are some mean values of the height and the velocity of the particle above the plane. Combine that with a mean gravitational field of an infinite uniform disk that is proportional to the disk surface density, $\mu(\tau)$, and obtain in time $h \propto v^{-1} \propto \mu^{-1/3} \propto R^{2/3}$, where R is the expanding length scale in the plane, i.e., the pancake becomes flatter like $h/R \propto R^{-1/3}$.

The flattening in the numerical models is found to resemble that of the LSC (Tully 1981 and references therein) near the focusing time τ=1, and it becomes much flatter later. The fit is insensitive to fine details in the models, such as the exact peculiar velocity of our galaxy toward Virgo or the degree of subclustering. A similar result is obtained for observed external SC's, which means that the nondissipative pancake scenario can account for the observed flattening of SC's in general and that dissipation is not required. The velocity field in the LSC (Dekel 1983b) and the subclustering (Dekel 1982; 1983c) are found to be consistent with this scenario also. Hence, galaxies may have formed before SC's, and not necessarily as a result of their collapse which could have been very recent (z<1) and nondissipative.

REFERENCES:

Dekel, A. 1982, Ap. J. Lett., 261 (Oct. 1); 1983a, Ap. J., 269 (Jan. 15);
     1983b, in preparation; 1983c, in preparation.
Tully, R. B. 1981, Ap. J., 255, 1.
Zeldovich, Ya. B. 1970, Astr. Ap., 5, 84.

THICK DISCS AND THE FORMATION OF DISC GALAXIES

Rosemary F.G. Wyse,
Institute of Astronomy, Cambridge, UK.
Bernard J.T. Jones,
Observatoire de Meudon, France.

The formation of disc galaxies is generally believed to involve the dissipative collapse of primordial gas clouds, moving in the potentials of dark, self-gravitating massive halos. Until now, most theories have considered only spherically symmetric collapse, with the luminous disc being formed out of a plane of cold gas which is in centrifugal equilibrium. However, it is unlikely that the halo potential will in fact be spherical.

We have investigated the fate of gas collapsing in the potential of an oblate halo, obtaining analytic results for various density profiles, as a function of halo axial ratio. For all but the most nearly spherical halos, analysis of the dynamical timescales in the vertical and radial directions (governed by the dark matter) indicate that the gas will be shock heated, and able to cool and form stars before angular momentum is important in the collapse. Thus the resultant stellar disc will be formed out of centrifugal equilibrium. The subsequent relaxation processes lead to a hot, pressure supported system with exponential light profiles in the outer regions. These are identified with the thick discs seen in external edge-on disc galaxies. The inner, denser regions will have suffered a higher rate of dissipation leading to a stellar distribution which is more centrally concentrated, and here rotation will be more important: the bulge.

The thin disc forms from the gas lost by stellar evolution in the thick disc and accretion of primordial gas from the outer regions of the halo which can cool only by this later time. The halo potential is now steady and this gas settles to form a cold thin disc that relaxes dissipatively into centrifugal equilibrium. The resultant disc follows the exponential profile of the thick disc.

The important model parameters are the axial ratio of the halo potential, and the slope and central value of the halo density profile. These govern the collapse dynamics, i.e. timescales, collapse factors in the radial direction and the temperature of the shocked gas.

E. Athanassoula (ed.), Internal Kinematics and Dynamics of Galaxies, 389–390.

If the cooling is by Compton scattering off background photons it is governed purely by the redshift Z_f assumed for galaxy formation. However, for this to be the dominant cooling mechanism, $Z_f \gtrsim 15$. Bremsstrahlung and line radiation are more important at later epochs, i.e. for larger masses, which for most reasonable power spectra of primordial fluctuations collapse later. The cooling times for these processes are determined by the temperature and density of the shocked gas itself. This last density depends in turn on the halo density profile and the fraction of the total halo mass which is initially gaseous, as opposed to in some dark form. A further model dependent parameter of relevance here is the enhancement of gas density due to the collapse itself. It should be noted that λ_H, the dimensionless angular momentum parameter for the halo, is not an important quantity in this theory.

The assumption of suitable scaling laws to describe, for example, the dependence of the power spectrum of primordial density fluctuations, and that of the fraction of matter in gaseous form, on total mass, together with the derived behaviour of the model parameters allow in principle a comparison of the theory predictions with the observations of other galaxies. It is unfortunate that there is not an overwhelming body of relevant observational data which is self consistent, but we can say that our models are in agreement with such seemingly fundamental observations as the Tully Fisher relation.

The observational situation concerning our own Galaxy is somewhat more clear cut. Our model predicts a distinct dichotomy between the thick and thin discs, both in kinematics and chemistry: since the thick disc relaxes after forming stars we expect there to be no metallicity gradient, whereas the thin disc forms once relaxation processes have been completed. We also prefer a rapid formation for the thick disc and a much longer formation time for the thin disc. These predictions are indeed upheld by the observations, if we identify the high-velocity stars of the solar neighbourhood as the thick disc population, and use an updated metallicity calibration for the RRlyrae stars.

A full description of the model is in Jones & Wyse (1982) to appear in Astronomy and Astrophysics.

GALAXY FORMATION: SOME COMPARISONS BETWEEN THEORY AND OBSERVATION

S. Michael Fall
Institute of Astronomy, Cambridge, England.

I. INTRODUCTION

Before theoretical ideas in this subject can be compared with observational data, it is necessary to consider the properties of galaxies that are likely to be relics of their formation. Most astronomers would agree that the list of important parameters should be headed by the total mass M, energy E and angular momentum J. Next on the list should probably be the relative contributions to these quantities from the disc and bulge components of galaxies and denoted D/B for the mass ratio. They can be estimated from the median (i.e. half-mass) radius R, velocity dispersion σ and rotation velocity v of each component, either through the virial theorem or through the luminosity L and an assumed value of M/L. As a first approximation, it is reasonable to suppose that galaxies of a given disc-to-bulge ratio or morphological type form a sequence with mass as the fundamental parameter. The comparison of theory with data is further simplified by considering the extreme cases of ellipticals, with D/B << 1, and late-type spirals, with D/B >> 1. The approach outlined below is to explore the consequences of relaxing in succession the constraints that E, J and M be conserved during the collapse of proto-galaxies. In this article I concentrate on theories that are based on some form of hierarchical clustering because the pancake and related theories are not yet refined enough for a detailed confrontation with observations.

II. A HIERARCHY WITH E, J AND M CONSERVED

The starting point for each of the theories discussed here is a primordial spectrum of density perturbations with a power-law form, $\delta\rho/\rho \propto M^{-\frac{1}{2}-n/6}$, and an index in the range $-1.5 \lesssim n \lesssim 0$. Such a distribution of matter would evolve by gravitational forces alone into a nested hierarchy that might resemble the distribution of galaxies on scales from about 0.1 to 10 Mpc (Fall 1979, Peebles 1980). In the process, individual structures would be given a small amount of rotation by the tidal torques of their neighbours, which can be quantified in

E. Athanassoula (ed.), Internal Kinematics and Dynamics of Galaxies, 391–399.

terms of the dimensionless spin parameter, $\lambda \equiv J|E|^{\frac{1}{2}} G^{-1} M^{-5/2}$
(Peebles 1969, Efstathiou & Jones 1979). The dependence of the spin
and median radius on mass have been difficult to predict analytically
as a result of the broad spectrum of structures in the hierarchy at a
fixed density contrast. Recently Barnes & Efstathiou (private communication) have analysed a 20,000 body simulation with white noise initial
conditions ($n \approx 0$) and a critical density parameter ($\Omega = 1.0$). By
fitting the results to power-laws over the range $0.1M* \lesssim M \lesssim 10 M*$, they
find

$$\lambda_H \propto M_H^{-0.03\pm0.02}, \qquad R_H \propto M_H^{0.52\pm0.02}, \qquad \text{(H for hierarchy)} \qquad (1)$$

and a median spin $\lambda_H = 0.07$ near the characteristic mass M*. These
relations may depend weakly on n and Ω but the exact dependence will
not be known until more simulations are available. The important
feature of eqns. (1) is that the exponent in the radius-mass relation
differs significantly from the often-quoted value $(n+5)/6$.

Elliptical galaxies are usually taken to be the prototypes for
non-dissipative formation because their density profiles and velocity
anisotropies can be reproduced in N-body collapse models (Aarseth &
Binney 1978, van Albada 1982). It is therefore of interest to compare
the predictions above with the data on 44 ellipticals compiled by Davies
et al. (1982). By fitting the results to power-laws over the range
$0.1L* \lesssim L \lesssim 10L*$, they find

$$\lambda_E \propto M_E^{-0.33\pm0.09}, \qquad R_E \propto M_E^{0.50\pm0.03}. \qquad \text{(E for elliptical)} \qquad (2)$$

These relations have been derived from the formulae appropriate for
self-gravitating bodies with de Vaucouleurs profiles: $\lambda_E \approx 0.4(v_m/\overline{\sigma})$,
$M_E \approx 5.0(\overline{\sigma})^2 r_e/G$ and $R_E \approx 1.35 r_e$ where v_m is the maximum rotation
velocity along the major axis, $\overline{\sigma}$ is the average velocity dispersion
within $\frac{1}{2}r_e$ and r_e is the effective radius. A comparison of eqns. (1)
and (2) shows that the visible bodies of elliptical galaxies are not
typical of the structures that form in a clustering hierarchy with E,
J and M conserved. An even stronger objection to this idea is that the
luminosity densities of L* galaxies, both ellipticals and spirals, are
3 or 4 orders of magnitude higher than those implied by an extrapolation
of the pair-correlation function for galaxies down to their median radii
(Fall 1981). Thus, if ellipticals reached their present states by
violent relaxation, this must have been preceeded by some dissipation.
One possibility along these lines is that elliptical galaxies formed by
the merging of spiral galaxies or subgalactic structures.

III. A HIERARCHY WITH E DISSIPATED AND J AND M CONSERVED

A currently popular notion is that the visible bodies of galaxies
formed by the dissipative collapse of residual gas in a hierarchy of
dark haloes (White & Rees 1978). Several recent investigations have
shown that this provides a natural explanation for the masses, radii

and spins of disc galaxies (Fall & Efstathiou 1980, Silk & Norman 1981, Burstein & Sarazin 1982, Faber 1982, Gunn 1982). As an illustration, consider a spherical halo with a constant circular velocity v_c out to some truncation radius r_t. Without having to specify the rotation curve of the dark material, its angular momentum and mass can be expressed as

$$J_H = \sqrt{2} \, \lambda_H \, v_c^3 \, r_t^2/G, \qquad M_H = v_c^2 \, r_t/G, \qquad \text{(H for halo)} \qquad (3)$$

when terms of order λ_H^2 are neglected in the energy $E_H = -v_c^4 \, r_t/2G$. Now suppose the gas that collapses in such a halo arranges itself into an exponential disc with a scale-radius α^{-1} determined by the circular velocity v_c and the specific angular momentum $J_D/M_D = 2v_c/\alpha$. Since the disc material would have experienced the same tidal torques as the halo material before dissipating any energy it seems reasonable to set $J_D/M_D = J_H/M_H$; this implies

$$R_H/R_D = 0.30(\alpha r_t) = 0.42/\lambda_H \approx 6. \qquad \text{(D for disc)} \qquad (4)$$

The first equality follows from the expressions $R_H = \frac{1}{2} r_t$ and $R_D = 1.67 \, \alpha^{-1}$ for the median radii of the halo and disc and the last equality is appropriate for $\lambda_H \approx 0.07$. These results agree nicely with the more exact calculations of Fall & Efstathiou (1980), which include the effects of a bulge and the gravity of the disc.

The picture outlined above has several interesting consequences. For a bright spiral such as the Milky Way, with $\alpha^{-1} \approx 4$ kpc, the extent of the surrounding halo is predicted to be $r_t \approx 80$ kpc. This is comparable with the radius of the proto-galaxy deduced by Eggen, Lynden-Bell & Sandage (1962) from the motions of old stars in the solar neighbourhood. When eqns. (3) and (4) are combined with eqns. (1) from the Barnes-Efstathiou simulations, the results are

$$\mu_o \propto (M_D/M_H)^{0.98} M_D^{0.02}, \qquad v_c \propto (M_D/M_H)^{-0.24} M_D^{0.24}, \qquad (5)$$

where $\mu_o = \alpha^2 M_D/2\pi$ is the central surface density of an exponential disc. For relatively isolated disc-halo systems, it seems natural to expect $M_D/M_H \approx$ const because this should be roughly the ratio of mass in primordial gas to that in dark matter. In this case, eqns. (5) take forms that are reminiscent of Freeman's (1970) law and the Tully-Fisher (1977) relation. The coefficients of proportionality in these expressions are determined by eqns. (1) and therefore by the scaling of the N-body simulations to astronomical units. A simple way to do this is by means of the empirical relation $v_c^2 \approx \alpha \, M_D G$, which implies $M_H/M_D \approx 1.4/\lambda_H$ when combined with eqns. (3) and (4). For $\lambda_H \approx 0.07$, the result is $M_H/M_D \approx 20$ and this agrees with current estimates of the average ratio of dark to luminous material, $\Omega(\text{dark})/\Omega(\text{lum}) \sim 10$ (Faber 1982). Thus, a hierarchy with E dissipated but J and M conserved appears to account for the most important properties of disc galaxies.

If elliptical and spiral galaxies formed in the same kind of haloes,

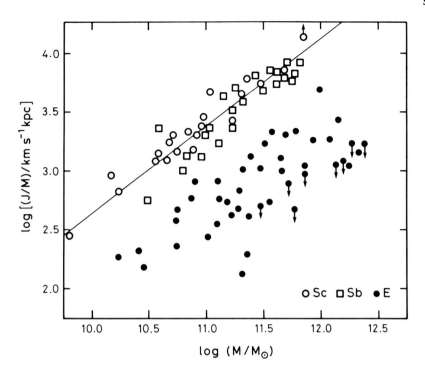

Figure 1. Specific angular momentum against luminous mass for galaxies of different morphological types. Spirals: $J/M = 2v_c\alpha^{-1}$ with $v_c = V(R_{25}^{ib})$ and $\alpha^{-1} = 0.32\ R_{25}^{ib}$; $M = (M/L)L(B_T^{ib})$ with $M/L = 3.0$ for Sb and $M/L = 1.5$ for Sc; data from Rubin et al. (1980, 1982). Ellipticals: $J/M = 2.5\ v_m\ r_e$ as appropriate for a de Vaucouleurs profile, a flat rotation curve and random orientations; $M = (M/L)L(B_T^b)$ with $M/L = 6.0$; data from Davies et al. (1982). A Hubble constant of 50 km s^{-1} Mpc^{-1} is adopted but the relative positions of points do not depend on this distance-scale.

they would occupy similar positions in the J/M - M plane, irrespective of how much energy they dissipated. As Fig. 1 shows, however, the visible bodies of spirals have about 6 times as much angular momentum as ellipticals of the same mass. The values of J/M in this diagram were computed from rotation velocities and photometric radii and the values of M were computed from total luminosities and M/L ratios for the mean B-V colour of each type in the population synthesis models of Tinsley (1981). This procedure avoids the awkward question of whether the luminous parts of galaxies are self-gravitating and the spurious correlations that may arise when v_c is used to compute both J/M and M for disc galaxies (Freeman 1970). It is not yet possible to plot galaxies with D/B ≈ 1 on the diagram because of the lack of coordinated photometric and kinematic data needed to estimate the contributions of each component to J/M and M. Nevertheless, the few data that are available indicate that intermediate types fill the gap between Sb/c

and E galaxies and that J/M increases with D/B at each value of M. This
is a refinement of the suggestion by Sandage, Freeman and Stokes (1970)
that the relative prominence of the disc and bulge components of galaxies
should reflect the proportions of material with high and low angular
momentum. The immportant point about Fig. 1 in the present context is
that it may contain useful information on the haloes of galaxies.

IV. A HIERARCHY WITH E DISSIPATED AND J AND M STRIPPED

The simplest explanation for the different angular momenta of
elliptical and spiral galaxies would be in terms of a spread in the
tidal spins of their haloes (Kashlinski 1982). In this case, low λ
haloes must be produced preferentially in regions of high density and
high λ haloes in regions of low density to account for the strong
correlation between the clustering environments of galaxies and their
morphological types (Dressler 1980). Such a coupling of long and short
wavelength modes seems unlikely in the linear phase of growth when tidal
torques are induced. Instead, some sort of non-linear process is probably needed if the initial spectrum of density perturbations had random
phases. One suggestion along these lines is that the outer envelopes
of haloes in clusters were stripped off by tidal interactions with
their neighbours whereas isolated haloes were not disturbed in this way.
The morphologies of galaxies might then be a reflection of the degree
to which the gas of high specific angular momentum was dispersed before
it could collapse (Binney & Silk 1978, Larson, Tinsley & Caldwell 1980).
This scheme requires that the stripping and collapse time-scales be
comparable in the outer parts of proto-galaxies and that the mass of
the diffuse gas and the visible bodies of galaxies be comparable in
rich clusters. To see how the second condition has a bearing on the
properties of galactic haloes, it is instructive to consider the following simple model.

The spatial and projected density profiles of the 'isothermal' halo
discussed in Section III are

$$\rho_H(s) = (M_H/4\pi r_t^3)(r_t/s)^2, \quad \mu_H(r) = (M_H/2\pi r_t^2)(r_t/r)\cos^{-1}(r/r_t), \quad (6)$$

where s denotes distance from the centre and r denotes distance from
the rotation axis. Prior to stripping, the fraction of mass with
specific angular momentum in the interval (h, h+dh) is assumed to be
$f_H(h)dh$ with

$$f_H(h) = (\beta/h_t)(h/h_t)^{\beta-1}\cos^{-1}(h/h_t)^{\beta} \quad \text{for } h \leq h_t. \quad (7)$$

Here β is an adjustable parameter and h_t is related to r_t and λ_H by
eqns. (3) and the normalization

$$J_H/M_H = \int_o^{h_t} dh\, f_H(h)h = \tfrac{1}{2}\sqrt{\pi}\beta h_t(1+\beta)^{-2}\Gamma(\frac{1}{2\beta})/\Gamma(\frac{1}{2} + \frac{1}{2\beta}), \quad (8)$$

where Γ denotes the usual gamma function. If the halo rotates on cylinders, its velocity is determined by the condition

$$M_H f_H(h)dh = 2\pi r \mu_H(r)dr \qquad \text{with} \quad h = r\, v_H(r), \tag{9}$$

hence
$$v_H(r) = (h_t/r_t)(r/r_t)^{1/\beta-1}. \tag{10}$$

The surface density of a disc with the same distribution of specific angular momentum is determined by the condition

$$M_D f_H(h)dh = 2\pi r \mu_D(r)dr \qquad \text{with} \quad h = r\, v_c, \tag{11}$$

hence
$$\mu_D(r) = (\beta M_D/2\pi r_o^2)(r/r_o)^{\beta-2}\cos^{-1}(r/r_o)^{\beta}, \tag{12}$$

where $r_o = h_t/v_c$ is the corresponding truncation radius. It is straightforward to show that the angular momentum of the halo within the radius s is

$$J(<s) = 4\pi\int_o^s dr v_H(r)r^2 \int_r^s ds'\rho_H(s')s'(s'^2-r^2)^{-\frac{1}{2}} = J_H(s/r_t)^{1/\beta+1} \tag{13}$$

If the halo is stripped on spherical shells, the angular momentum and mass of the remaining material are related by the simple power-law

$$J(<s)/J_H = [M(<s)/M_H]^{1/\beta+1} \qquad \text{or} \qquad M \propto (J/M)^{\beta} \tag{14}$$

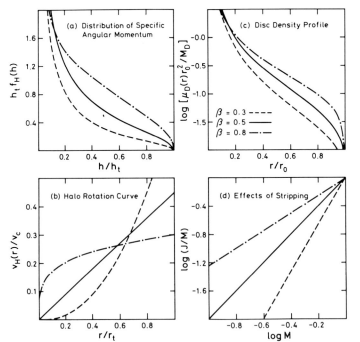

Figure 2. Illustrative model of a disc-halo system with three values of the parameter β. (a) from eqn. (7); (b) from eqns. (3), (8) and (10) with $\lambda_H = 0.07$; (c) from eqn. (12); (d) from eqn. (14).

Several properties of this model are shown in Fig. 2 for $\beta = 0.3$, 0.5 and 0.8. In each case, a disc that is roughly exponential over 3.5 scale-radii could form in the halo if no stripping occurred and each element of gas conserved its angular momentum during the collapse. Smaller values of β imply steeper rotation curves in the halo and therefore more angular momentum lost when a given fraction of mass is stripped. The significance of this effect can be seen by comparing Fig. 2(d) with Fig. 1. To convert the haloes of spirals into the haloes of ellipticals requires that their masses be reduced by factors of about $(6\pm2)^{\beta}$. In the core of the Coma cluster, the ratio of the mass in X-ray emitting gas to the mass in optically visible material probably lies in the range 1 to 4 depending on the value of the Hubble constant and other uncertainties (Faber 1982). This implies $0.4 \lesssim \beta \lesssim 0.9$, which should be compared with the rotation curves induced by tidal torques. In the absence of any detailed calculations, a reasonable model for a proto-halo during the linear phase of growth might be a uniform density sphere with solid body rotation (Gunn 1982). This corresponds to an 'isothermal' halo with a nearly flat rotation curve and therefore to $\beta \approx 1$ if $f_H(h)$ is conserved in the approach to dynamical equilibrium. The slow rotation of ellipticals then requires the stripping of so much material that it becomes hard to understand why they are more massive than giant spirals.

V. CONCLUSIONS

The preceeding arguments are summarized in the table below.

Hierarchy	Spirals	Ellipticals
E conserved J conserved M conserved	Radii too small. Spins too high. Etc.	Radii too small. Spin-mass rela- tion too steep.
E dissipated J conserved M conserved	Radii, masses and. spins about right	Location in clusters hard to explain.
E dissipated J stripped M stripped	As above in small groups and the field.	Small spins imply large mass-loss.

The strength of the picture outlined here is its success with spirals and the weakness is its difficulty with ellipticals. This is not, however, a firm conclusion because some combination of effects might explain the differences between early and late types and their correlation with environment. For example, the factor of 6 spread in the angular momenta of galaxies at a fixed mass might reflect both the distribution of tidal spins in the linear regime and some tidal stripping in the non-linear regime. Finally, it must be emphasized that all of these remarks pertain to the general framework of hierarchical clustering

with scale-free initial conditions. A test of this hypothesis is not
likely to come from studies of the internal properties of galaxies
because the theoretical interpretation is usually ambiguous. For more
direct tests, we will probably have to rely on studies of the large-
scale distribution of galaxies and fluctuations in the cosmic background radiation.

REFERENCES

Aarseth, S.J. & Binney, J., 1978. Mon. Not. Roy. Astr. Soc., 185, 227.
Binney, J. & Silk, J., 1978, Comm. Astr. Astrophys., 7, 139.
Burstein, D. & Sarazin, C.L., 1982, Astrophys. J., in press.
Davies, R.L., Efstathiou, G., Fall, S.M., Illingworth, G. & Schechter,
      P.L., 1982. Astrophys. J., in press.
Dressler, A., 1980. Astrophys. J., 236, 351.
Efstathiou, G. & Jones, B.J.T., 1979. Mon. Not. Roy. Astr. Soc., 186,133.
Eggen, O.J., Lynden-Bell, D. & Sandage, A., 1962. Astrophys. J., 136,748.
Faber, S.M., 1982. in "Astrophysical Cosmology", eds. H.A. Brück, G.V.
      Coyne & M.S. Longair, Pontificia Academia Scientarium, Vatican,
      p.191.
Fall, S.M., 1979. Rev. Mod. Phys., 51, 21.
Fall, S.M., 1981. in "The Structure and Evolution of Normal Galaxies",
      eds. S.M. Fall & D. Lynden-Bell, Cambridge University Press, p.1.
Fall, S.M. & Efstathiou, G., 1980. Mon. Not. Roy. Astr. Soc., 193, 189.
Freeman, K.C., 1970. Astrophys. J., 160, 811.
Gunn, J.E., 1982. in "Astrophysical Cosmology", eds. H.A. Brück, G.V.
      Coyne & M.S. Longair, Pontificia Academia Scientarium, Vatican,
      p.233.
Kashlinski, A., 1982. Mon. Not. Roy. Astr. Soc., 200, 585.
Larson, R.B., Tinsley, B.M. & Caldwell, C.N., 1980. Astrophys. J., 237,
      692.
Peebles, P.J.E., 1969. Astrophys. J., 155, 393.
Peebles, P.J.E., 1980. "The Large-Scale Structure of the Universe",
      Princeton University Press.
Rubin, V.C., Ford, W.K. & Thonnard, N., 1980. Astrophys. J., 238, 471.
Rubin, V.C., Ford, W.K., Thonnard, N. & Burstein, D., 1982. Astrophys.
      J., in press.
Sandage, A., Freeman, K.C. & Stokes, N.R., 1970. Astrophys. J., 160,831.
Silk, J. & Norman, C.A., 1981. Astrophys. J., 247, 59.
Tinsley, B.M., 1981. Mon. Not. Roy. Astr. Soc., 194, 63.
Tully, R.B. & Fisher, J.R., 1977. Astron. Astrophys., 54, 661.
van Albada, T.S., 1982. Mon. Not. Roy. Astr. Soc., in press.
White, S.D.M. & Rees, M.J., 1978. Mon. Not. Roy. Astr. Soc., 183, 341.

DISCUSSION

DEKEL : With regard to the idea of spin-loss : there are indications
that the rotation curves of ellipticals are flat. If so, the specific
angular momentum indeed rises with radius. If the gas was contracted from
a r^{-2} distribution, the original halo rotation curve must have been rising
exponentially and to obtain a factor of 3 decrease in total spin, 30 %
mass loss may have to occur. Tidal encounters in clusters might easily
account for this. The spin-loss can be naturally combined with a scenario
in which all galaxy types are formed by gas contraction in dark halos.
The structure of the halo can determine the type : halos that are puffed-
up by slow tidal encounters in protoclusters allow star formation at larger radii, i.e. more bulge.

FALL : As you say, the main constraint on this scheme is that the rotation curves of the haloes must rise steeply to avoid the removal of too
much mass as the angular momentum of proto-galaxies is reduced by stripping. The density profiles and rotation curves of the visible bodies of
galaxies might be used to infer the corresponding properties of the haloes
if one has reason to suppose that each element of mass conserved its angular momentum during the collapse. But why would you use ellipticals for
this purpose when the hypothesis being tested is that their masses and
angular momenta were not conserved ? When these arguments are applied to
relatively isolated spirals on the grounds that their haloes are more
likely to have remained intact, the inferred rotation curves are fairly
flat. In any case, the real question is whether tidal torques would endow
the structures that form in a clustering hierarchy with this distribution
of specific angular momentum. To my knowledge, the necessary calculations
have not yet been made but simple arguments suggest that exponentially
rising rotation curves are too steep,especially after virial equilibrium
is reached. This is one of the reasons I am sceptical about halo stripping
as the sole explanation for all of the morphological and environmental
differences between galaxies.

VIII

SUMMARIES

SUMMARY : OBSERVATIONAL VIEWPOINT

W.B. Burton
Leiden Observatory, Leiden, the Netherlands

The organizers took advantage of my position as an outsider to much
of the work which has been discussed at this symposium when they assigned
me this task, thinking that my ignorance would keep me unaware of the in-
digestion which would result from trying to summarize everything on the
menu which we have had in front of us for the past 5 days. I am more
familiar with work on our own galaxy, but it does seem to me that IAU
Symposium No. 100 fits perfectly into the series of meetings which the
IAU has sponsored in encouragement of work on the Milky Way. This series
began in 1953 with IAU Symposium No. 1 held in Groningen and entitled
"Co-ordination of Galactic Research". I expect that this symposium will
serve to co-ordinate extragalactic research in similar ways: I say this
because of the impression formed during this week that the study of
normal galaxies is experiencing a period of consolidation, when there is
a willingness to agree on issues, and when these issues are mature enough
that co-ordination of work is possible. Do I correctly perceive that
consolidation means that the field is not experiencing the chaos associated with breakthrough? For example, I don't think that this symposium
experienced an analogue to the introduction of the brand new Westerbork
HI maps experienced at the Besançon meeting of 8 years ago, and it is a
year or two too soon to expect comprehensive work from the VLA line
system.

The tone of consolidation was set on the first morning by Rubin's
plaintive hope that the non-luminous mass would turn out to be "something
we can be comfortable with", such as low-mass stars. Her talk, and Bosma's
which followed it, stressed several points which recurred throughout the
meeting, and which established recurring themes: 1) rotation curves are
flat (or rise) over the optical image, and 2) this is a general property
shared by field and cluster galaxies. Bosma extended these general, and
by now familiar, conclusions with other, also familiar, conclusions:
1) HI rotation curves extend to greater radii than optical curves, and
2) they are also generally flat over the entire observed extent.

Bosma carefully described the HI asymmetries which have to be
accounted for before one can trust these generalizations. Asymmetries

403

E. Athanassoula (ed.), Internal Kinematics and Dynamics of Galaxies, 403–410.

abound, due to spiral arm motions, bar or oval distortions, warps, and the sort of structural asymmetries and kinks that led us to be reminded often during the meeting that spiral rectification in the manner of Danver rarely works well. These asymmetries plague even such familiar galaxies as M31 and M33, both of which are probably warped, and M101 which is lopsided, as is the inner part of M33, and as is, quite possibly, our own galaxy. Simkin illustrated a different kinematic asymmetry with NGC 2903, whose optical center is displaced from the rotational center.

The structural and kinematic asymmetries made Bosma caution us to be careful with HI rotation curves: the envelope of the measured velocities might represent some characteristic of the galaxy other than rotation in circular orbits. Nevertheless a question from the audience forced Bosma to admit that there is "no solid evidence yet for a falling HI rotation curve". He mentioned curves which appear to fall: NGC 891 is asymmetric, and NGC 5907 is warped such that simple geometry probably does not suffice when deriving the kinematics. Warps were established as a theme of this meeting when Kennicutt provocatively claimed that the number of warped and distorted galaxies is equal to the number of galaxies studied. Bosma stressed that the HI warps are kinematic as well as structural. Although they occur in a variety of complicated forms, these forms often share the property of starting at the optical edge of the galaxy. He presented a counter example of his own, however, in UGC 2885 where the HI follows distorted optical arms to large distance.

Thonnard summarized work by him, Roberts, Rubin, and Ford on some 60 Sa,b,c spirals. Asymmetries are shared by $H\alpha$ and HI and commonly occur over large portions of the disks. Shostak discussed a kinematic study of the warp in IC 10, an apparently messy, nearby, irregular barred magellanic galaxy. The kinematics of that galaxy show enough systematics when observed with 8 km s^{-1} resolution at Westerbork to allow study of a face-on warp.

IC 10 also shows a hole in the HI distribution towards its center. Central deficiences of HI in the cores of galaxies have become familiar. Such holes - our galaxy and M31 are prototypes - are being put into a new context by the studies of comparative gas morphology reported by Solomon and Young. These new studies are centered on extensive radio observations of the CO line. Their groups have independently mapped the CO distribution in several dozen galaxies. The data show that 1) annuli similar to the one found in our galaxy are found in many other galaxies; 2) CO does not follow the HI distribution, although in the region of overlap the HI and CO kinematics are similar; 3) CO does follow the blue light distribution over 2 orders of magnitude in CO and luminosity; 4) the surface density of H_2 dominates HI in the central regions of high luminosity galaxies; and 5) the surface density of H_2 is approximately equal to that of HI in the central regions of low luminosity galaxies.

Young pointed out that high luminosity Sc's have no CO central holes, but that Sb's with large bulges do have holes. NGC 253 has a hole like our galaxy's. Young indeed suggested that NGC 253 is a galaxy similar to ours. Both galaxies might well be barred, a point van Albada had in mind when he asked if oval distortions might be responsible for the central CO deficiencies.

de Vaucouleurs asked that someone present soon the convolution which would show what the Milky Way's HI and CO radial distribution would look like from some Mpc distance. For the HI, that would be easy to do, but for the CO, it would require careful consideration of antenna beam characteristics and beam filling factors. We can imagine circumstances where the hole in our galaxy would be overlooked. Results from our galaxy, particularly on the distribution parameters of the molecular cloud ensemble, can be very useful in interpreting the CO data now becoming available for other galaxies.

These distribution parameters are bound to be important in a different context. Numerous discrete objects of some 10^6 $M_\odot$ embedded in the ISM would govern many aspects of the medium. The same can be said for the holes in the HI report often this week, for example by Shostak in IC 10 and by Bajaja in M31, and for the Hα bubbles reported by Marcelin in NGC 5128. Masses of 10^6 are involved, and expansion motions of some 20 km s^{-1}. These constituents must be important even on a global scale. Norman emphasized this in his review of galactic theory. Perhaps at this meeting it would have been useful to have had a review specifically on the morphology of the interstellar medium in spirals.

Interpretation of the correlation of the radial distribution of CO with that of blue luminosity in late-type galaxies generated much animated discussion. Does this correlation mean that H_2 density follows blue light, and that the star formation rate follows the 1st power of H_2 density? This was suggested, but Blitz countered that the CO-to-H_2 conversion involves many uncertainties. In M31, the blue light follows the metallicity, so that relation of CO to blue light might be a metallicity dependence, not a density one. Liszt remarked that the correlation might also reflect temperature because of the expected heating of CO in the interior of high luminosity galaxies. More study in our galaxy could guide us on this topic, too. Kennicutt remarked that in any case blue luminosity is not a good measure of the star formation rate.

Manchester presented the impressive results of the first southern CO survey. Arguments raging amongst members of our own galaxy's CO guild regarding the degree to which the CO is confined to spiral arms might be guided by work in nearby galaxies. Rydbeck said there is no arm/interarm contrast in M51.

Solomon expressed some scepticism that Kutner's work correctly shows the amount of CO emission beyond the Sun. But there can be no doubt that the HI dominates in the outer galaxy. Blitz discussed this for our own galaxy, and Sancisi for other galaxies. In general, HI extends furthest out of all observed components. The poster of Wever showing the Palomar-Westerbork survey of radial distributions of light, color, and HI mass was very relevant to this conclusion.

Sancisi compared outer-galaxy optical and HI profiles. Often (e.g. NGC 4564, 5907, 628) he finds an exponential fall-off to a sharp truncation of the optical disk, just where the HI shoulder starts and continues to much larger radii (the HI in Mkn 348 extends to 5 times the optical extent). Sancisi speculated that a critical radius separates the region of star formation from the HI galaxy. Also, the warp or unusual HI kinematics would begin at this critical radius. Casertano presented such a model for NGC 5907. Verification of Sancisi's idea requires searches for warps in the outer disks composed of old stars. Wamsteker reports optical counterparts of the warp in M83, extending to some 40 kpc, although this might be an encounter distortion. Trully remarked that our own galaxy's warp begins inside the Holmberg radius; I find it curious that we still are not sure if the stars in the outer disk of our own galaxy follow the HI warp.

Blitz gave a general review of the HI distribution in the outer Milky Way. According to his work, the rotation curve rises above 300 km s^{-1} at R $\sim$ 20 kpc, with the most important source of error residing in the possibility of motions in non-circular orbits. To me the most valuable uses of the outer galaxy HI lie in the availability of the $<z_{\frac{1}{2}}>$-distribution as a function of R, and, although it also has not yet been derived, of the variation of σ_v with R. Blitz concludes from the z-distribution that the HI cannot be confined by the observed mass: the non-luminous mass (80% of it) must lie in a spheroidal distribution.

van der Kruit emphasized the need for measurements of $<z_{\frac{1}{2}}>(R)$ and $\sigma_v(R)$ in other galaxies. His work with Shostak on several galaxies finds $<\sigma_v^2>^{\frac{1}{2}} \sim 7\text{-}10$ km s^{-1} and $<z_{\frac{1}{2}}> \sim 300$ pc, with the force term requiring in NGC 891 that one third of the mass reside in the disk interior to optical termination, and two thirds in a halo of necessarily dark mass.

If the symposium were to have voted, I suspect a majority would have resolved that 1) flat rotation curves imply dark mass, 2) $<z_{\frac{1}{2}}>$ and σ_v imply that the dark mass resides in the halo, and 3) M/L rises in the outer regions of spiral galaxies. But the vote would not have been unanimous, and the arguments of the dissidents need to be given very careful attention. For example, Tully remarked that accretion of high angular momentum material might give the apparently flat curves, without much dark mass. Observations should be able to resolve this; the outer boundary of our own galaxy is remarkably well defined, with no obvious important perturbations. Kalnajs' remark that he can model optical curves to large distance with a constant M/L was made casually, but was heard by all. I do not know how to respond to it. Certainly I do not know how to respond to Bekenstein's suggestion of non-Newtonian dynamics at low accelerations. But I feel that no rebuttal has been made to the flat HI curves at very large distance, or, as Blitz reminded us, to the globular cluster distribution in our galaxy at very large distance. Until we know better, the problem of non-luminous mass remains.

It might seem trite in a symposium on kinematics to say that the principal observational need centers on more measures of velocity fields and dispersions. It is nevertheless true. Kalnajs emphasized the importance of σ_v measures not only to determination of the mass distribution but to theories of spiral stability. Collisions of clouds might preserve spiral structure larger than in a gas composed of stars only. Carlberg's age v. σ_v theory could be tested. According to the theory, star formation leads to patterns of low σ_v which provide instabilities giving a short-lived spiral.

We heard rather less than I would have expected on observational predictions of the stochastic star formation theory, or for that matter, of the density-wave theory. Feitzinger claimed that stochastic star formation can generate realistic morphological sequences. Binney recalled us of Schweizer's demonstration of a smooth underlying red component, which the stochastic theory might find awkward. Norman concluded similarly that the coherence length is not long enough. Lynden-Bell reminded us of repeating structures, and Tohlin reminded us of bars: both are difficult for the stochastic theory. In favor of the theory, van Woerden mentioned large complexes in M101.

More observational work on the cross-sectional distribution of spiral arm tracers seems called for. The need for such work was emphasized by Allen, who speculated that Taurus will show that the kinematics of HII will differ, as predicted by the density-wave theory, from the kinematics of HI.

Three themes kept returning: warps, halo-mass, and bars. Toomre in his review of theories of warps asked if warped galaxies are also flare-edged (a propos of a flag-flapping instability) or if they have bars. These are questions which observations should soon be able to answer. Sandage remarked that M31 and M33 have warps but no halos.

Kormendy's review emphasized also the need for velocity fields in bars. His work to get the stellar velocity dispersion in bars represents a major and difficult program. Apparently the velocity dispersion in the disk around bars is hot, and this region of transition shows a dip in the rotation curve; clearly the bar potential rotates differently from the galaxy-at-large and will stir the galaxy up. The dust lanes on the leading sides of only pure bars are not yet understood, but as Kormendy emphasized, must be a key property.

Lindblad reported on work on NGC 1365 which showed that the velocities are perpendicular to the bar. In a poster on NGC 1365, van de Hulst showed in VLA data concentrations of HI on the edge of the bar but not on dust lanes.

Illingworth reviewed a major observational effort on elliptical galaxies. Some of his points: 1) some ellipticals have rising rotation curves, 2) some have dispersions approximately constant, meaning M/L probably rises outwards, 3) many ellipticals show dust and gas on the minor axis, implying that more than one kinematic system is present (this is a general problem, also for our galaxy, and was confronted in

this symposium by theories of response to a triaxial potential), and
4) galactic bulges are different from ellipticals (photometry required),
because bulge rotation velocities stay constant with z, whereas in elliptical the rotation velocity falls slowly into halo.

Dressler has studied the luminous bulges of SO's: they rotate rapidly, not as equivalent E galaxies. Although most SO's occupy the same
region of the Fisher-Tully diagram as spirals, some 10-20% lie far outside that region, implying that those SO's are not disk systems like
spirals. Dressler urged caution in inferring kinematics from the optical
morphology of many SO galaxies. Nevertheless, Sandage argued that SO's,
and smooth armed Sa's, are part of the Hubble sequence. SO's have a
high disk surface brightness; Sc's have a low disk brightness. According
to Sandage, SO's can not be formed by stripping Sc's of gas. I found the
Westerbork results presented by van Woerden showing HI in enormous rings
enveloping some SO's among the most important results presented at the
symposium. The HI distribution extends to some three optical diameters.

Tully presented an extragalactic analogue to the HR diagram, based
on hopefully distance-independent parameters of color and magnitude.
Rubin wondered if M does not depend on luminosity, however, not just on
velocity.

Knapp's discussion of gas in ellipticals emphasized the new result
that the gas is distributed in clumps, and includes cold HI and molecular
material. The VLA results on Cen A reported by van Gorkum show absorption
lines very near the nucleus, with an accretion rate of a few .1 $M_\odot$ yr^{-1}.
Do we see the radio galaxy being fed? This result, and also Gottesman's
results on the elliptical companions of M31 which show HI offset from
the symmetry of the distribution light, make it hard to imagine that the
gas does not come from outside the galaxy.

Ulrich's discussion of the ionized extended gas in normal E's was
also relevant in this regard, especially the result that there is no
correlation between the optical axis and the velocity field of gas. She
likewise concluded that some of the gas must come from outside.

We had a spirited review of mergers by Schweizer, backed up by a
handsome collection of photographs of ellipticals. Schweizer has searched
for prototypes of mergers in advanced stages; he chooses to reject as
prototypes pairs of galaxies, like the antennae, which represent encounters rather than mergers, and also chooses not to include multiple-
nuclei cD's, claiming evidence based on their color is unreliable and
that the components of cD nuclei may differ in their velocities by 100
km s^{-1}. In Schweizer's view the best examples for mergers are Fornax A
(which he describes as an elliptical in a very extended envelope), Cen A
(described as a dust disk in an elliptical), and NGC 7252 (described as
two disks, each with independent kinematics).

The signature for these and other Schweizer mergers includes ripples
and tails. The tails we are to understand according to the Toomre gospel.

The ripples were remarkably well simulated in Quinn's movie showing the response of a disk to capture by a central potential representing an elliptical.

Schweizer would have about one quarter of all field ellipticals have ripples and would have many (all?) field ellipticals have accreted disks. Several points were made in the discussion and in subsequent contributions which are relevant to, and difficult for, this point of view. Sandage challenged the view that ellipticals commonly formed by mergers on the basis of observations that 1) the density of field galaxies is too low, 2) the velocity dispersion of the galaxy ensemble is too low, and 3) ellipticals exist over a range of 12 magnitudes in Virgo, implying a range 10^6 in mass, and making too extreme demands on the Schweizer scenario. Sandage prefers to consider E's as part of the Hubble sequence.

Special problems of merger in the specific case of Cen A were pointed out by Marcelin, based on Hα interferometry, and by Taylor, on the basis of Taurus interferometry. Both investigations have impressive velocity resolution and both maps show a very regular velocity field. The Taurus data were modelled by a rotating disk, perhaps warped but otherwise well-behaved in ways that a merging disk might not be. Rubin remarked, however, that she has observed a pair of galaxies whose rotation curves fall through zero, indicating a measure of interaction.

Freeman's discussion on globular cluster systems held some surprises for me. The metal-rich clusters in M31 rotate as rapidly as the HI! In our own galaxy the rotation of the cluster system is much slower, and in the LMC the system is highly flattened.

The Scientific Organizing Committee of Symposium 100 put together a program which admirably incorporated the current activity in studies of the kinematics and dynamics of normal galaxies. It is tempting to speculate on the advances which we might anticipate hearing discussed at a meeting on this topic held a few years hence. It seems to me that we might well hear of substantial advances based on new instrumentation, on mature instrumentation, and on persistence in following current major studies. Regarding new instrumentation: I found myself especially impressed with the potential described by Atherton of the Taurus Fabry-Perot interferometer. This instrument will map out Hα spectral line profiles channel by channel with a 1/3 Å filter bank. The velocity resolution of 10 km s^{-1} will provide extremely valuable kinematic fields for galaxies, bars, planetary nebulae, and supernovae remnants. Regarding mature instrumentation: I think especially of HI line work which we may expect from the VLA. At this meeting such work was represented only in posters but included the remarkable maps of M81 by van der Hulst and van Gorkum, and of M31's satellites made by Gottesmann and Hunter. Regarding advances due to persistence: I was especially struck by the enormous wealth of details contained in the Westerbork HI survey of M31 carried out by Brinks and reported on by him and by Bajaja. A program of this magnitude is a long-term undertaking, but it will have provided

in the near future for M31 the same sort of spatial and kinematic resolution as is available now only in the standard surveys of our own
galaxy. An evolutionary development will thus be brought full circle,
with problems formulated but never solved in our own galaxy now available
for study under many more favorable conditions in other galaxies.

DISCUSSION

DJORGOVSKI : In your talk you asked whether our galaxy has a stellar
warp following the hydrogen one. I have looked into this question
following suggestions from Harold WEAVER and Leo BLITZ. The aim was to
see if it would be possible to detect the stellar warp in the infrared.
I have made a simple galaxy and stellar population model, and integrated
the direction of the HI warp. The results indicate that it may be (marginally) possible to detect an excess of IR light there, but the uncertainties are very high. I then looked at NASA IR survey maps, and indeed there
seems to be something there (galactic longitude about 80, latitude $5°$ –
$10°$), but it is highly uncertain whether this is due to foreground sources
or to the warp.

SUMMARY: THEORETICAL VIEWPOINT

Scott Tremaine
Massachusetts Institute of Technology

Rather than give a comprehensive review, I want to concentrate on a few areas where I was particularly impressed by the theoretical results discussed at this meeting.

The first such area is spiral structure theory. Kalnajs presented a clear and thoughtful review (and Lin gave us a fast Fourier transform of the review), but I mainly want to stress a very important point first made by Colin Norman.

Density wave theory was introduced in the early 1960's and quickly became the standard theory of spiral structure. It was based on the idea that the spiral arms were a wave pattern or normal mode of the galactic disk. For the past twenty years a number of theorists, primarily Kalnajs, Lin and Toomre, have been trying to understand these modes in detail. They have travelled a long and winding road, and now seem to be near their goal. In fact, this is the first meeting at which these three major protagonists all agree that they qualitatively understand the dominant normal modes in galactic disks.

Despite the sophisticated mathematics, the behaviour of these modes can be simply understood. Any normal mode has a corotation radius, where the wave pattern rotates at the same angular speed as the disk. Waves outside or inside corotation have positive or negative angular momentum respectively (this result is messy to prove but easy to rationalize: outside [inside] corotation the wave pattern angular speed is greater [less] than the disk angular speed so that the presence of the wave effectively increases [decreases] the angular momentum of the disk).

Consider a wave inside the corotation radius which propagates outward. It cannot pass corotation, since to do so its angular momentum would have to change sign. Hence the wave effectively reflects off the corotation circle and begins to propagate inward. The inward directed wave may reflect off the central bulge or propagate through the center of the galaxy; in either case a resonant cavity is set up whose

E. Athanassoula (ed.), Internal Kinematics and Dynamics of Galaxies, 411–416.
Copyright © 1983 by the IAU.

wall is the corotation circle. Moreover, tunneling through the wall
amplifies the standing wave inside the cavity, since the generation
of positive angular momentum waves outside corotation removes angular
momentum from the cavity and thus strengthens the negative angular
momentum waves inside. This is the amplification process variously
called the "WASER" or "swing amplifier."

This argument (and most of the analytic work in spiral structure
theory) is based on a "local" approximation, i.e. an approximation that
the separation of the spiral arms is small compared to the galactic
radius. Thus its relevance to realistic galaxies is limited. However,
there are now a number of numerical codes which exactly calculate the
linear normal modes of both gaseous disks (Bardeen, Haass and Iye) and
stellar disks (Athanassoula, Kalnajs and Zang). Remarkably, all of the
codes seem to show that these simple analytic arguments based on the
local approximation work quite well, even for large scale waves, and
can be used to predict fairly accurately the growth rates and shapes of
unstable normal modes.

Thus, we now largely understand the linear normal modes of galactic
disks, and in a sense the fundamental problem of density wave theory has
therefore been solved. To put this accomplishment in perspective I want
to use the mountain climbing analogy suggested several years ago by
Toomre (see diagram).

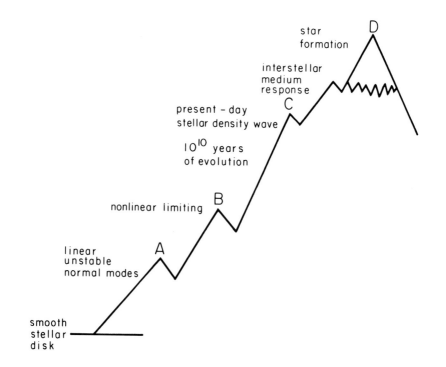

We have now reached peak A. Next we must understand what limits the growth of the linear normal modes (saturation of the Lindblad resonances? shock damping? nonlinear effects in the stellar disk?). This is the climb to peak B. After a steady spiral pattern is established we must follow it for 10^{10} yr, as it evolves due to angular momentum transfer, star formation, dynamical friction from the bulge and halo, infall and a variety of other effects. The importance of angular momentum transfer was pointed out ten years ago by Lynden-Bell and Kalnajs (1972) and again by Kalnajs at this meeting; the importance of infall was also stressed here by Gunn, and Kormendy has discussed morphological evidence for evolution in disk galaxies. Nevertheless, this part of the climb, from B to C, has not received the attention it deserves. It is difficult but rewarding, since the present structure of spiral galaxies may largely be determined during this phase.

Finally, we must understand star formation in sufficient detail to relate the density and potential perturbations in the stellar disk to the bright stars that define the optical arms (the climb to D).

An important cause for optimism is the recent development of efficient and accurate N-body codes, by Aarseth, Hohl, James, Miller, Sellwood, Wilkinson, Zang and others. We can now follow realistic galaxy models for $\sim 10^{10}$ yr, and can thus check our understanding of galaxy dynamics at all stages of the climb up to peak C (Sellwood described some checks of linear normal mode calculations in his review).

To summarize, we should congratulate the density wave theorists, who are approaching the successful completion of twenty years of work on the stability of galactic disks, and wish them equal success in the study of the origin and evolution of spiral structure.

The second area in which I feel that real progress has been made is the theory of triaxial stellar systems. Of course, we have known that such systems exist ever since Hubble defined the class of barred spiral galaxies, and some specialized triaxial models were constructed by Freeman in the 1960's. But the first real clue to the importance of these systems came from N-body experiments conducted by Hohl and Miller (the best review of these was given at the last Besançon meeting on dynamics [Hohl 1975]). They found that rapidly rotating axisymmetric stellar systems are always strongly unstable to the formation of a bar-like or triaxial subsystem.

The second clue came from experiments on the collapse of non-rotating triaxial systems. Aarseth and Binney, Miller, and especially Wilkinson and James (1982) showed that triaxiality is conserved in a collapse, so that triaxial systems are naturally formed in a collapse from irregular initial conditions. The Wilkinson and James calculation also shows conclusively that the equilibrium system preserves its tri-axiality, showing virtually no evolution over a large fraction of a Hubble time.

The conclusion is that triaxiality is a natural state for stellar systems, whether rapidly or slowly rotating, and that axisymmetry should be regarded as the exception, not the rule.

Accompanying these calculations has been Schwarzschild's (1979) development of a general algorithm for the construction of stellar systems. The application of this method to axisymmetric systems has been described by Richstone and de Zeeuw at this meeting, and Schwarzschild has used it to construct a non-rotating triaxial stellar system (which might be called the Schwarzschild ellipsoid). The density distribution in the Schwarzschild ellipsoid was chosen a priori to resemble the emissivity distribution in a real galaxy; hence the potential has no special features such as integrability or separability. Thus it is significant that Schwarzschild's algorithm rapidly converged on a self-consistent solution; this result lends support to the notion that tri-axial stellar systems are easy to form in nature. In addition, Schwarz-schild's work showed that the proper initial approach to the study of triaxial systems is morphological: one studies and classifies the fami-lies of orbits in order to get a qualitative grasp of how to combine these orbits to make up a self-consistent stellar system. We have heard about studies of this kind by Mulder and de Zeeuw.

In summary, we obviously do not understand triaxial systems yet, but we do understand that they are important; we have some preliminary theoretical results; and we have all the tools we need to finish the job. Ironically, many of the observations discussed at this meeting argue against strongly triaxial galaxies: Illingworth showed that faint ellipticals rotate like oblate spheroids with isotropic pressure tensors; Schweizer stressed that isophotal twists are often due to recent mergers; and studies of apparent axis ratio distributions are either inconclusive or weakly favour oblate systems.

The last topic that I shall discuss is the evidence for massive halos. At the last Besançon meeting, in 1974, there was a spirited controversy over whether the HI rotation curve of M31 was flat, and at this meeting there were similar controversies. However, over the last eight years Bosma, Rubin and others have accumulated impressive data showing that most spiral galaxies have flat rotation curves extending to at least one Holmberg radius, and hence that these galaxies have heavy halos containing at least 1-2 disk masses. The rotation curves provide the strongest evidence that such halos exist. However, there are many other clues.

One can combine M/L estimates for the disk and bulge with their photometric profiles and thus ask whether they contain enough mass to produce the observed rotation speed at radii $\leq$10 kpc. The answer is generally "no" (van der Kruit, at this meeting; also Bahcall and Soneira 1980).

Dynamical models of the Magellanic Stream require that our Galaxy has a massive halo out to $\geq$70 kpc if there are no non-gravitational forces on the Stream (e.g. Lin and Lynden-Bell 1982; also Lynden-Bell at this meeting).

The Local Group timing (e.g. Gunn 1975) strongly suggests that the Galaxy and M31 have a total mass of $\sim 3\times10^{12}$ $M_\odot$, far larger than their combined disk and bulge masses.

Studies of the dynamics of the Galactic globular cluster system (Frenk and White 1980) and the ellipticity of the Galactic bulge (Monet, Richstone and Schechter 1981) also suggest that an extended massive halo is present in the Galaxy.

Ostriker and Peebles (1973; see Sellwood's review) argued that either a halo or a hot disk component with mass comparable to the observed disk mass was needed for stability. This argument still lends strong support to the heavy halo hypothesis, although a number of speakers here have suggested that the second alternative, the hot disk, should be considered seriously.

There are also tests of the heavy halo hypothesis which are useful in principle but weak in practice. The selection effects in studies of binary galaxy dynamics (White et al. 1982) are so large that they offer no strong evidence for or against heavy halos. Similarly, satellite galaxy tidal radii cannot be used to constrain halo masses, since the observed radii are uncertain, the satellite galaxy M/L's are unknown, and the theory of tidal radii is poorly developed (although progress on the latter two problems was reported here by Illingworth and Freeman).

To summarize, the evidence for halos containing ~ 1-2 disk masses is very strong but not conclusive. A number of new arguments support the heavy halo hypothesis and there is still no substantial evidence against it. On the negative side, the evidence for very heavy halos extending to several hundred kpc is still very slim, and we have made virtually no progress in understanding what the composition of the halo could be (10^6 $M_\odot$ black holes? 1 $M_\odot$ black holes? Jupiters? massive neutrinos? etc.).

To close, may I say that I hope we can look forward to a third Besançon symposium on dynamics in another eight years, and I hope that that meeting will be as exciting and enjoyable as this one has been.

REFERENCES

Bahcall, J.N. and Soneira, R.M.: 1980, Astrophys. J. (Lett.) 238, L17.
Frenk, C.S. and White, S.D.M.: 1980, Monthly Notices Roy. Astron. Soc. 193, 295.
Gunn, J.E.: 1975, Comm. Astrophys. Sp. Phys. 6, 7.
Hohl, F.: 1975, in "Dynamics of Stellar Systems", ed. A. Hayli (Dordrecht: Reidel), 349.
Lin, D.N.C. and Lynden-Bell, D.: 1982, Monthly Notices Roy. Astron. Soc. 198, 707.

Lynden-Bell, D. and Kalnajs, A.J.: 1972, Monthly Notices Roy. Astron.
      Soc. 157, 1.
Monet, D.G., Richstone, D.O. and Schechter, P.L.: 1981, Astrophys. J.
      245, 454.
Ostriker, J.P. and Peebles, P.J.E.: 1973, Astrophys. J. 186, 407.
Schwarzschild, M.: 1979, Astrophys. J. 232, 236.
White, S.D.M., Huchra, J., Latham, D., and Davis, M.: 1982, preprint.
Wilkinson, A. and James, R.A.: 1982, Monthly Notices Roy. Astron. Soc.
      199, 171.

DISCUSSION

OORT : Our chairman, Dr. Toomre, asked for my reaction to the symposium.
I have been impressed - and to some extent confused - by the wonderful
account of the large progress made not only in observations of the structure of galaxies, but also in the understanding of these structures. A
particularly impressive example was the discovery of the triaxiality of
elliptical galaxies. My attention was first drawn to it in an introduc-
tory report by Freeman at the IAU General Assembly in Sydney, when I
found it very surprising. For there seemed to be such good grounds for
supposing that the apparent flattening was caused by rotation; until
Illingworth found that there was no rotation, and Binney showed how the
apparent flattening could be understood without rotation.
A second totally unforeseen development has been the realization of the
enormous influence of mergers on the evolution process of galaxies which
our chairman of this afternoon has had the boldness to introduce.
What has impressed me most is the boldness and the success with which one
is now attempting to understand the formation as well as the evolution
of galaxies.
It is wonderful to have such adventurous discoverers in the astronomical
family.

NAME INDEX

<u>nnn</u>, Name is referred to in several pages of the same paper

Aaronson, M. 6, 9, 373
Aarseth, S.J. 280, 343, 352, 373, 413
Ables, H.D. 16, 50
Adams, M.T. 352
Ahmad, A. 338
Albada, G.D. van 16, 61, 114, 169, 215, 221, 225, 227, 405
Albada, T.S. van 75, 171, 223, 245, 275, 286, 288, 334, 339 392
Alladin, S.M. 339, 349
Allen, D.A. 347
Allen, R.J. 14, 35, 56, 73, 147, 149, 170, 215, 407
Alloin, D. 322
Anderson, S. 13
Aoki, S. 125
Appleton, P.N. 95
Arnett, W.D. 109
Arnold, V. 245
Arp, H.C. 234, 320, 347
Athanassoula, E. 12, 56, 155, 170, 203, 209, 211, 229, 239, 243, 254, 412
Atherton, P.D. 14, 31, 147, 331, 409
Avez, A. 245

Baade, W. 367
Bahcall, J.N. 60, 86, 94, 414
Bajaja, E. 22, 26, 27, 139, 405
Baldwin, J.E. 323
Balick, B. 303, 323
Ball, J.R. 93, 199, 235

Bardeen, J.M. 122, 412
Barnes 392
Bash, F.N. 133, 147, 169
Battaner, E. 157
Beck, R. 159
Beck, S.C. 67
Bekenstein, J.D. 205, 406
Bergh, S. van den 200, 320, 361
Bergvall, N. 326, 367
Berman, R.H. 207, 215
Bertin, G. 117, 119, 165, 180
Bertola, F. 189, 258, 277, 297, 311, 316
Bettoni, D. 311
Bienaymé, O. 209
Binggeli, B. 328
Bingham, R.G. 68
Binney, J.J. 17, 179, 189, 275, 285, 290, 295, 297, 343, 392, 407, 413
Blaauw, A. 166
Blackman, C.P. 300, 311
Blandford, R. 301
Blitz, L. 37, 43, 56, 172, 405
Bodenheimer, P.H. 279
Bohnenstengel, H.D. 65
Boksenberg, A. 309
Bond, H.E. 359
Borne, K. 327
Boroson, T.A. 52, 72, 253
Bosma, A. 11, 55, 69, 81, 172, 188 234, 245, 253, 329, 403, 414
Bottinelli, L. 33, 305
Boulesteix, J. 238

OBJECT INDEX

Object are listed in various categories. Page numbers given refer to the first page of the paper discussing the object.

LOCAL GROUP GALAXIES

Our Galaxy   35, 43, 45, 47, 49, 55,
    69, 77, 81, 89, 133, 135, 137,
    139, 143, 159, 161, 163, 177,
    189, 197, 285, 359, 379, 403,
    411
LMC   89, 139, 141, 177, 239, 359,
    403
SMC   89, 359
Magellanic Stream   89, 177, 411
Carina 89
Draco 89
Fornax 89
Leo I 89
Leo II 89
Pal 1,3,4,5,12,13   89
Sculptor 89
Ursa Minor 89
M31 = NGC 224   11, 23, 27, 35, 55,
    135, 139, 155, 159, 163, 177, 275,
    307, 335, 359, 403, 411
M32 81
M33 = NGC 598   11, 49, 87, 155, 159,
    177, 335, 365, 403
N206 in M31   139
N185 297, 307
N205 297, 307
IC 10 33, 403

RUBIN'S SAMPLE OF SPIRALS

Sa's

NGC 1024 3
NGC 1357 3
NGC 2639 3
NGC 2775 3
NGC 2844 3
NGC 3281 3
NGC 3593 3
NGC 3898 3
NGC 4378 3, 87
NGC 4419 3
NGC 4594 = M104   3, 81, 257, 275

Sa's (continued)

NGC 4698 3
NGC 4845 3
NGC 6314 3
IC   724 3
UGC 10205 3

Sb's

NGC 1085 3
NGC 1325 3
NGC 1353 3
NGC 1417 3
NGC 1515 3
NGC 1620 3, 29
NGC 2590 3
NGC 2708 3
NGC 2815 3
NGC 3054 3
NGC 3067 3
NGC 3145 3
NGC 3200 3
NGC 3223 3
NGC 4448 3
NGC 4800 3
NGC 7083 3
NGC 7171 3
NGC 7217 3, 87
NGC 7537 3
NGC 7606 3, 29
UGC 11810 3
UGC 12810 3

Sc's

NGC 701 3
NGC 753 3
NGC 801 3
NGC 1035 3
NGC 1087 3, 29
NGC 1421 3
NGC 2608 3
NGC 2715 3
NGC 2742 3
NGC 2998 3

Sc's (continued)

NGC 3495   3
NGC 3672   3
NGC 4062   3
NGC 4321 = M100   3, 11, 49, 153, 163
NGC 4605   3, 29
NGC 4682   3
NGC 7541   3
NGC 7664   3
IC   467   3
UGC 2885   3, 11, 29, 403
UGC 3691   3

OTHER SPIRALS

NGC   253   159, 403
NGC   300   11
NGC   488   257
NGC   628 = M74   11, 35,55,69,177,403
NGC   891   11, 35, 55, 69, 403
NGC   925   335
NGC 1058   11, 69
NGC 1068   49
NGC 1097   233
NGC 1300   215, 227
NGC 1313   11, 239, 335
NGC 1317   319
NGC 1365   11, 31, 231, 233, 403
NGC 1398   11,
NGC 1566   151, 163
NGC 2090   253
NGC 2403   11, 49
NGC 2523   193
NGC 2805   11
NGC 2841   11, 49, 55, 163
NGC 2903   11, 403
NGC 2997   11, 155, 163
NGC 3031 = M81   11, 81, 115, 133,
    147, 159, 403
NGC 3198   11, 55
NGC 3359   11
NGC 3504   215
NGC 3938   69
NGC 3992   11, 87, 93, 235
NGC 4027   11
NGC 4151   11
NGC 4244   77
NGC 4254   153
NGC 4258   11
NGC 4303   153

OTHER SPIRALS (continued)

NGC 4340   153
NGC 4501   153
NGC 4535   153
NGC 4565   11, 55, 63, 257, 403
NGC 4579   153
NGC 4618   239
NGC 4731   11, 235
NGC 4736   11, 55, 243
NGC 5033   11
NGC 5055   11, 55
NGC 5194 = M51   11, 35, 49, 53,
    109, 135, 155, 159, 163, 403
NGC 5236 = M83   11, 49, 55, 65,
    147, 177, 403
NGC 5383   11, 147, 215, 229, 233,
    243
NGC 5457= M101   11, 35, 49, 147, 403
NGC 5907   11, 55, 63, 77, 177, 187,
    197, 403
NGC 5908   11
NGC 5963   253
NGC 6503   11
NGC 6946   11, 35, 49, 159
NGC 7331   11, 49
NGC 7479   215
NGC 7723   215
NGC 7741   237
NGC 7793   87
IC   342   11, 35, 49, 159
Mk   348   55

SO'S AND EARLY Sa's

NGC   16   271
NGC   128   257, 271, 275
NGC   584   271
NGC   821   271
NGC   890   271
NGC   936   163, 193
NGC 1023   99, 271
NGC 1175   271
NGC 1209   271
NGC 1332   271
NGC 1380   271
NGC 1381   257, 271
NGC 1386   271
NGC 1461   271
NGC 1553   243
NGC 1726   271

SUBJECT INDEX

Page numbers given refer to the first page of the paper.